Two week loan
Benthyciad pythefnos

Please return on or before the due date to avoid overdue charges
*A wnewch chi ddychwelyd ar neu cyn y dyddiad a nodir ar eich llyfr os
gwelwch yn dda, er mwyn osgoi taliadau*

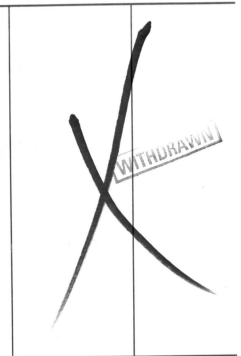

http://library.cardiff.ac.uk
http://llyfrgell.caerdydd.ac.uk

MUCUS IN HEALTH
AND DISEASE–II

ADVANCES IN EXPERIMENTAL MEDICINE AND BIOLOGY

Recent Volumes in this Series

MUCUS IN HEALTH AND DISEASE—II

Edited by

Eric N. Chantler
Department of Obstetrics and Gynecology
University Hospital of South Manchester
Manchester, England

James B. Elder
University of Department of Surgery
Manchester Royal Infirmary
Manchester, England

and

Max Elstein
Department of Obstetrics and Gynecology
University Hospital of South Manchester
Manchester, England

PLENUM PRESS • NEW YORK AND LONDON

Library of Congress Cataloging in Publication Data

International Symposium on Mucus in Health and Disease (2nd: 1981: Manchester
  University)
  Mucus in health and disease II.

  Advances in experimental medicine and biology; v. 144)
  "Proceedings of the Second International Symposium on Mucus in Health and Dis-
ease, held September 1-4, 1981, at Manchester University, Manchester, England" — T.p.
verso.
  Includes bibliographies and index.
  1. Mucus — Congresses. I. Chantler, Eric N., 1947-   II. Elder, James B.,
1928-   . III. Elstein, Max. IV. Title. V. Series. [DNIM: 1. Mucus — Congresses. 2.
Glycoproteins — Congresses. Wl AD559 v.144 / QS 532.5.M8 I62 1981m]
QP215.I57 1981   612'.4   82-424
ISBN 0-306-40906-2   3/6/82   AACR2

Proceedings of the Second International Symposium on Mucus in Health and Disease,
held September 1-4, 1981, at Manchester University, Manchester, England

© 1982 Plenum Press, New York
A Division of Plenum Publishing Corporation
233 Spring Street, New York, N.Y. 10013

Printed in the United States of America

FOREWORD

This second International Symposium on Mucus in Health and Disease once again brings together basic scientists such as Biochemists, Anatomists, Biologists and Clinicians who are dealing with aspects of mucus in the various tracts of the body where it is of such great functional importance. It is fitting that the meeting should take place at Manchester University where there is so much activity in this field and our grateful thanks are due to Dr Eric Chantler for his untiring efforts in organising this meeting.

At the first Mucus meeting, Sir Francis Avery Jones stated "this is a subject which will justify further Symposia, both local and international". As he predicted, this meeting succeeds the first and adds further to our progress in understanding the complex and unique structure and function of the mucus secretion in its various sites of the body. Much was learned from the first meeting and it is hoped that the second will be an appropriate successor to it. The emphasis in this meeting has been to encourage discussion and the presentation of research material. In this respect, review articles have been kept to a minimum. The structure of the Conference has been organised around eight keynote addresses: one on the biosynthesis of the general mucus glycoproteins and another on its physical properties. Other keynote papers are on the biochemical and clinical aspects of mucus in the respiratory, gastrointestinal and urogenital tracts by recognised authorities in these subjects.

To maintain the vital aspect of the proceedings, it is intended that they will be published, with the co-operation of Plenum Press, for dissemination shortly after the meeting. In order to achieve this, we have chosen not to include the discussions in favour of speed of production.

The review papers provide a valuable statement of the current status of the subject. From this base an exciting collection of research papers have been submitted which contain many significant contributions to our knowledge. The design of the meeting and the presentation of the scientific material has been planned to ensure a stimulating exchange of ideas.

This volumne, together with the proceedings of the first Symposium on Mucus in Health and Disease, provides the most complete statement of current knowledge in this fascinating interdisciplinary field of study. The quality and diversity of the work presented demonstrates the enthusiasm of the participants upon whom the success of any conference finally rests.

                              Max Elstein
                              Professor of Obstetrics & Gynaecology
                              University Hospital of South Manchester

ACKNOWLEDGEMENTS

The organising committee wish to express their thanks
for the financial support of the Symposium to the following:

Abbott Laboratories Ltd

Ayrst Laboratories Ltd

Beecham Research Laboratories

Boehringer Ingelheim Limited

Brocades (Great Britain) Limited

Haigh and Hochland

Imperial Chemical Industries

L.K.B.

Organon Laboratories Limited

Roche Products Limited

Searle Pharmaceuticals

Smith, Kline and French Laboratories Limited

Syntex Limited

Tobacco Advisory Council

Winthrop Laboratories

Wyeth Laboratories

CONTENTS

MUCUS BIOSYNTHESIS: IMMUNOLOGICAL AND RHEOLOGICAL PROPERTIES

MUCUS BIOSYNTHESIS: IMMUNOLOGICAL AND RHEOLOGICAL PROPERTIES

BIOSYNTHESIS OF MUCUS GLYCOPROTEINS

Harry Schachter and David Williams

Research Institute, Hospital for Sick Children, and
Department of Biochemistry, University of Toronto
Toronto, Canada M5G 1X8

INTRODUCTION

Mucus in higher organisms is usually defined as the viscous
fluid lining the epithelium of the gastro-intestinal, respiratory
and genito-urinary tracts. Mucus is a complex mixture containing
large glycoproteins (mucins), water, electrolytes, sloughed epithelial cells, enzymes and various other materials, including bacteria
and bacterial products depending on the source and location of the
mucus. This discussion will be limited to a survey of the biosynthesis of the carbohydrate portion of mucins, the major components
of mucus. Mucins are large glycoproteins (molecular weight is
usually over 1×10^6) containing from 50 to 80% or more by weight
of carbohydrate. As will be discussed in more detail below, the
carbohydrate is linked to the polypeptide by an O-glycosidic bond
between GalNAc and Ser or Thr. Mucins are polydisperse with respect
to molecular size and oligosaccharide sequences and chain lengths.
Thus structural information on mucin oligosaccharides must be obtained by cleaving the Ser(Thr)-GalNAc bond, usually by alkali-
catalyzed β-elimination, and purifying individual oligosaccharides.
By analogy, our discussion of biosynthesis will deal with the assembly of a single oligosaccharide at a time.

Mucins are synthesized either by goblet cells lining the mucous
epithelium or by special exocrine glands with mucous cell acini such
as some of the salivary glands. The peptide backbone is assembled
on membrane-bound ribosomes and sugars are added subsequently as the
mucin moves through the cell from the rough endoplasmic reticulum
towards the Golgi apparatus, as has been shown by various autoradiographic studies (1-5). Mucin-containing vesicles bud off the Golgi
apparatus and mature into secretory granules. These granules migrate

3

towards the apex of the cell and eventually discharge their contents
out of the cell by a process called exocytosis (4, 6); secretion can
be slow and continuous or can occur in a large burst of material in
response to a secretory stimulus. The dramatic response of a mucous
cell to a secretory stimulus will not be discussed here but the process has interesting implications for the cell's glycoprotein synthesis machinery. The secretory process often results in a large
increase in the cell's surface membrane; this material is brought
back into the cell by endocytosis and it is not clear whether the
membrane-bound glycoproteins of this material are degraded and re-
assembled or are re-utilized intact. Also, secretion depletes the
cell of secretory granules; in some systems it has been shown that
this results in an increased rate of glycosylation but the mechanism
of this control is unknown (7).

OLIGOSACCHARIDE STRUCTURES IN MUCINS

     Mucins contain only oligosaccharides linked by a Ser(Thr)-GalNAc
bond to the polypeptide chain. There are of course various other
types of oligosaccharides, e.g., the very common N-glycosidic type
(Asn-GlcNAc linkage) found in many mammalian glycoproteins, the
Xyl-Ser type found in mucopolysaccharides and the hydroxylysine-Gal
type found in collagens and basement membranes (8). Mucins do not
appear to carry such oligosaccharides; Ser(Thr)-GalNAc oligosaccharides do, however, occur in glycoproteins that are not mucins, e.g.,
glycophorin A, fetuin and IgA. Mucin oligosaccharides show a great
diversity; they can contain anywhere from 1 to 15 or more sugars and
may be linear or branched. Commonly occuring sugars are GalNAc,
Gal, Fuc, GlcNAc and sialic acid; Man, Glc and uronic acids are not
found but many mucins are sulphated either at a Gal or GlcNAc residue
(6). Mucins frequently carry at the non-reducing termini of their
oligosaccharides antigenic determinants for one or more of the human
blood groups (ABO, Lewis and Ii systems) and our discussion will
therefore have to include some comments on the biosynthesis of
human blood groups.

     We have found it convenient to classify mucin oligosaccharide
structures into four categories depending on the type of core present in the oligosaccharide (Figure 1). Core 1 (Galβ1-3GalNAc-)
and, to a lesser extent, Core 2 (Galβ1-3[GlcNAcβ1-6]GalNAc) have,
to date, been found most frequently whereas Core 4 (GlcNAcβ1-3-
[GlcNAcβ1-6]GalNAc) has been only recently described in ovine gastric
mucin (9). The classification is not comprehensive; for example,
both GalNAc-Ser(Thr)- and sialylα2-6GalNAc-Ser(Thr)- occur in ovine
submaxillary mucin yet do not fit into any of the four categories
(Figure 1).

     The biosynthesis of these various oligosaccharides is discussed
in the following sections. Again, however, the discussion is not

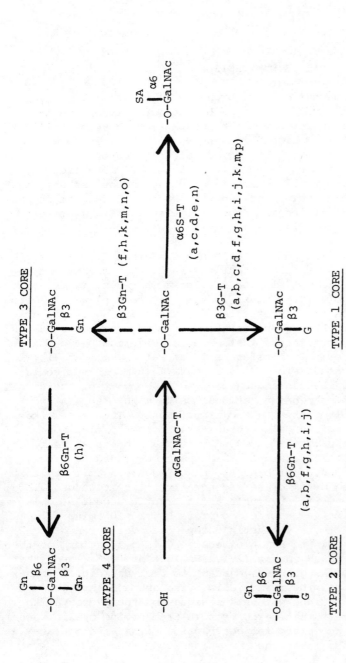

Figure 1. Assembly of Ser(Thr)-GalNAc oligosaccharides; four basic cores are made.
Abbreviations: -OH, hydroxyl of Ser or Thr in polypeptide backbone; SA, sialic acid; Gn, D-GlcNAc; G, D-Gal, F, L-Fuc; T, transferase; continuous arrows, enzymes established by in vitro experiments; discontinuous arrows, enzymes postulated on the basis of known mucin structures; letters under arrows indicate organs of origin with appropriate references in brackets, as follows: a, canine submaxillary (12); b, human submaxillary (13); c, ovine submaxillary (10); d, porcine submaxillary (11); e, bovine submaxillary (22); f, equine gastric (15); g, human gastric (16); h, ovine gastric (9 and personal communication, E.F.Hounsell); i, porcine gastric (14); j, human ovarian cyst (17,56); k, human bronchial (18); m, rat small intestinal (19); n, rat colonic (20); o, rat sublingual (21); p, armadillo submandibular (61).

comprehensive. Several relatively large mucin oligosaccharide structures have been described in the literature (12-17, 20, 21) and no
attempt is made to include the largest of these in the schemes to be
presented below.

INITIATION OF OLIGOSACCHARIDE SYNTHESIS

    All mucin oligosaccharides are probably initiated in the same
way, i.e., by the action of the enzyme UDP-GalNAc:polypeptide α-N-
acetylgalactosaminyltransferase on the polypeptide backbone. This
enzyme was first described by McGuire and Roseman (23) in 1967 in
sheep submaxillary gland; it transferred GalNAc to carbohydrate-free
ovine submaxillary mucin but was unable to act on 24 low molecular
weight sugars, 31 glycoproteins and glycolipids, and on carbohydrate-
free ovine submaxillary mucin digested with Pronase. The enzyme was
subsequently discovered in bovine and porcine submaxillary glands
and in a variety of other tissues (24-26).

    The strict specificity of the GalNAc-transferase for glycose
acceptor has remained a topic of great interest. Since this enzyme
determines the initiation of all mucin oligosaccharides, it is not
surprising that the polypeptide must in some way control which serine
and threonine residues are glycosylated and which are not. The original experiments with Pronase-digested mucin have been verified by
others and it is now established that a certain minimum polypeptide
length is essential for the transferase to operate. Hill et al. (26)
carried out sequence studies on ovine submaxillary mucin but could
find no unique primary sequences adjacent to 28 known 0-glycosidically
substituted serine and threonine residues which might explain the
specificity of the polypeptide GalNAc-transferase. The same workers
prepared a 30-fold purified GalNAc-transferase from porcine submaxillary glands (specific activity was 0.06 units/mg protein) by chromatography on UDP-hexanolamine-agarose, SE-Sephadex C-50 and Sephadex
G-100 (26); the enzyme did not bind to the UDP-agarose affinity column.
The partially purified transferase was equally active towards carbohydrate-free ovine submaxillary mucin and a tryptic digest of this
mucin; the smallest active peptide acceptors had 16-18 amino acid
residues. More recently, a synthetic octapeptide Val-Thr-Pro-Arg-
Thr-Pro-Pro-Pro was shown to be an effective GalNAc acceptor for the
porcine submaxillary gland enzyme (27). Hill et al. (26) found that
20 of the 28 amino acid sequences adjacent to 0-glycosidically substituted serine or threonine residues in ovine submaxillary mucin
had at least one Pro residue within 4 residues of the hydroxyamino
acid. It is of interest that basic protein from bovine myelin, one
of the few non-mucin acceptors for polypeptide GalNAc-transferase
(24), has 3 Pro residues near the Thr which accepts GalNAc.

The above findings can be compared with analogous studies on the synthesis of the Asn-GlcNAc bond (25, 28). It has been established that this bond is made by the transfer of oligosaccharide from dolichol pyrophosphate oligosaccharide to polypeptide and that transferase specificity is determined by the sequence -Asn-X-Ser(Thr)- where X can be almost any amino acid. However, the tripeptide sequence must have both its N- and C-terminal ends blocked and the sequence must be sterically available to the transferase. There is no apparent equivalent primary amino acid sequence requirement for the polypeptide GalNAc-transferase but the sequences around the acceptor hydroxyamino acid residues suggest that these residues are in regions with little secondary structure (26); thus the GalNAc-transferase seems to require accessibility and a minimum polypeptide length rather than a specific amino acid sequence.

Synthesis of the Asn-GlcNAc linkage is a highly specialized process requiring pre-assembly of a large oligosaccharide as a lipid-bound intermediate (28); this oligosaccharide is transferred to an Asn residue and is then processed by removal and re-incorporation of sugars to the final glycoprotein product (29). Hanover et al. (30) have investigated the possibility that a similar process may be involved in the assembly of the Ser(Thr)-GalNAc linkage. They found that a partially purified preparation of porcine submaxillary gland UDP-GalNAc:polypeptide GalNAc-transferase could transfer GalNAc to Val-Thr-Pro-Arg-Thr-Pro-Pro-Pro without the formation of either dolichol monophosphate GalNAc or dolichol pyrophosphate oligosaccharide. Further, neither tunicamycin nor exogenous dolichol phosphate had any appreciable effect on the rate of this reaction; tunicamycin completely inhibits and dolichol phosphate greatly stimulates the incorporation of radioactive GlcNAc from UDP-GlcNAc into Asn-linked oligosaccharides. They then examined oviduct microsomes for the ability to transfer radioactive GalNAc from UDP-GalNAc to exogenous octapeptide and to transfer radioactive Man from GDP-Man, in the presence of UDP-GlcNAc, to exogenous S-carboxymethylated α-lactalbumin. The microsomes carried out both reactions effectively, i.e., they were capable of both O- and N-glycosylation. The O-glycosylation reaction mixture did not accumulate either dolichol monophosphate monosaccharide nor dolichol pyrophosphate oligosaccharide and the reaction was insensitive to both tunicamycin and dolichol phosphate addition. In contrast, the N-glycosylation reaction mixture accumulated the expected lipid intermediates, and tunicamycin addition caused an 86% inhibition of Man incorporation into exogenous acceptor while dolichol phosphate addition stimulated this reaction 26%. Babczinski (31) found that bovine submaxillary gland membranes could transfer radioactive GalNAc from UDP-GalNAc to endogenous high molecular weight acceptor without the formation of GalNAc-containing lipid intermediates, whether or not dolichol phosphate was added to the incubation. The conclusion drawn by these workers is that formation of the Ser(Thr)-GalNAc linkage does not require oligosaccharide pre-assembly nor lipid intermediates but occurs by the addition of a

single GalNAc residue. It is perhaps of interest to point out that
0-mannosylation of Ser(Thr) residues in fungal glycoprotein synthesis
does require the formation of lipid intermediates (32, 33).

The subcellular. sites of oligosaccharide initiation have also
been investigated by various groups. Three general approaches have
been applied to this problem: (i) Radioautography following injection
of radioactive precursors into living animals (ii) Subcellular
fractionation of tissues from animals previously injected with radioactive precursors and (iii) Glycosyltransferase assays of subcellular
fractions. These 3 approaches have established quite clearly that
N-glycosylation is initiated within the rough endoplasmic reticulum,
either on nascent ribosome-bound polypeptide or on post-ribosomal
polypeptide, and that subsequent processing and elongation reactions
occur within the endoplasmic reticulum and Golgi apparatus (see 25
and 29 for reviews of this literature). The subcellular site of 0-
glycosylation has proven to be more difficult to establish, primarily
because subcellular fractionation methods have been difficult to
apply to tissues rich in mucins. Nevertheless, the concensus appears
to be that the UDP-GalNAc:polypeptide GalNAc-transferase is located
in smooth-surfaced membranes in rat small intestinal mucosa (34), in
brain (35), in hen oviduct (30), and in HeLa cells (36). There have,
however, been reports of 0-glycosylation in the rough endoplasmic
reticulum, e.g., Strous (37) isolated membrane-bound polyribosomes
from rat stomach, purified peptidyl-tRNA from these by DEAE-Sephadex
chromatography and treated the peptidyl-tRNA with alkaline borotri-
tide. He thus showed the release of tritiated N-acetylgalactosamini-
tol from nascent peptide. He further showed that the nascent peptides
had to have a molecular weight of at least 4000 for glycosylation to
occur. The difficulty with this type of experiment is to be sure
that the peptidyl-tRNA preparation does not contain small amounts of
non-nascent 0-glycosidically linked material. On the other hand, it
is possible that in some systems, varying amounts of 0-glycosylation
may occur in the rough endoplasmic reticulum. The evidence presently
available indicates that the bulk of 0-glycosylation probably occurs
in the smooth endoplasmic reticulum and/or Golgi apparatus after release of nascent peptide from the ribosome.

Once the first sugar has been incorporated into the peptide
chain, subsequent elongation can occur along various paths (Figure
1). The main controlling factors determining the path of elongation
appear to be the glycosyltransferases available to act on the growing
glycoprotein oligosaccharides. These paths will now be discussed.

ASSEMBLY OF OLIGOSACCHARIDES WITH TYPE 1 CORE

The key enzyme required for routing the biosynthetic machinery
towards oligosaccharides with a type 1 core is UDP-Gal:GalNAc-mucin
β3-galactosyltransferase, first described in 1971 by Schachter et al

(38) in porcine submaxillary gland, and subsequently studied by
others in a variety of tissues (25, 39-42).

The porcine submaxillary gland β3-Gal-transferase (38) is a membrane-bound enzyme requiring Mn2+ and Triton X-100 for activity.
Attempts at solubilizing and purifying this enzyme have been unsuc-
cessful to date. The crude enzyme has a specific activity of about
5×10^{-4} units per mg protein and requires GalNAc-mucin (sialidase-
treated ovine or porcine submaxillary mucin) for full activity.
The crude enzyme can transfer Gal to GalNAc or glycosides of GalNAc
but the K_m is much higher than for mucin substrates (180 mM as
opposed to 0.7 mM). Competition experiments have shown (38) that
the mucin Gal-transferase is a different enzyme from UDP-Gal:β-N-
acetylglucosaminide β-Gal-transferase and UDP-Gal:Tay-Sachs ganglioside β-Gal-transferase but enzyme purification is required to establish that the mucin and ganglioside enzymes are indeed different.

The β3-galactosyltransferase competes with at least 2 other
enzymes for a common substrate (Figure 1); one of these is CMP-sialic
acid:GalNAc-mucin α6-sialyltransferase, which is discussed in a
later section, and the other enzyme, not yet demonstrated experimentally, is suggested to be UDP-GlcNAc:GalNAc-mucin β3-N-acetyl-
glucosaminyltransferase. The product of the porcine submaxillary
gland Gal-transferase was originally identified as Galβ1-3GalNAc-
mucin by paper chromatography of Galβ1-3GalNAcOH following alkaline
borohydride treatment (38). Further, similar analysis of the product obtained by the action of the Gal-transferase on native ovine
submaxillary mucin yielded Galβ1-3GalNAcOH; no sialic acid-containing
oligosaccharide was detected showing that the prior action of the
CMP-sialic acid:GalNAc-mucin α6-sialyltransferase (Figure 1) prevented subsequent action of the Gal-transferase. It has also been
shown that ovine submaxillary glands have a preponderance of the
α6-sialyltransferase over the β3-galactosyltransferase whereas the
reverse is true for porcine submaxillary glands (38, 43). These
facts explain why ovine submaxillary mucin contains primarily sialyl-
α2-6GalNAc oligosaccharides and very few larger Gal-containing
oligosaccharides, whereas porcine submaxillary mucin is rich in
more complex oligosaccharides containing the Galβ1-3GalNAc core
structure (Figure 2).

The product formed by the action of porcine submaxillary gland
β3-galactosyltransferase on sialic acid-free ovine submaxillary
mucin was recently examined by more definitive methods, including
high resolution proton nuclear magnetic resonance spectroscopy, and
shown to be Galβ1-3GalNAc-mucin (39), as had previously been shown
by less reliable methods (38). It should be pointed out that the
enzyme is required for the biosynthesis not only of mucins but also
of other non-mucin glycoproteins carrying Galβ1-3GalNAc core structures, e.g., glycophorin A (44), antifreeze glycoprotein from
Antarctic fish (41), fetuin, IgA, etc. These proteins can, in fact,

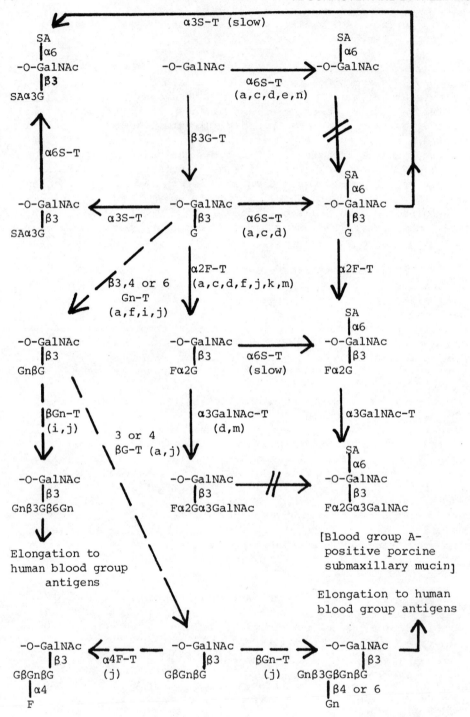

<u>Figure 2.</u>   Synthesis of type 1 core oligosaccharides.   See legend to
          Figure 1 for abbreviations.

be used as acceptors for the transferase, e.g., derivatives of
Antarctic antifreeze glycoprotein (41). It is therefore not surprising that the enzyme has been demonstrated in such diverse tissues
as rat pancreas (40), rat liver (42), rat intestine and human serum
(see 25 and 39 for references).

The subcellular localizations of most transferases involved in
mucin biosynthesis have not been firmly established due to the difficulty of fractionating mucin-secreting tissues. However, since
some of the transferases involved in mucin synthesis are found in
tissues which do not secrete appreciable amounts of mucin, it is
possible to study the subcellular locations of these enzymes in
such tissues since they are more suitable for subfractionation
procedures. For example, the UDP-Gal:GalNAc-mucin β3-galactosyl-
transferase has been studied in rat liver (42) and shown to be
enriched 23-fold in a Golgi subfraction but only 1.3-fold in a rough
microsome fraction. The authors found evidence that the microsomal
and Golgi Gal-transferases might be separate enzymes. However,
their data indicates that the Golgi apparatus is the major site in
rat liver for the synthesis of the Galβ1-3GalNAc linkage. This
suggests that this organelle is involved in the elongation of both
O- and N-glycosidically linked oligosaccharides. Incubation of rat
gastric mucosa with [^{3}H]-galactose followed by autoradiography showed
the Golgi complex to be the almost exclusive site of galactose incorporation (4, 45); these experiments suggest that galactose incorporation into rat gastric glycoproteins occurs within the Golgi
apparatus but the autoradiographic approach cannot determine what
linkages are being synthesized.

Figure 2 summarizes the pathways that branch off from the
Galβ1-3GalNAc core. Three major pathways can be defined.

The Porcine Submaxillary Mucin Pathway

Porcine submaxillary mucin was shown by Carlson in 1968 (11)
to contain a series of oligosaccharides the largest of which was a
pentasaccharide carrying the antigenic determinant for human blood
group A (GalNAcα1-3[Fucα1-2]Galβ-); this structure is shown in
Figure 2. All five glycosyltransferases required for the assembly
of this pentasaccharide have been demonstrated in vitro (23-27, 29,
38, 39, 43, 46-53). Three of the five transferases have been purified from porcine submaxillary gland (47-52), and a preferential
pathway of synthesis has been worked out (38, 53). Although the
pathway has been established with porcine submaxillary gland enzymes,
it is likely from a study of mucin oligosaccharide structures that
various parts of the pathway are applicable also to the synthesis
of canine (12) and ovine (10) submaxillary mucins, rat intestinal
mucin (19), human bronchial mucin (18), equine gastric mucin (15)

and human ovarian cyst mucin (17); the relevant parts of the pathway
are indicated in Figure 2.

The porcine submaxillary mucin synthetic path has several alternative routes. The core structure Galβ1-3GalNAc-mucin is a good
substrate for either CMP-sialic acid:α-N-acetylgalactosaminide
α6-sialyltransferase or GDP-Fuc:β-galactoside α2-fucosyltransferase.
Action of the α6-sialyltransferase leads to sialylated oligosaccharides whereas action of the α2-fucosyltransferase results mainly in
neutral oligosaccharides (Figure 2). The reason for this is that
the presence of the Fucα1-2Gal- group tends to inhibit the rate of
catalysis by the α6-sialyltransferase (53).

The α6-sialyltransferase was first studied by Carlson et al.
(46) in ovine submaxillary glands. It catalyzed incorporation of
sialic acid into GalNAc-mucin to form sialyl2-6GalNAc mucin; the
product was identified by chromatographic analysis of sialyl-N-
acetylgalactosaminitol released by alkaline borohydride and the 2-6
linkage was determined by periodate oxidation. The anomeric linkage
of the incorporated sialic acid was not determined nor was Galβ1-3-
GalNAc-mucin tested as a substrate. The enzyme was purified about
44-fold from ovine submaxillary glands and was also found in porcine,
bovine and canine submaxillary glands; ovine heart, liver, kidney,
duodenum, stomach, lung, spleen and brain showed minimal enzyme
levels. The transferase utilized either CMP-N-acetylneuraminic acid
or CMP-N-glycollylneuraminic acid as sialyl donor. For this reason
we have retained the term "sialyl" in naming these enzymes.

Sadler et al. (49, 50) carried out further studies on the porcine submaxillary gland α6-sialyltransferase and eventually succeeded
in purifying this enzyme to homogeneity. An early problem in their
work was the finding that highly purified asialo-ovine submaxillary
mucin contained small amounts of Fucα1-2Galβ1-3GalNAc- (2.2 mol% of
total GalNAc) and Galβ1-3GalNAc- (0.5 mol% of total GalNAc) oligosaccharides; the latter disaccharide, but not the trisaccharide
(53), was an excellent substrate for yet another porcine submaxillary gland enzyme, CMP-sialic acid:Galβ1-3GalNAc α3-sialyltransferase,
to be discussed below. They prepared a specific acceptor for the
α6-sialyltransferase by treating asialo-ovine submaxillary mucin
with an endo-α-N-acetylgalactosaminidase which removed Galβ1-3GalNAc
but did not affect terminal GalNAc residues. Using this assay, the
two sialyltransferases were both purified to homogeneity from
Triton X-100 extracts of submaxillary gland. The key steps in the
purification of both transferases were adsorptions to CDP-hexanola-
mine-Sepharose columns; the α6-sialyl-transferase adheres to the
affinity column more weakly than the α3-sialyltransferase and the
enzymes can be separated by repeated affinity chromatography.
Highly purified α6-sialyltransferase was obtained after 5 CDP-

agarose steps followed by gel filtration on Sephadex G-200; the
final preparation was obtained in 2% yield, had a purification
factor of 117,000-fold over homogenate, a specific activity of
44.6 units/mg, and was free of α3-sialyltransferase (50). The pure
enzyme shows no metal requirement.

Highly purified α6-sialyltransferase transfers sialic acid to
GalNAc-α-Ser(Thr)-proteins like asialo-ovine submaxillary mucin, as
well as to Galβl-3GalNAc-α-Ser(Thr)-proteins like antifreeze glycoprotein; in fact, the Galβl-3GalNAc-protein is a better substrate.
The product of α6-sialyltransferase action on asialo-ovine submaxillary mucin was identified primarily by periodate oxidation; anomeric
configuration was assumed. No product identification was carried
out with antifreeze glycoprotein as substrate. However, the product
formed by the action of purified α6-sialyltransferase on antifreeze
glycoprotein is an excellent substrate for highly purified α3-
sialyltransferase and α2-fucosyltransferase, both known to act on
terminal Gal residues (53); this indicates that α6-sialyltransferase
incorporates sialic acid into linkage with GalNAc as depicted in
Figure 2.

The α6-sialyltransferase is subject to controls in porcine submaxillary glands (53). The enzyme rate is decreased if substrate
has a Fucα1-2Gal sequence and is inhibited completely if substrate
has a GalNAcα1-3Gal sequence (Figure 2).

The GDP-Fuc:β-galactoside α2-fucosyltransferase from porcine
submaxillary gland resembles the human blood group H-gene specified
α2-fucosyltransferase; porcine submaxillary mucins rich in terminal
Fucα1-2Gal- groups are indeed excellent blood group H antigens.
The subject of human blood groups in general and of the ABO-Lewis
blood group system in particular is too large to be discussed in
detail in this review. There are many comprehensive reviews available on the subject (for example, 25, 54). The human ABO-Lewis
antigenic determinants are carbohydrate in nature and occur on the
red cell membrane as glycosphingolipids and as protein-bound oligosaccharides of both the Asn-GlcNAc and Ser(Thr)-GalNAc types.
Similar membrane-bound substances occur on the surfaces of cells
other than the red cell. Finally, these antigenic determinants
also occur on mucins secreted by human epithelial cells as well as
by epithelial cells from certain other species such as the pig.
Associated with the ABO-Lewis antigenic system is yet another human
blood group system, Ii. The I and i determinants are now also believed to be carbohydrate in nature, occuring internal to the ABO-
Lewis determinants; these determinants have also been shown to occur
in mucins (54, 55).

Fucosyltransferases capable of synthesizing the H-determinant,
Fucα1-2Gal-, are widely distributed in human tissues, e.g. milk,

submaxillary gland, stomach mucosa, serum, and bone marrow. The enzyme is dependent on the presence of the H gene; rare individuals
with the genotype hh lack this enzyme. The α2-fucosyltransferase
acts on secreted glycoproteins only in individuals with a secretory
gene Se; the mechanism of action of the Se/se system is not known.
The substrate for the α2-fucosyltransferase can be large or small
and the only requirement appears to be a terminal β-D-Gal residue;
even phenyl-β-D-galactoside is a substrate and is in fact very useful
since it is more specific for the α2-fucosyltransferase than larger
oligosaccharides which can act as substrates for other fucosyltransferases (54).

The porcine submaxillary gland α2-fucosyltransferase has been
purified to homogeneity from a Triton X-100 extract by chromatography
on SP-Sephadex, GDP-hexanolamine-agarose and Sephadex G-200 (51,52).
The enzyme was purified 124,000-fold with a 6% yield and had a
specific activity of 20 units per mg protein. It acted on both
large acceptors such as Galβ1-3GalNAc-α-Ser(Thr)-protein and disaccharides with a terminal β-D-Gal residue. The porcine enzyme pre-
fers Galβ1-3GalNAc and Galβ1-3GlcNAc as acceptors compared to the
1-4 and 1-6 linked disaccharides.

The final enzyme in the porcine submaxillary mucin pathway is
also similar to a human blood group enzyme, the blood group A-
dependent UDP-GalNAc:Fucα1-2Gal α3-N-acetylgalactosaminyltransferase.
The ABO gene system segregates independently of the H/h, Le/le, Se/se
and I/i systems. The A determinant is GalNAcα1-3(Fucα1-2)Galβ-;
the A-dependent transferase can only act on the H determinant,
i.e., the A antigen cannot be expressed unless the H antigen has
first been synthesized (25, 54). The A-dependent transferase is
widely distributed in human tissues and has been purified from
human serum (54). The porcine submaxillary gland enzyme has been
purified 38,000-fold in 13% yield by repeated chromatography of
Triton X-100 extracts on UDP-hexanolamine-agarose columns (47).
The purified enzyme had a specific activity of 30 units per mg
protein, showed an absolute requirement for divalent metal ions,
and acted only on acceptors with a terminal Fucα1-2Gal disaccharide
(48). Product identification was essentially based on conversion
of H antigen to A antigen (48); no product characterization was
carried out using chemical as opposed to purely immunological
methods.

The Macromolecular Pathways

We have termed the second pathway the "macromolecular" pathway for want of a better name. We have arbitrarily called oligosaccharides containing over 5 monosaccharides macromolecules since
the largest oligosaccharides in the other two type 1 core paths contain 5 and 4 monosaccharides respectively. The key transferases

leading to elongated oligosaccharides with a type 1 core appear to
be N-acetylglucosaminyltransferases adding GlcNAc in β1-3, β1-4 or
β1-6 linkages to the Gal residue of the Galβ1-3GalNAc core (Figure 2).
None of the enzymes shown in Figure 2 as being components of the
macromolecular pathways have as yet been demonstrated in vitro.
They are postulated to occur as shown in the Figure on the basis of
published mucin oligosaccharide structures. Thus, for example,
GlcNAc addition to the Gal residue of the core should occur during
synthesis of canine submaxillary mucin (12), porcine gastric mucin
(14), and certain human ovarian cyst mucins (17, 56). It is interesting that Ziderman et al. (62) reported in 1967 the ability of
rabbit, human and baboon stomach mucosal extracts to catalyze incorporation of GlcNAc from UDP-GlcNAc into β-methylgalactoside and
β-Gal-terminal disaccharides. The linkage synthesized was not determined but it is likely the enzyme discovered by Ziderman et al. (62)
is involved in elongation of mucins as shown in Figures 2, 3 and 4.
The usual GlcNAc-Gal linkage is either β1,3 or β1,6; the β1,4 linkage
has been reported in only one oligosaccharide isolated from equine
gastric mucin (15).

The various paths shown in Figure 2 are not meant to be comprehensive but serve to indicate the intricate variety of the mucin
synthesis machinery. Frequently one or more of the arms of a mucin
oligosaccharide is terminated at its non-reducing end by a human
blood group determinant, e.g., A, GalNAcα1-3(Fucα1-2)Gal; B, Galα1-3-
(Fucα1-2)Gal; H, Fucα1-2Gal; Le[a], Galβ1-3(Fucα1-4)GlcNAc; Le[b], Fucα1-2-
Galβ1-3(Fucα1-4)GlcNAc. The repeating structure (Galβ1-4GlcNAcβ1-3)$_n$
is found in certain large mucin structures; this structure is now
known to be associated with human i antigenic activity (54, 55).
When branching occurs on this structure to form a GlcNAcβ1-3(GlcNAc
β1-6)Galβ1-4GlcNAcβ1-3- branch point, human blood group I activity
is created (54, 55); such a structure is shown in the macromolecular
pathway in Figure 2. Mucin oligosaccharides with a type 1 core and
terminating in human blood group determinants have been isolated
primarily from human ovarian cyst fluids (17, 56).

The Glycophorin Pathway

The third pathway shown in Figure 2 leads to the tetrasaccharide
sialylα2-3Galβ1-3(sialylα2-6)GalNAc-Ser(Thr)-. This tetrasaccharide
has been demonstrated in glycophorin A (44) and other glycoproteins
but has not as yet been shown to occur in a typical mucin. It is
included in this discussion to illustrate the variety of known type
1 core O-glycosidically linked structures (Figure 2). Tetrasaccharide synthesis requires the action of two sialyltransferases on the
type 1 core, i.e., CMP-sialic acid:α-N-acetylgalactosaminide α6-
sialyltransferase (discussed above) and CMP-sialic acid:β-galactoside
α3-sialyltransferase. The latter enzyme is preferably termed

CMP-sialic acid:Galβ1-3GalNAc α3-sialyltransferase to distinguish
it from yet another transferase acting on N-glycosidically linked
oligosaccharides, i.e., CMP-sialic acid:Galβ1-4GlcNAc α3-sialyl-
transferase.

The α3-sialyltransferase can be assayed (49, 57) by measuring
incorporation of sialic acid into Galβ1-3GalNAc-mucin (e.g., porcine
submaxillary mucin after removal of sialic acid residues), antifreeze glycoprotein, or disaccharides such as Galβ1-3GalNAc (K_m =
0.2 mM), Galβ1-3GlcNAc (K_m = 85 mM) and Galβ1-4Glc (K_m = 130 mM).
Although lactose is a relatively poor acceptor for the α3-sialyl-
transferase, it is useful for routine assays in tissues such as
porcine submaxillary gland which lack CMP-sialic acid:Galβ1-4GlcNAc
α6-sialyltransferase (49); the latter enzyme acts on lactose to
form 6'sialyllactose (sialylα2-6Galβ1-4Glc). It is of interest that
the ability of rat liver to synthesize 3'sialyl-lactose (sialylα2-
3Galβ1-4Glc) is due to the presence of the CMP-sialic acid:Galβ1-
3GalNAc α3-sialyltransferase since rat liver has no detectable
CMP-sialic acid:Galβ1-4GlcNAc α3-sialyltransferase (29), an enzyme
acting on N-glycosidically linked oligosaccharides.

The α3-sialyltransferase has been purified to homogeneity from
Triton X-100 extracts of porcine submaxillary gland (49) by chromatography on SP-Sephadex, CDP-hexanolamine-agarose (3 separate columns)
and Sephadex G-200. The final preparation was purified 92,000-fold
relative to the crude homogenate and was obtained in 5% yield. The
specific activity of the pure enzyme was 10.6 units per mg protein.
The purified enzyme formed a sialylα2-3Gal bond with a variety of
Gal-terminal compounds. The best acceptors were Galβ1-3GalNAc and
glycoproteins with O-glycosidically linked Galβ1-3GalNAc such as
glycosidase-treated mucins, antifreeze glycoprotein, and asialofetuin. Ganglioside G_{M1}, i.e., Galβ1-3GalNAcβ1-4(sialylα2-3)Galβ1-
4Glc-ceramide, is also a good acceptor for the highly purified
α3-sialyltransferase (57) resulting in the synthesis of ganglioside
G_{D1a}; further, ganglioside G_{M1} and antifreeze glycoprotein compete
for active sites on the purified enzyme. Other glycolipids lacking
a terminal Galβ1-3GalNAc sequence are not acceptors. The conclusion
can be drawn that porcine submaxillary gland CMP-sialic acid:
Galβ1-3GalNAc α3-sialyltransferase is responsible for synthesis of
the sialylα2-3Gal linkage in mucins and ganglioside GD_{1a}; it is
also the enzyme in rat liver which makes 3'sialyllactose. This
enzyme, however, cannot account for the occurence of sialylα2-3Gal-
β1-4GlcNAc sequences in N-glycosidically linked oligosaccharides
nor for the sialylα2-3Galβ1-4Glc sequence in hematoside; these conclusions are based on the findings that Galβ1-4GlcNAc and Galβ1-4-
Glc-ceramide are very poor acceptors. Table I summarizes the sialyltransferases presently known to act on glycoproteins and glycolipids.

Since neither porcine submaxillary mucin nor any other mucin
has to date been shown to contain a sialylα2-3Galβ1-3GalNAc sequence,

TABLE I

MAMMALIAN SIALYLTRANSFERASES

Substrates Acted On	Linkage Synthesized	Tissue Sources[a]
O-glycosidic oligo-saccharides in mucins & other glycoproteins	sialylα2-6GalNAc-Ser(Thr)-	Salivary glands; intestinal mucous cells[b]; human bone marrow[b]
O-glycosidic oligo-saccharides (excluding mucins) & gang-liosides	sialylα2-3Galβl-3GalNAc-	Porcine submaxil-lary glands & liver; human bone marrow;[b] chick embryo brain
N-glycosidic oligosaccharides	sialylα2-6Galβl-4GlcNAc-	Goat, bovine & human colostrum; rat, porcine, bovine & human liver; porcine & human serum
N-glycosidic oligosaccharides	sialylα2-3Galβl-4GlcNAc-	Has been difficult to demonstrate but must exist in fetal calf liver[b]
Gangliosides	sialylα2-3Galβl-4Glc-Cer	Chick embryo brain
Gangliosides	sialylα2-8sialylα2-3Gal-	Chick embryo brain

[a]See references (25, 29) for references to these transferases.

[b]The transferase has not been detected in this tissue but a complex carbohydrate isolated from this tissue has been shown to contain the appropriate linkage.

the question of why porcine submaxillary glands have the α3-sialyl-
transferase must be briefly considered. The in vivo function of
this enzyme may be restricted to ganglioside synthesis; this does
not explain why the enzyme does not act on mucin-bound Galβl-3GalNAc
sequences. Figure 2 shows that there are at least 3 glycosyltransferases competing for Galβl-3GalNAc termini; the action of the
α2-fucosyltransferase and α6-sialyltransferase (Figure 2) in porcine submaxillary glands might deprive the α3-sialyltransferase of
substrate. Or it is possible that the α3-sialyltransferase is
simply restricted to cell types in the gland which are not involved
in mucin synthesis.

Beyer et al. (53) have studied the control of N- and O-glycosidically linked oligosaccharides. They have provided several
examples for preferential pathways of synthesis in which the prior
action of one transferase prevents or slows down the subsequent
action of another transferase. The synthesis of the tetrasaccharide
sialylα2-3Galβl-3(sialylα2-6)GalNAc- is under such control. It was
found that synthesis was 30 times faster if the α3-sialyltransferase
acts before the α6-sialyltransferase rather than the other way
around (Figure 2).

The sialylα2-3Gal- sequence may be important at cell surfaces.
It is an important constituent of components found solely or
primarily in the plasma membrane, i.e., glycophorin A, epiglycanin
and gangliosides G_{Dla} and G_{Tlb}. A recent paper by Markwell and
Paulson (58) suggests a role for this oligosaccharide sequence in
Sendai virus infection. Bovine kidney cells become resistant to
infection by Sendai virus when surface sialic acid is removed by
neuraminidase; infectivity can be restored by treating the cells
with CMP-sialic acid and α3-sialyltransferase to restore cell
surface sialylα2-3Galβl-3GalNAc- sequences. The interesting finding is that if the neuraminidase-treated cells are treated with
CMP-sialic acid and CMP-sialic acid:Galβl-4GlcNAc α6-sialyltrans-
ferase, no restoration of infectivity is observed. An earlier
study had shown that hemagglutination of erythrocytes by Sendai
virus is abolished by neuraminidase treatment and restored by
sialylation with the purified α3-sialyltransferase but not by
treatment with either CMP-sialic acid:Galβl-4GlcNAc α6-sialyltrans-
ferase or CMP-sialic acid:α-N-acetylgalactosaminide α6-sialyltrans-
ferase. It appears that Sendai virus requires the sialylα2-3Gal
sequence as a host cell receptor determinant (58); it is not known
whether glycoprotein, or glycolipid, or both, is the physiological
virus receptor in the cell membrane.

ASSEMBLY OF OLIGOSACCHARIDES WITH TYPE 2 CORE

The type 2 core is made from the type 1 core by an enzyme
recently described in canine submaxillary glands, UDP-GlcNAc:Galβl-

3GalNAc (GlcNAc→GalNAc) β6-N-acetylglucosaminyltransferase (Figure 1 and refs. 59, 60). This core structure has been described in human submaxillary mucin (13), in porcine (14), ovine (9), equine (15) and human (16) gastric mucins, and in human ovarian cyst glycoproteins (17). It is curious that canine submaxillary mucin (12) does not appear to have this core type although canine submaxillary glands have the β6-GlcNAc-transferase (59, 60); a possible explanation is that Lombart and Winzler (12) looked only at acidic components of canine submaxillary mucin and type 2 core may be present only in neutral oligosaccharides, but this hypothesis has no experimental support at this time.

The β6-GlcNAc-transferase has not been purified and crude canine submaxillary gland membranes have a specific activity of about 0.005 units per mg protein. The enzyme shows an absolute specificity for acceptors with a Galβ1-3GalNAc- terminus; however, the enzyme will work with either disaccharides or with Galβ1-3GalNAc-Ser(Thr)-proteins such as antifreeze glycoprotein and porcine submaxillary mucin which has been hydrolyzed with mild acid to remove sialic acid and fucose residues. The product of the β6-GlcNAc-transferase was identified as Galβ1-3(GlcNAcβ1-6)GalNAc- by methylation analysis, high resolution proton NMR spectroscopy and comparison with synthetic trisaccharide (60).

On studying in detail the substrate specificity of this transferase, at least two control phenomena were uncovered. Asialo-ovine submaxillary mucin (GalNAc-Ser(Thr)-protein) was found to be an ineffective acceptor indicating that the UDP-Gal:GalNAc-mucin β3-galactosyltransferase must act before the β6-GlcNAc-transferase. Also, asialo-, afuco-porcine submaxillary mucin was an appreciably better acceptor than asialo-porcine submaxillary mucin (60). The latter result showed that the presence of the Fucα1-2Gal moiety in the porcine submaxillary mucin oligosaccharides inhibited transferase activity. This conclusion was recently verified by the demonstration that synthetic Fucα1-2Galβ1-3GalNAcαphenyl was a poor acceptor when compared to non-fucosylated disaccharides (unpublished observations). Thus, biosynthesis of the tetrasaccharide Fucα1-2-Galβ1-3(GlcNAcβ1-6)GalNAc follows the preferential pathway: α-GalNAc-transferase to β3-Gal-transferase to β6-GlcNAc-transferase to α2-Fuc-transferase (Figure 3).

Reactions subsequent to the synthesis of the type 2 core (Figure 3) have not been carefully investigated. We have shown (60) that Galβ1-3(GlcNAcβ1-6)GalNAc-benzyl will accept L-fucose from GDP-Fuc in the presence of crude canine submaxillary gland membranes; this Fuc residue is probably incorporated in α2 linkage to the terminal Gal residue as shown in Figure 3 but definitive product identification has not as yet been carried out. From a consideration of the various structures that have been found in mucins (Figure 3), it is evident that at least 5, and possibly more, path-

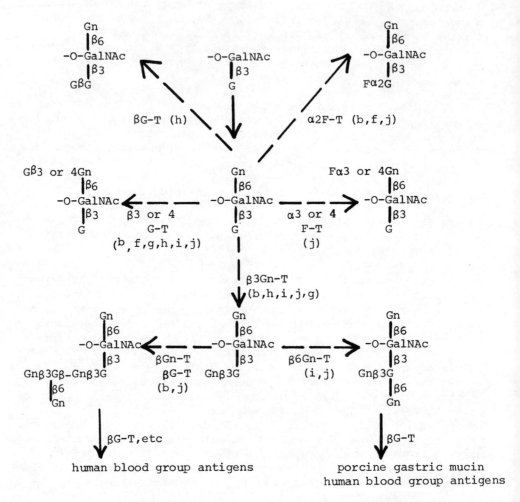

Figure 3.   Synthesis of type 2 core oligosaccharides. See legend
             to Figure 1 for abbreviations.

ways can branch off the type 2 core structure. Human blood group
determinants with A, B or H antigenicity can be added directly to
the Gal residue of the type 2 core, or to a Gal residue attached to
the GlcNAc residue of the core, or to the macromolecule described by
Kabat's group in human ovarian cyst fluids (17), as indicated in
Figure 3. The reader is referred to reviews on blood group synthesis (25, 54) for a detailed consideration of this complex topic.

ASSEMBLY OF OLIGOSACCHARIDES WITH TYPE 3 and 4 CORES

     Neither of the two GlcNAc-transferases involved in the assembly
fo type 3 and 4 cores (Figure 1) have been described. It is presumed
that there must be a β3-GlcNAc-transferase capable of making a type 3
core but nothing is known about its occurrence or substrate specificity. There must also be a β6-GlcNAc-transferase for the assembly of
the type 4 core structure. This enzyme may be identical to the β6-
GlcNAc-transferase described above for the assembly of type 2 core;
however, GlcNAcβ1-3GalNAc- terminal substrates were not available for
testing when this β6-GlcNAc-transferase was first studied (59, 60).

     Figure 4 summarizes some of the structures that have been des-
cribed containing type 3 and 4 cores. None of these enzymes have
been demonstrated in vitro. As is indicated in the legend to the
figure, type 3 cores have been described in rat mucins from the
small intestine (19), colon (20) and sublingual gland (21), in human
bronchial mucin (18) and in equine (15) and ovine (9) gastric mucins.
Type 4 core is a relatively new structure which has to date been
found only in ovine gastric mucin (9). The type 3 core can be extended by addition of sialic acid residues to GalNAc (rat colon and
sublingual gland) in α6 linkage; this may be catalyzed by the same
CMP-sialic acid: α-N-acetylgalactosaminine α6-sialyltransferase that
has been purified from porcine submaxillary glands (see above) but
the enzyme has not been tested with type 3 core substrates. Alter-
nately, type 3 core extension may begin by addition of Gal residues
to the GlcNAc residue of the core (Figure 4). The relationship of
this Gal-transferase to the well-known UDP-Gal GlcNAc β4-transferase
acting on N-glycosidic oligosaccharides (25, 29) remains to be studied. Type 4 core may also be extended by addition of β4-linked Gal
residues to either or both of its GlcNAc residues (Figure 4) and here
again very little is known about the enzymes involved.

CONCLUDING REMARKS CONCERNING THE CONTROL OF MUCIN SYNTHESIS

     The above summary is obviously no more than a progress report
on the biosynthesis of mucins. Much more information is available
on the assembly of N-glycosidically linked oligosaccharides (25,
28, 29) but many of the conclusions drawn from this area of study
may be applicable also to mucins. It was postulated many years ago
that oligosaccharide assembly required one glycosyltransferase for

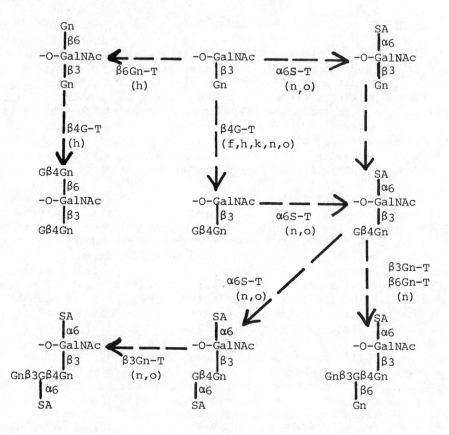

Figure 4. Synthesis of type 3 and 4 core oligosaccharides.
See legend to Figure 1 for abbreviations.

every novel linkage made, i.e., there were separate transferases
for every sugar transferred, for the type of linkage formed (i.e.,
α or β, and position of attachment to the penultimate sugar), and
for the nature of the penultimate sugar. Recent work by R.L.Hill's
group (47-53, 57) has supported this concept by showing that highly
purified glycosyltransferases do indeed have great specificity for
the nature of the linkage made. The "one linkage-one glycosyltransferase" theory is therefore a sound working hypothesis although
there may be occasional variations and exceptions.

Many of the steps shown in Figures 1 to 4, therefore, are
probably catalyzed by separate enzymes although it is likely that
some of the more peripheral and recurring linkages, e.g., Galβ1-4-
GlcNAc, GlcNAcβ1-3Gal, the various blood group determinants and so
on, are synthesized by the same transferases. The control of oligosaccharide assembly therefore resides in the presence of a particular
glycosyltransferase and its substrate specificity. In other words
a structure is made in a cell because it has a certain battery of
glycosyltransferases (multiglycosyltransferase system) and the substrate specificities of these transferases determine the oligosaccharide sequence. The situation becomes complex when a mucin contains different oligosaccharides that are obviously not assembled
in the same pathway. The only conclusion we can draw is that
whenever there is a branch point at which a single substrate can
be acted on by more than one transferase, the relative amounts
and activities of these enzymes control the distributions along
the different paths, e.g., the distribution of "ovine-like" and
"porcine-like" structures in ovine and porcine submaxillary mucins
depends on the relative activities of CMP-sialic acid: α-N-acetyl-
galactosaminide α6-sialyltransferase and UDP-Gal:GalNAc-mucin
β3-galactosyltransferase (Figure 1).

Competing pathways for a common substrate is one source of
microheterogeneity in glycoproteins. Another common source is
lack of completion of an oligosaccharide such as occurs, for example, in porcine submaxillary mucin. The latter may be due to a
lack of balance between the rate of transit of the glycoprotein
through the endomembrane system and the ability of the endomembrane
system to add sugars to the glycoprotein.

The endomembrane system undoubtedly plays a role in control
of mucin synthesis but little is known about this. It is clear
that glycosylation of mucins occurs mainly in smooth-surfaced
membranes, probably in the Golgi apparatus. Whether different
mucins are compartmentalized in different subcellular organ-
elles or in different cells remains to be investigated.

A final cautionary word is in order. The present stage of our
knowledge forces us to pool data from different organs and different

species into uniform schemes. It is possible that variations between organs and species may add even further complexities to an
already complicated field of research.

REFERENCES

1.   M. Neutra and C.P. Leblond. Synthesis of the carbohydrate of
       mucus in the Golgi complex as shown by electron microscope
       radioautography of goblet cells from rats injected with
       glucose-H^3, J.Cell Biol. 30:119-136 (1966).
2.   G. Bennett, C.P. Leblond and A. Haddad. Migration of glycopro-
       tein from the Golgi apparatus to the surface of various
       cell types as shown by radioautography after labeled
       fucose injection into rats. J.Cell Biol. 60:258-284 (1974).
3.   B. Mayrick and L. Reid. In vitro incorporation of ^{3}H-threonine
       and ^{3}H-glucose by the mucous and serous cells of the human
       bronchial submucosal gland. J.Cell Biol. 67:320-344 (1975).
4.   M.F.Kramer, J.J. Geuze and G.J.A.M. Strous. Site of synthesis,
       intracellular transport and secretion of glycoprotein in
       exocrine cells, in "Respiratory Tract Mucus", Ciba Founda-
       tion Symposium 54, pp. 25-51, Elsevier, Amsterdam (1978).
5.   N.B.Berg and B.P. Austin. Intracellular transport of sulfated
       macromolecules in parotid acinar cells. Cell Tiss.Res.
       165:215-225 (1976).
6.   J.F.Forstner. Intestinal mucins in health and disease,
       Digestion 17:234-263 (1978).
7.   J.J.Geuze and J.W. Slot. Synthesis and secretion of glyco-
       proteins in rat bulbourethral (Cowper's) gland. 1. The
       effect of copulation on the glandular content and on the
       incorporation of galactose and leucine. Biol. Reprod.
       15:118-125 (1976).
8.   A.R.Zinn, J.J. Plantner and D.M.Carlson. Nature of linkages
       between protein core and oligosaccharides, in: "The Glyco-
       conjugates", Vol. 1, M.I. Horowitz and W. Pigman, eds.,
       pp. 69-85, Academic Press, New York (1977).
9.   E.F.Hounsell, M.Fukuda, M.E. Powell, T. Feizi and S. Hakomori.
       A new O-glycosidically linked tri-hexosamine core structure
       in sheep gastric mucin: a preliminary note. Biochem. Biophys.
       Res. Communs. 92:1143-1150 (1980).
10. H.D.Hill, Jr., J.A. Reynolds, and R.L. Hill. Purification,
       composition, molecular weight, and subunit structure of
       ovine submaxillary mucin. J.Biol.Chem. 252:3791-3798 (1977).
11. D.M. Carlson. Structures and immunochemical properties of
       oligosaccharides isolated from pig submaxillary mucins.
       J.Biol.Chem. 243:616-626 (1968).

12. C.G.Lombart and R.J. Winzler. Isolation and characterization
     of oligosaccharides from canine submaxillary mucin. Eur.J.
     Biochem. 49:77-86 (1974).
13. M.S.Reddy, R.H. Shah and O.P. Bahl. Structures of the carbo-
     hydrate units of mucins from normal and fibrocystic human
     submaxillary secretions. Proc.Ann.Meeting Soc.Complex
     Carbohydrates, Toronto, abstract No. 40 (1979).
14. V.A.Derevitskaya, N.P. Arbatsky and N.K. Kochetkov. The
     structure of carbohydrate chains of blood-group substance.
     Isolation and elucidation of the structure of higher
     oligosaccharides from blood-group substance H. Eur.J.
     Biochem. 86:423-437 (1978)
15. W. Newman and E.A. Kabat. Immunochemical studies on blood
     groups. Structures and immunochemical properties of nine
     oligosaccharides from B-active and non-B-active blood
     group substances of horse gastric mucosae. Arch.Biochem.
     Biophys. 172:535-550 (1976).
16. M.D.G.Oates, A.C. Rosbottom and J. Schrager. Further in-
     vestigations into the structure of human gastric mucin:
     the structural configuration of the oligosaccharide chains.
     Carbohyd.Res. 34:115-137 (1974).
17. L. Rovis, B. Anderson, E.A. Kabat, F. Gruezo and J. Liao.
     Structures of oligosaccharides produced by base-borohydride
     degradation of human ovarian cyst blood group H, Le[b] and Le[a]
     active glycoproteins. Biochemistry 12:5340-5354 (1973).
18. G. Lamblin, M. Lhermitte, A. Boersma, P. Roussel and
     V. Reinhold. Oligosaccharides of human bronchial glyco-
     proteins. Neutral di- and tri-saccharides isolated from
     a patient suffering from chronic bronchitis. J.Biol.Chem.
     255:4595-4598 (1980).
19. H.E.Carlsson, G. Sundblad, A. Hammarström and J. Lonngren.
     Structure of some oligosaccharides derived from rat in-
     testinal glycoproteins. Carbohyd.Res. 64:181-188 (1978).
20. B.L.Slomiany, V.L.N. Murty and A. Slomiany. Isolation and
     characterization of oligosaccharides from rat colonic
     mucus glycoprotein. J.Biol.Chem. 255:9719-9723 (1980).
21. A. Slomiany and B.L.Slomiany. Structures of the acidic
     oligosaccharides isolated from rat sublingual glyco-
     protein. J.Biol.Chem. 253:7301-7306 (1978).
22. A. Herp, A.M. Wu and J. Moschera. Current concepts of the
     structure and nature of mammalian salivary mucous glyco-
     proteins. Mol.Cell Biochem. 23:27-43 (1979).
23. E.J.McGuire and S. Roseman. Enzymatic synthesis of the
     protein-hexosamine linkage in sheep submaxillary mucin
     J.Biol.Chem. 242:3745-3747 (1967).
24. A. Hagopian, F.C. Westall, J.S. Whitehead and E.H. Eylar.
     Glycosylation of the Al protein from myelin by a polypep-
     tide N-acetylgalactosaminyltransferase. Identification
     of the receptor sequence. J.Biol.Chem. 246:2519-2523
     (1971).

25. H. Schachter. Glycoprotein Biosynthesis, in: "The Glyco-
    conjugates", Vol. II, M.I. Horowitz and W. Pigman, eds.,
    pp. 87-181, Academic Press, New York (1978).
26. H.D.Hill, Jr., M. Schwyzer, H.M. Steinman and R.L. Hill.
    Ovine submaxillary mucin. Primary structure and peptide
    substrates of UDP-N-acetylgalactosamine: mucin transferase.
    J.Biol.Chem. 252:3799-3804 (1977).
27. J.D.Young, D. Tsuchiya, D.E. Sandlin and M.J. Holroyde.
    Enzymic O-glycosylation of synthetic peptides from sequences
    in basic myelin protein. Biochemistry 18:4444-4448 (1979).
28. D.K.Struck and W.J. Lennarz. The function of saccharide-
    lipids in synthesis of glycoproteins, in: "The Biochem-
    istry of Glycoproteins and Proteoglycans", W.J. Lennarz,
    ed., pp. 35-83, Plenum Press, New York (1980).
29. H. Schachter and S. Roseman. Mammalian Glycosyltransferases.
    Their role in the synthesis and function of complex carbo-
    hydrates and glycolipids, in: "The Biochemistry of Glyco-
    proteins and Proteoglycans", W.J. Lennarz, ed., pp. 85-
    160, Plenum Press, New York (1980).
30. J.A.Hanover, W.J. Lennarz and J.D. Young. Synthesis of N-
    and O-linked glycopeptides in oviduct membrane preparations.
    J.Biol.Chem. 255:6713-6716 (1980).
31. P. Babczinski. Evidence against the participation of lipid
    intermediates in the in vitro biosynthesis of serine-
    (threonine)-N-acetyl-D-galactosamine linkages in sub-
    maxillary mucin. FEBS Letters 117:207-211 (1980).
32. P. Babczinski and W. Tanner. Involvement of dolichol mono-
    phosphate in the formation of specific mannosyl linkages
    in yeast glycoproteins. Biochem.Biophys.Res.Communs.
    54:1119-1124 (1973).
33. L. Lehle and W. Tanner. Biosynthesis and characterization
    of large dolichyl diphosphate-linked oligosaccharides in
    saccharomyces cervisiae, Biochim.Biophy.Acta 539:218-
    229 (1978).
34. Y.S.Kim, J. Perdomo and J. Nordberg. Glycoprotein biosyn-
    thesis in small intestinal mucosa. I. A study of glycosyl-
    transferases in microsomal subfractions. J.Biol.Chem.
    246:5466-5476 (1971).
35. G.K.W. Ko and E. Raghupathy. Glycoprotein biosynthesis in the
    developing rat brain. II. Microsomal galactosaminyltrans-
    ferase utilizing endogenous and exogenous protein acceptors.
    Biochim.Biophy.Acta 264:129-143 (1972).
36. A. Hagopian, H.B. Bosmann and E.H. Eylar. Glycoprotein bio-
    synthesis: the localization of polypeptidyl: N-acetylgala-
    ctosaminyl, Collagen:glycosyl, and Glycoprotein:galactosyl
    Transferase in HeLa cell membrane fractions. Arch.Biochem.
    Biophys. 128:387-396 (1968).
37. G.JA.M. Strous. Initial glycosylation of proteins with acetyl-
    galactosaminylserine linkages. Proc.Nat.Acad.Sci.USA 76.
    2694-2698 (1979).

38. H. Schachter, E.J. McGuire and S. Roseman. Sialic acids. XIII.
    A uridine diphosphate D-galactose:mucin galactosyltrans-
    ferase from porcine submaxillary gland. J.Biol.Chem.
    246:5321-5328 (1971).
39. D.H.Van den Eijnden, R.A. Barneveld and W.E.C.M. Schiphorst.
    Structure of the disaccharide chain of galactosyl-N-
    acetylgalactosaminyl-protein synthesized in vitro.
    Eur.J.Biochem. 95:629-637 (1979).
40. D.M. Carlson, J. David and W.J. Rutter. Galactosyltransferase
    activities in pancreas, liver and gut of the developing
    rat. Arch.Biochem.Biophys. 157:605-612 (1973).
41. W.T. Shier and G.J. Roloson. Preparation and galactosyltrans-
    ferase acceptor activities of derivatives of anti-freeze
    glycoproteins of an antarctic fish. Can.J.Biochem.
    55:886-893 (1977).
42. G.N. Andersson and L.C. Eriksson. Studies on the latency of
    UDP-galactose:asialo-mucin galactosyltransferase activity
    in microsomal and Golgi subfractions from rat liver.
    Biochim.Biophys.Acta 600:571-576 (1980).
43. E.J. McGuire. Biosynthesis of submaxillary mucins, in:
    "Blood and Tissue Antigens", D. Aminoff, ed., pp. 461-478,
    Academic Press, New York (1970).
44. D.B. Thomas and R. J. Winzler. Structural studies on human
    erythrocyte glycoproteins. Alkali-liable oligosaccharides.
    J.Biol.Chem. 244:5943-5946 (1969).
45. M.F. Kramer and J.J. Geuze. Comparison of various methods to
    localize a source of radioactivity in ultrastructural
    autoradiographs. The site of (^{3}H)-galactose incorporation
    in surface mucous cells of the rat stomach. J.Histochem.
    Cytochem. 28:381-387 (1980).
46. D.M. Carlson, E.J. McGuire, G.W. Jourdian and S. Roseman.
    The Sialic Acids. XVI. Isolation of a mucin sialytrans-
    ferase from sheep submaxillary gland. J.Biol.Chem.
    248:5763-5773 (1973).
47. M. Schwyzer and R.L. Hill. Porcine A blood group-specific
    N-acetylgalactosaminyltransferase I. Purification from
    porcine submaxillary glands. J.Biol.Chem. 252:2338-2345
    (1977).
48. M. Schwyzer and R.L. Hill. Porcine A blood group-specific
    N-acetylgalactosaminyltransferase II. Enzymatic properties.
    J.Biol.Chem. 252:2346-2355 (1977).
49. J.E. Sadler, J.I. Rearick, J.C. Paulson and R.L. Hill. Puri-
    fication to homogeneity of a β-galactoside α2-3 sialyl-
    transferase and partial purification of an α-N-acetylgala-
    ctosaminide α2-6 sialyltransferase from porcine submaxillary
    glands.J.Biol.Chem. 254:4434-4442 (1979).
50. J.E. Sadler, J.I. Rearick and R.L. Hill. Purification to
    homogeneity and enzymatic characterization of an α-N-acetyl-
    galactosaminide α2-6 sialyltransferase from porcine sub-
    maxillary glands. J.Biol.Chem. 254:5934-5941 (1979).

51. T.A. Beyer, J.E. Sadler and R.L. Hill. Purification to homo-
     geneity of the H blood group β-galactoside α1-2 fucosyl-
     transferase from porcine submaxillary gland. J.Biol.Chem.
     255:5364-5372 (1980).
52. T.A. Beyer and R.L. Hill. Enzymic properties of the β-
     galactoside α1-2 fucosyltransferase from porcine submaxil-
     lary gland. J.Biol.Chem. 255:5373-5379 (1980).
53. T.A. Beyer, J.I. Rearick, J.C. Paulson, J-P. Prieels,
     J.E. Sadler and R.L. Hill. Biosynthesis of mammalian
     glycoproteins. Glycosylation pathways in the synthesis
     of the nonreducing terminal sequences. J.Biol.Chem.
     254:12531-12541 (1979).
54. W.M. Watkins. Biochemistry and genetics of the ABO, Lewis and
     P blood group systems, in: "Advances in Human Genetics",
     Vol. 10, H. Harris and K. Hirschhorn, eds., pp. 1-136, 379-
     385, Plenum Publishing Corp., New York and London (1980).
55. E. Wood, E.F. Hounsell, J. Langhorne and T. Feizi. Sheep
     gastric mucins as a source of blood group-I and -i
     antigens. Biochem.J. 187:711-718 (1980).
56. F. Maisonrouge-McAuliffe and E.A. Kabat. Immunochemical
     studies on blood groups. Structures and immunochemical
     properties of oligosaccharides from two fractions of blood
     group substance from human ovarian cyst fluid differing
     in B, I and i activities and reactivity toward Concanavalin
     A. Arch.Biochem.Biophys. 175:90-113 (1976).
57. J.I. Rearick, J.E. Sadler, J.C. Paulson and R.L. Hill.
     Enzymatic characterization of β-D-galactoside α2-3 sialyl-
     transferase from porcine submaxillary gland. J.Biol.Chem.
     254:4444-4451 (1979).
58. M.A.K. Markwell and J.C. Paulson. Sendai virus utilizes
     specific sialyloligosaccharides as host cell receptor
     determinants. Proc.Nat.Acad.Sci.USA 77:5693-5697 (1980).
59. D. Williams and H. Schachter. Mucin synthesis. I. Detection
     in canine submaxillary glands of an N-acetylglucosaminyl-
     transferase which acts on mucin substrates. J.Biol.Chem.
     255:11247-11252 (1980).
60. D. Williams, G. Longmore, K.L. Matta and H. Schachter.
     Mucin synthesis. II. Substrate specificity and product
     identification studies on canine submaxillary gland UDP-
     GlcNAc:Galβ1-3GalNAc (GlcNAc→GalNAc) β6-N-acetylglucosa-
     minyltransferase. J.Biol.Chem. 255:11253-11261 (1980).
61. A.M. Wu, A. Slomiany, A. Herp and B.L. Slomiany. Structural
     studies on the carbohydrate units of armadillo submandi-
     bular glycoprotein. Biochim.Biophys.Acta 578:297-304
     (1979).
62. D. Ziderman, S. Gompertz, Z.G. Smith and W.M. Watkins.
     Glycosyltransferases in mammalian gastric mucosal linings.
     Biochem.Biophys.Res.Communs. 29:56-61 (1967).

ANTIGENICITIES OF MUCINS - THEIR RELEVANCE TO TUMOUR ASSOCIATED AND STAGE SPECIFIC EMBRYONIC ANTIGENS

Ten Feizi

Applied Immunochemistry Research Group, Division
of Communicable Diseases, Clinical Research Centre
Harrow, Middlesex, England

SECRETED GLYCOPROTEINS (MUCINS) HAVE PROVIDED FUNDAMENTAL INFORMATION ON THE NATURE OF THE MAJOR BLOOD GROUP ANTIGENS AND THEIR PRECURSORS

Because of their relative abundance compared with membrane associated glycoconjugates of cells, secreted glycoproteins (mucins) have been the major source of oligosaccharides for defining the chemical nature of the blood group A, B, H, Lewis[a] and Lewis[b] antigens (reviewed in refs 1 and 2). It is now known that blood group antigenic determinants expressed on membrane-associated glycosphingolipids are the same as those on mucins (3). The patterns of antigen expression on these two categories of macromolecule are determined by the same blood group genes although their presence on secreted glycoproteins may in addition, be subject to the action of regulator genes such as the secretor gene (1).

Similarly, the first indication that the oligosaccharide backbones (precursor chains) bearing these blood group structures may be antigens in their own right, was derived from studies of secreted glycoproteins which were partially degraded to remove their outermost (non-reducing) monosaccharide residues. Glycoproteins thus treated react with horse antisera to Type XIV pneumococcal polysaccharide (1). These precursor chains consist of alternating galactose and N-acetylglucosamine residues joined by one of two types of linkage:

$$Gal\beta1 \longrightarrow 3GlcNAc \qquad Type\ 1$$

$$or \qquad Gal\beta1 \longrightarrow 4GlcNAc \qquad Type\ 2$$

29

Both types of precursor chain exist in human ovarian and gastric
mucosal glycoproteins and antisera to Type XIV pneumococcal
polysaccharide react more strongly with the Type 2 precursor
chains (1).

BLOOD GROUP I AND i ANTIGENS ARE EXPRESSED ON TYPE 2 PRECURSOR CHAINS AND THE EXPRESSION OF I ANTIGEN OF 'MA TYPE' IN GASTRIC MUCINS IS GENETICALLY DETERMINED

Studies with the monoclonal autoantibodies which occur in
the sera of patients with cold agglutinin disease (anti-I and
anti-i cold agglutinins) have shown that the I and i antigens
are expressed on oligosaccharide precursors of the blood group
ABH antigens (4). By antigenic analyses of ovarian cyst
glycoproteins from persons who are secretors and non-secretors
of blood group A, B and H (4), oligosaccharides derived from
these (4) and, more recently, glycosphingolipids (hexa-octagly-
cosylceramides) obtained from erythrocyte membranes (5-7), it
has been established that the I and i antigenic determinants are
built of Type 2 precursor chains; i antigen is expressed on the
repeating linear sequence:

$Gal\beta1 \longrightarrow 4GlcNAc\beta1 \longrightarrow 3Gal\beta1 \longrightarrow 4GlcNAc\beta1 \longrightarrow 3Gal....$

and I on the related branched structure

$Gal\beta1 \longrightarrow 4GlcNAc\beta1$

$\searrow 6$

$_{3}Gal\beta1 \longrightarrow 4GlcNAc\beta1 \longrightarrow 3Gal....$

$\nearrow$

$Gal\beta1 \longrightarrow 4GlcNAc\beta1$

Gastric mucosal glycoproteins of different animal species
differ in their I and i antigen activities. These differences
are in part due to differences in (a) the proportions of Type 1
and Type 2 precursor chains and (b) their length and degree of
branching (reviewed in ref 8). Furthermore, when Ii structures
are additionally glycosylated in their peripheral regions and
converted into the blood group H, A or B active structures as
shown below, the I and i activities become masked (4-7).

H
$$\begin{array}{c} Fuc\alpha \\ \downarrow 1,2 \\ Gal\beta1 \longrightarrow 4GlcNAc\beta1 \longrightarrow \end{array}$$

A
$$\begin{array}{c} Fuc\alpha \\ \downarrow 1,2 \\ GalNAc\alpha1 \longrightarrow 3Gal\beta1 \longrightarrow 4GlcNAc\beta1 \longrightarrow \end{array}$$

B
$$\begin{array}{c} Fuc\alpha \\ \downarrow 1,2 \\ Gal\alpha1 \longrightarrow 3Gal\beta1 \longrightarrow 4GlcNAc\beta1 \longrightarrow \end{array}$$

Thus in a given individual, the blood group antigen and
regulator genes such as the secretor gene are among factors
which determine to what extent the Ii structures are accessible
for reaction with anti-I and anti-i antibodies.

It has been shown by this laboratory that the expression of
one of the I antigenic determinants (Galβ1⟶4GlcNAcβ1⟶6Gal)
on gastric mucosal glycoproteins is genetically determined (9)
(Fig. 1A). This antigenic determinant, termed I(Ma), is recognized by the monoclonal anti-I antibody of a patient (Ma), and
it is usually strongly expressed in the gastric mucosal glycoproteins of non-secretors.   In secretors we envisage that this
determinant is masked by the blood group ABH monosaccharides.

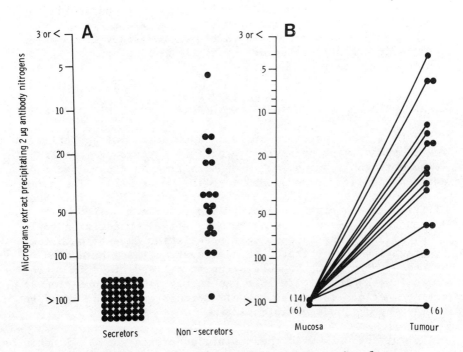

Fig. 1 Blood group I (Ma) activities of gastric glycoproteins determined by quantitative precipitin assays.
A, gastric mucosal glycoproteins of healthy persons;
B, uninvolved mucosae and tumours of 20'secretors' with
gastric cancer. Taken from refs 9 and 10 by permission.

CHANGES IN ANTIGENICITY OF GASTRIC MUCOSAL CLYCOPROTEINS
ASSOCIATED WITH CANCER

In contrast to the glycoprotein extracts of non-neoplastic
tissues, those obtained from gastric cancer tissues of secretors
usually have strong I antigen activity of Ma type (9,10)
(Fig. 1B). This is presumably due to incomplete biosynthesis of
the blood group ABH structures with resulting exposure of
precursor chains. Thus there is a change in 'antigenic
phenotype' for the glycoprotein extracts obtained from tumour
tissues of secretors resemble the mucosal extracts of
non-secretors.

By immunofluorescence studies it has been confirmed that
the tumour cells are the source of the I(Ma) antigen activities
in the extracts (11). However, the immunofluorescence studies
have revealed focal areas of abnormal antigen expression in
histologically normal looking areas of the gastric mucosa, not
only of patients with gastric cancer, but also of certain
patients with chronic benign peptic ulceration. Further
investigations are required to determine whether these focal
areas of antigenic atypia reflect precancerous changes in the
mucosal cells. We are currently comparing the antigenicity of
gastric juice glycoproteins obtained at endoscopy from patients
with various gastric disorders with those from apparently
healthy volunteers.

Ii AND RELATED ANTIGENS AS STAGE SPECIFIC EMBRYONIC ANTIGENS

The I and i antigens are well known as developmentally
regulated antigens on human erythrocytes (reviewed in ref. 6).
We have recently shown that they also rank as stage specific
antigens in early embryogenesis in the mouse (12). The earliest
of mouse embryos express I antigen but not i (Fig. 2). When
differentiation first occurs, primary endoderm cells can be
discerned and these cells express i antigen in addition to I
(Fig. 3). Later stages of differentiation have been followed
by observing mouse teratocarcinoma systems which have been
induced to mimic embryogenesis and to differentiate in vitro for
up to 7 weeks in culture. It was observed that expression of I
antigen became restricted to only certain epithelial areas and i
antigen became undetectable (12).

Simultaneously with the restricted expression of Ii antigens
in the developing embryos, there occurs increased expression of
blood group H antigen (12). It is possible that the
'disappearance' of the Ii antigens is in part due to their being
masked by the blood group H associated fucose residues.

Fig. 2. Indirect
immunofluorescence
staining with anti-I
(Ma) of the surface
of two 2-cell mouse
embryos (A) and
corresponding phase
contrast (B); pb
denotes polar body
corresponding to each
embryo. Taken from
ref 12 by permission.

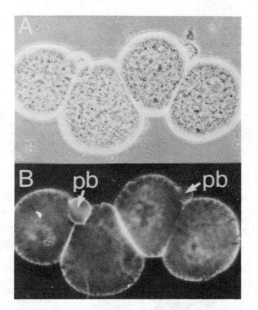

Recent studies from this laboratory (15) have shown that the
stage specific embryonic antigen, SSEA-1, defined by a hybridoma
antibody (16), is also closely related to Ii antigens. This
antigenic determinant appears at the 8-cell stage of mouse
embryos and during differentiation its expression becomes restricted, being confined to only a minority of adult cells (17).

We have shown that anti-SSEA-1 binds to a glycoprotein-rich
extract of human meconium (15). In elucidating the carbohydrate
sequence recognized by this antibody, inhibition of binding
assays were performed using ovarian cyst glycoproteins with
defined ABH, Ii, Le^a and Le^b activities. The SSEA-1
determinant was expressed most strongly in ovarian cyst
glycoproteins lacking ABH activities but was shown to involve a
fucose residue with a linkage other than those associated with
ABH, Le^a and Le^b activities (15). The most potent inhibitor
was an oligosaccharide derived from an ovarian cyst glycoprotein
having the terminal sequence:

$$Gal\beta \longrightarrow 4GlcNAc \longrightarrow$$
$$\uparrow \alpha 1,3$$
$$Fuc$$

Thus it can be envisaged that I and i active structures can be
converted into SSEA-1 by a single fucosylation step.

In Figure 4 we present schematically the way single glycosylation steps can result in changing antigenicities at
different stages of development and differentiation.

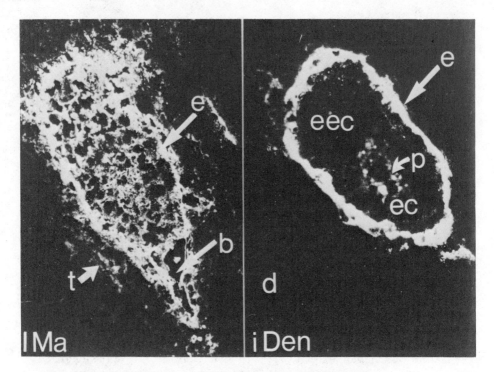

Fig. 3. Indirect immunofluorescence staining of
cryostat sections of a 5-day mouse embryo surrounded
by maternal decidua. Consecutive sections have been
stained with anti-I (Ma) and anti-i (Den) respectively.
The entire embryo including trophoblast is stained with
anti-I. In contrast only primary endoderm and the luminal
aspects of ectoderm cells lining the pro-amniotic cavity
are stained with anti-i. Symbols, d: maternal decidua; t:
trophoblast; e: endoderm; ec: embryonic ectoderm; eec:
extra-embryonic ectoderm; p: cells lining the pro-amniotic
cavity; b: blastocoelic cavity. Magnification X 270 taken
from ref 14 with permission.

Antigen activity

	I	i	SSEA	H	Other

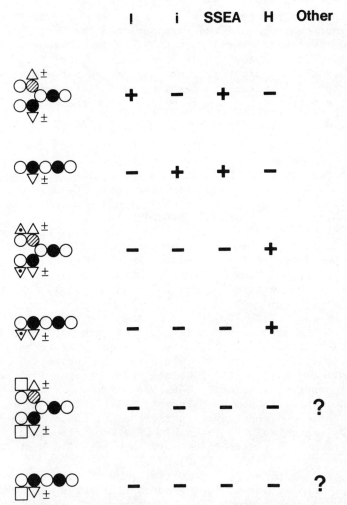

Fig. 4 Schematic presentation of the proposed structure for the SSEA-1 antigenic determinant and its relationship to blood group, I, i or H active oligosaccharides and to sialylated structures of undetermined antigenic specificity. Symbols: ○ , β1—→4 linked galactose; ● , β1—→3 linked N-acetyl glucosamine; ⊘ , β1—→6 linked N-acetyl glucosamine; △ , α1—→3, linked fucose; ⊿ , α1—→2 linked fucose; □ , α2—→3 linked sialic acid; + implies optional substitution with α1—→3 linked fucose, such that in the absence of other peripheral monosaccharides, either the I, i or the SSEA-1 antigenic determinants are expressed. It is not known whether the α1—→2 fucosylation or the sialyla-lation steps precede or follow the α1—→3 fucosylation of these precursor chains. Taken from ref. 15 by permission.

SUMMARY

     It can now be envisaged that there exists a distinct set of differentiation antigens formed by linear (i type) or branched (I-type) oligosaccharides and their fucosylated (and perhaps sialylated) derivatives. Several of these antigens are expressed in secreted glycoproteins (mucins) which are relatively abundantly available. Thus mucins can serve as valuable sources of oligosaccharide haptens of differentiation antigens just as they have done in the past with the major blood group antigens.

REFERENCES

1. W. M. Watkins, Biochemistry and genetics of the ABO, Lewis and P blood group systems, in "Advances in Human Genetics", H. Harris, K. Hirschhorn, eds. Vol.10. Plenum Publishing Co. pp 1 and 379 (1980).
2. E. A. K. Kabat, The carbohydrate moiety of the water-soluble human A, B, H, Le^a and Le^b substances, in: "Blood and Tissue Antigens" p 187. Academic Press, New York (1970).
3. S. -I. Hakomori, A. Kobata, Blood Group Antigens. In The Antigens. M. Sela, ed Academic Press, New York. Vol. II; p 79 (1974)
4. T. Feizi, E. A. Kabat, G. Vicari, B. Anderson and W. L. Marsh. Immunochemical studies on blood groups XLIX. The I antigen complex : specificity differences among anti-I sera revealed by quantitative precipitin studies; partial structure of the I determinant specific for one anti-I serum. J. Immunol. 106:1578 (1971).
5. H. Niemann, K. Watanabe, S. Hakomori, R.A. Childs, and T. Feizi. Blood group i and I activities of "Lacto-N-nor-hexaosyl ceramide" and its analogues: the structural requirements for i-specificities. Biochem. Biophys. Res. Commun. 81, 1286-1293 (1978).
6. K. Watanabe, S. Hakomori, R.A. Childs, and T. Feizi, Characterization of a blood group I-active ganglioside. Structural requirements for I and i specificities. J. Biol. Chem. 254:3221 (1979).
7. T. Feizi, R.A. Childs, K. Watanabe, and S. Hakomori, Three types of blood group I specificity among monoclonal anti-I autoantibodies revealed by analogues of a branched erythrocyte glycolipid. J. Exp. Med. 149:975 (1979).
8. E. F. Hounsell, E. Wood, T. Feizi, M. Fukuda, M. E. Powell and S. Hakomori, Structural analysis of hexa-octasaccharide fractions isolated from blood group I and i active sheep gastric glycoproteins. Carbohyd. Res. 90:283 (1981) .
9. J. Picard, D. Waldron-Edward and T. Feizi, Changes in the expression of the blood group A,B,H, Le^a and Le^b antigens and the blood group precursor associated I (Ma)

antigen in glycoprotein-rich extracts of gastric
carcinomas. J. Clin. & Lab. Immunol. 1:119 (1978).

10. T. Feizi, J. Picard, A. Kapadia and G. Slavin. Changes in
the expression of the major blood group antigens ABH and
their precursor antigens Ii in human gastri cancer
tissues. In: Protides of Biological Fluids 27th
Colloquium. Ed. H. Peeters Pergamon Press Oxford and New
York. pp. 221. (1979).

11. A. Kapadia, T. Feizi, D. Jewell, J. Keeling, and G. Slavin.
Immunocytochemical studies of blood group A,H,I and i
antigens in gastric mucosae of infants with normal
gastric histology and of patients with gastric carcinoma
and chronic benign peptic ulceration. J. Clin Pathol.
34, 320 (1981).

12. A. Kapadia, T. Feizi, and M.J. Evans, Changes in the expression and polarization of blood group I and i antigens in
post-implantation embryos and teratocarcinomas of mouse
associated with cell differentiation. Exp. Cell. Res.
131:185 (1981).

13. T. Feizi, A. Kapadia, and H.C. Gooi. Human monoclonal
autoantibodies detect changes in expression and polarization of the Ii antigens during cell differentiation in
early mouse embryos and teratocarcinomas. In:
Teratocarcinoma and Cell Surface. Eds. Muramatsu, T. and
Ikawa, Y. (Japan Scientific Socities Press.) In press.

14. T. Feizi. Structural and biological aspects of blood group
I and i antigens on glycolipids and glycoproteins. Blood
Transfusion and Immunohaematology, 23:563. (1980).

15. H. C. Gooi, T. Feizi, A. Kapadia, B. B. Solter, D. and M.J.
Evans. Stage specific embryonic antigen (SSEA-1)
involves 3-fucosylated Type 2 precursor chain. Nature.
In press.

16. D. Solter and B.B. Knowles. Monoclonal antibody defining a
stage-specific mouse embryonic antigen (SSEA-1). Proc.
Natl. Acad. Sci. USA 75:5565. (1978).

17. N. Fox, I. Damjanov, and A. Martinez-Hernandez. Immuno-
histochemical localization of the early embryonic antigen
(SSEA-1) in postimplantation mouse embryos, and fetal and
adult tissues. Develop. Biol. In press.

STRUCTURAL AND ANTIGENIC DIVERSITY IN MUCIN CARBOHYDRATE CHAINS

Elizabeth F. Hounsell, Alexander M. Lawson and
Ten Feizi

Divisions of Communicable Diseases and Clinical
Chemistry, Clinical Research Centre, Harrow
Middlesex, England

There is considerable information on the structure and genetics of the peripheral regions of the complex carbohydrate chains of mucins which vary depending on blood group isotype specificity (1). However, much less is known of the detailed structures and biosynthetic control within the backbone and core regions of these chains. The backbones consist of either Type 1,

$$Gal\beta 1 \longrightarrow 3GlcNAc\beta 1 \longrightarrow, \text{ or Type 2, } Gal\beta 1 \longrightarrow 4GlcNAc\beta 1 \longrightarrow,$$

sequences (1) and they vary in length and branching. Type 2 sequences in the liner repeating structure

$$Gal\beta 1 \longrightarrow 4GlcNAc\beta 1 \longrightarrow 3Gal\beta 1 \longrightarrow 4GlcNAc\beta 1 \longrightarrow$$

are recognized by anti-blood group i antibodies (2). Addition of a $Gal\beta \longrightarrow 4GlcNAc\beta 1 \longrightarrow 6$ branch onto the internal galactose of this structure confers blood group I activity (3). Type 1 sequences are apparently not involved in the Ii antigenic system. Species and individual differences in blood group I and i activities may therefore reflect differences in branching patterns and the proportions of Type 1 and Type 2 chain sequences. Ii activities may also be masked by further glycosylation, for example by the addition of sugar residues confering blood group ABH activities. We have examined these various structural parameters in mucous glycoprotein preparations of different species and of varying blood group I, i, A, B and H activities in order to compare their structural and antigenic diversities.

The glycoprotein preparations analysed were (a) sheep gastric mucins; two fractions were obtained enriched for and depleted of Ii activities by affinity chromatography, (b) blood group H active hog gastric mucin which is known to express some I but no i activity, (c) a pool of human blood group B-rich meconium, and (d) an H-active human ovarian cyst glycoprotein. The glycoprotein preparations were subjected to mild alkaline borohydride degradation (4) to release the alkali labile chains intact. The dialysable oligosaccahrides were analysed by gas-liquid chromatography/mass spectrometry of partially O-methyl-ated alditol acetate derivatives. This analysis procedure gave information on the type of linkage of each monosaccharide in the carbohydrate chains as follows.

The N-acetylgalactosamine residue (GalNAc) originally attached to protein is readily identified. Some of these GalNAc residues are not further substituted indicating the presence of very short chains: GalNAc⟶ protein. Others are substituted at their C3 position and still others are di-substituted at C3 and C6. These substitutions probably correspond to the following core region structures so far described (reviewed in ref.5).

$$
\text{Gal}\beta1\longrightarrow 3\text{GalNAc,}
\qquad
\begin{array}{l}
\text{Gal}\beta1\searrow_3 \\
\text{GalNAc,} \\
\text{GlcNAc}\beta1\nearrow^6
\end{array}
\qquad
\begin{array}{l}
\text{GlcNAc}\beta1\searrow_6 \\
\text{GalNAc} \\
\text{GlcNAc}\beta1\nearrow^3
\end{array}
$$

Branching can also occur at intra-chain galactose residues. Most commonly galactose disubstituted at C3 and C6 is observed which corresponds to the following branching patterns:

$$
\begin{array}{l}
\text{GlcNAc}\beta1\searrow_3 \\
\text{Gal}\beta1\longrightarrow \\
\text{GlcNAc}\beta1\nearrow^6
\end{array}
\qquad
\begin{array}{l}
\text{Gal}\beta1\searrow_3 \\
\text{Gal}\beta1\longrightarrow \\
\text{GlcNAc}\beta1\nearrow^6
\end{array}
\qquad
\begin{array}{l}
\text{Gal}\beta1\searrow_3 \\
\text{Gal}\beta1\longrightarrow \\
\text{Gal}\beta1\nearrow^6
\end{array}
$$

     (ref.6)   (ref.7)   (ref.8)

Galactose in the chain periphery disubstituted at C2 and C3 is found as expected in blood group A or B active oligosaccharides. Monosubstituted galactose from the following chain terminating structures: Fucα⟶2Galβ1⟶ (blood group H active), GlcNAc(α or β)1⟶2 or 4Galβ1⟶ (unknown antigenic activities), are also observed. Galactose residues in the backbone regions are substituted at C3 and they alternate with N-acetyl-glucosamine residues. The ratio of Type 1 to Type 2 sequences can be determined by the relative proportions of N-acetyl-glucosamine monosubstituted at C3 and C4 respectively.

The average length of chains can also be determined by the amount of monosubstituted galactose or N-acetylglucosamine and the relative amount of core region N-acetylgalactosamine. However, such estimates of chain length, like those of total carbohydrate composition, give little comparative information when whole mucous glycoproteins are examined. The observed differences in branching patterns and in the ratio of Type 1 / Type 2 sequences corresponded to antigenic differences in the mucosal glycoproteins studied. Such analyses complement detailed structural characterisation of purified oligosaccharide chains. Continued integration of the various analytical methods discussed will give valuable information on the structural diversity of mucins and facilitate the interpretation of antigenic changes found in disease (9).

1. W.M. Watkins, Biochemistry and genetics of the ABO, Lewis
    and P blood group systems, in: Advances in Human Genetics,
    Hirschhorn K, Harris, H., eds. Vol. 10. Plenum Publishing
    Co. pp 1 and 379 (1980).
2. H. Niemann, K. Watanabe, S. Hakomori, R.A. Childs, and
    T. Feizi, Blood group i and I activities of "Lacto-N-nor-
    hexaosyl-ceramide" and its analogues: the structural
    requirements for i-specificities. Biochem. Biophys. Res.
    Commun. 81: 1286 (1978).
3. T. Feizi, R.A. Childs, K. Watanabe, and S. Hakomori, Three
    types of blood group I specificity among monoclonal anti-I
    autoantibodies revealed by analogues of a branched
    erythrocyte glycolipid. J. Exp. Med. 149: 975 (1979).
4. R.N. Iyer and D.M. Carlson. Alkaline borohydride
    degradation of blood group H substance. Arch. Biochem.
    Biophys. 142: 101 (1971).
5. E.F. Hounsell, M. Fukuda, M.E. Powell, T. Feizi, and
    S. Hakomori. A new O-glycosidically linked tri-hexosamine
    core structure in sheep gastric mucin: a preliminary note.
    Biochem. Biophys. Res. Commun. 92: 1143 (1980).
6. V.A. Derevitskaya, N.P. Arbatsky, and N.K. Kochetkov. The
    structure of carbohydrate chains of blood-group substance.
    Eur. J. Biochem. 86: 423 (1978).
7. E.F. Hounsell, E. Wood, T. Feizi, M. Fukuda, M.E. Powell,
    and S. Hakomori. Hexa-octasaccharide fractions isolated
    from blood group I and i active sheep gastric
    glycoproteins Carbohyd. Res. 90: 283 (1981).
8. B.L. Slomiany and K. Meyer. Oligosaccharide produced by
    acetolysis of blood group active (A + H) sulfated
    glycoproteins from hog gastric mucin. J. Biol. Chem. 248:
    2290 (1973).
9. J. Picard, D. Waldron-Edward, and T. Feizi. Changes in the
    expression of the blood group A,B,H, Le[a] and Le[b]
    antigens and the blood group precursor associated I(Ma)
    antigen in glycoprotein-rich extracts of gastric
    carcinomas. J. Clin. & Lab. Immunol. 1: 119 (1978).

THE USE OF RADIOIMMUNOASSAY AS A PROBE OF THE ANTIGENIC

DETERMINANTS AND STRUCTURE OF HUMAN SMALL INTESTINAL MUCIN

M. Mantle, A.W. Wesley, J.F. Forstner

Kinsmen Cystic Fibrosis Research Centre, Dept. of
Paediatrics and Biochemistry, The Hospital for Sick
Children and the University of Toronto, Toronto, Canada

Native undegraded mucins from a variety of epithelial tissues
in different animal species have been shown to be remarkably similar
in their overall chemical composition, containing the same complement
of sugars and spectrum of amino acids. Many have a high molecular
weight, similar rheological properties and are sensitive to thiol
reduction. It was considered possible, however, that subtle but significant differences in the structure and chemical features of mucins
might be detected by immunological techniques. To this end, a solid
phase radioimmunoassay (RIA) for human small intestinal (HSI) mucin
has been developed. The RIA was used to compare the antigenic
activity of various mucins and haptens, and to study the effects of
thiol reduction upon HSI mucin.

The experimental procedure involved binding a fixed amount (50
ng protein) of purified HSI mucin to the wells of a plastic micro-
titre plate. The remaining plastic was completely coated with bovine
serum albumin. HSI mucin antibody (1:6000 dilution of the IgG-rich
fraction of immune rabbit serum) was pre-incubated separately with
varying amounts of either standard HSI mucin or test material. The
antigen-antibody mixture was then added to the mucin antigen-coated
wells. At this stage, only the free antibody left in the pre-
incubation mixtures could bind to the antigen coating the wells.
^{125}I-protein A (from S. aureus) was then added (0.1 µCi/well) to
monitor the bound antibody. Bound ^{125}I radioactivity was plotted
against the concentration of mucin added to the pre-incubation mixtures and a standard curve was produced (Fig. 1) having a range of
5 to 100 ng mucin protein. For purposes of future comparison, the
slope of the standard curve was arbitrarily set at 1.0. The assay
required 24 hours, was reliable and sensitive, and less tedious to
perform than the previously reported double-antibody RIA for HSI

43

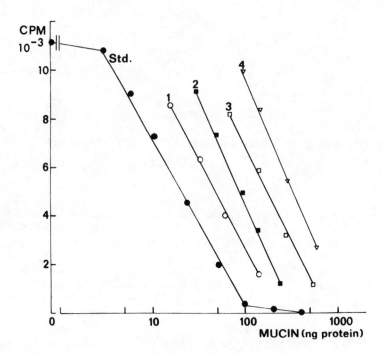

Fig. 1.   Solid phase radioimmunoassay of human small intestinal
           (HSI) mucin. "Std" is purified standard HSI mucin. Num-
           bers 1 to 4 refer to representative mucins purified from
           the intestinal tissue of four different human subjects.
           Note parallelism of the slopes with reference standard
           mucin. Antibody was a 1:6000 dilution of immune rabbit
           serum (IgG-rich fraction) developed against the standard
           reference mucin. Radioactivity represents ^{125}I-protein A
           bound to antibody in microtitre wells coated with the
           reference mucin.

mucin (Qureshi et al, 1976) since it did not involve the prolonged
incubation time necessary for immunoprecipitation.

     HSI mucin was purified (by a modification of the method reported
by Jabbal et al, 1976) from nine autopsy samples, obtained 2 to 12
hours post mortem. All gave a reaction of identity, (mean slope
0.95 ± 0.14) indicating identical antigenic determinants. The amount
of mucin protein required to give 50% binding ranged from 60-500 ng
protein (equivalent to 1-6 nmols total carbohydrate). This range
suggests little variation in mucin-antibody affinity despite known
differences in age, ABH and Lewis blood group specificity and terminal
illness of the individuals.

To investigate the specificity of the antibody for HSI mucin, various well-characterized mucins purified from other species and organs were assayed for their cross-reactivity. These included human lung mucin (HLM), pig gastric mucus glycoprotein, pig small intestinal and rat small intestinal mucin. Salivary mucins, some of which are known to contain less complex oligosaccharide side chains and to have a simpler polymeric structure, were also compared for their antigenic reactivity. Only two glycoproteins were found to be weakly cross-reactive - pig submaxillary mucin (slope = 0.3) and HLM (slope = 0.6), while ovine (desialated), bovine and canine submaxillary, rat sub-mandibular, pig gastric, and pig and rat small intestinal mucins were all non-reactive in the range of 0.2 to 25 µg dry weight.

Individual monosaccharides (Gal, Gluc, Man, GalNAC, GlcNAC, NANA, Fuc) either alone or together in equal proportions (3 to 200 nmols) did not interfere with the interaction of HSI mucin with its anti-body. This would seem to indicate that the antigenic determinants of HSI mucin (if carbohydrate in nature) may be complex, involving more than one sugar residue. Four oligosaccharides of known composition and structure: [NANA $\alpha 2 \rightarrow 6$ GalNAC-OL; Gal $\beta 1 \rightarrow 4$ GlcNAC $\beta 1 \rightarrow 3$ GalNAC-OL; GlcNAC $\beta 1 \rightarrow 3$ GalNAC-OL (6 $\leftarrow$ 2α NANA); Gal $\beta 1 \rightarrow 3$ GlcNAC $\beta 1 \rightarrow 3$ GalNAC-OL (6 $\leftarrow$ 2α NANA)] isolated from rat colonic mucin (Slomiany et al, 1980) also showed no cross-reactivity. Since these structures are common to many mucins and might be expected to occur in HSI mucin, their lack of reactivity suggests that HSI mucin determinants involve more complex oligosaccharides and/or peptide residues, uniquely spaced in a 3-dimensional structure.

Disulphide bridges have been shown to play a major role in maintaining the polymeric structure of many gastrointestinal mucus glycoproteins, including pig gastric (Snary et al, 1970), human gastric (Pearson et al, 1980), pig small intestinal (Mantle and Allen, 1981), and pig colonic mucin (Marshall and Allen, 1978). Thiol reduction of these glycoproteins resulted in breakdown of the high molecular weight native molecule (eluted in the excluded volume of a Sepharose 2B column) into smaller sized subunits (partially included by a Sepharose 2B gel). Exhaustive reduction of HSI mucin caused a similar change in its elution profile from a Sepharose 2B column. It was therefore considered of interest to know if conversion of the native molecule into smaller subunits induced changes in its antigenic determinants. One sample of purified HSI mucin was reduced in 0.2M 2-mercaptoethanol for 48 hours at 37°C and pH 8.6. Following exhaustive dialysis against distilled water (5 days, 2 changes per day) the resulting solution gave a slope of 0.2 in the RIA, while appropriate controls (incubated similarly but without mercaptoethanol) gave a slope of 1.2. Alkylated samples of the reduced and non-reduced mucins produced the same slopes in the RIA as those of corresponding non-alkylated samples (0.2 and 1.2 respectively). It would appear therefore that reduction of the thiol bridges of HSI mucin (with or without subsequent alkylation) resulted in a marked decrease in the number and/or affinity of the antigenic determinants. Neither alkylation alone nor incubation at $37^{\circ}C$ and pH 8.6 for 48 hours were responsible for these results. These findings imply that the overall structure and polymeric configuration of the HSI mucin molecule are important for its antigenic reactivity.

In conclusion, the antibody for HSI mucin shows a high degree of species and organ specificity as judged by RIA. Variation among HSI specimens was minimal. This was true despite known polydispersity of the oligosaccharide side chains of mucins, variable blood group activity, post mortem time before isolation and terminal illness. Thus, the antigenic determinants of HSI mucin appear to be common but unique structures in mucin macromolecules, having considerable stability as well as independence from terminal blood group sugars. Reduction of disulphide bonds in one sample of HSI mucin at $37^{\circ}C$ and pH 8.6 for 48 hours, caused a significant decrease in molecular weight and interaction of the mucin with its antibody. These findings suggest that the antigenic determinants of HSI mucin depend at least in part on the 3-dimensional configuration of the native molecule.

Acknowledgements

The authors are grateful to E. Katona (Dept. of Biochemistry, HSC, Toronto) for supplying pig, bovine and canine submaxillary mucins; N. Fleming (Dept. of Pathology, HSC, Toronto) for rat submandibular mucin; D. Williams (Dept. of Biochemistry, HSC, Toronto) for desialated ovine submaxillary mucin; M.C. Rose (Dept. of Biochemistry, Duke University Medical Center, North Carolina) for human lung mucin; A. Allen (Dept. of Physiology, Univeristy of Newcastle upon Tyne, U.K.) for pig gastric mucus glycoprotein; and A. and B.L. Slomiany (Dept. of Medicine, New York Medical College) for the four oligosaccharides from rat colonic mucin.

Financial support was provided by the Canadian Cystic Fibrosis Foundation.

References

Jabbal, I., Kells, D. I. C., Forstner, G. G., and Forstner, J. F., 1976, Human intestinal goblet cell mucin, Can. J. Biochem. 54:707.
Mantle, M., and Allen, A., 1981, Polymeric structure of pig small intestinal mucus glycoprotein, Biochem. J. (in press).
Marshall, T., and Allen, A., 1978, The isolation and characterization of the high-molecular-weight glycoprotein from pig colonic mucus, Biochem. J. 173:569.

Pearson, J. P., Allen, A., and Venables, C. W., 1980, Gastric mucus: isolation and polymeric structure of the undegraded glyco-protein, its breakdown by pepsin. Gastroent. 78:709.

Qureshi, R., Forstner, G. G., and Forstner, J. F., 1979, Radioimmuno-assay of human intestinal goblet cell mucin, J. Clin. Invest. 64:1149.

Slomiany, B. L., Murty, V. L. N., and Slomiany, A., 1980, Isolation and characterization of oligosaccharides from rat colonic mucus glycoprotein, J. Biol. Chem. 255:9619.

Snary, D., Allen, A., and Pain, R. H., 1970, Structural studies on gastric mucoproteins: lowering of molecular weight after reduction with 2-mercaptoethanol, Biochem. Biophys. Res. Comm. 40:844.

TRISACCHARIDES FROM BLOOD GROUP A_1 AND A_2 MUCOUS GLYCOPROTEINS

Alastair S.R. Donald

Division of Immunochemical Genetics

MRC Clinical Research Centre, Watford Road
Harrow, Middlesex HA1 3UJ, Great Britain

INTRODUCTION

The mucous glycoproteins of human origin may possess A, B, H and Lewis blood group activity and the structures of the blood group determinants responsible for these activities have been established (see ref.[1]). However, the nature of the difference between A_1 and A_2, the two main subgroups of A, is still unclear. One theory states that there are structural differences between the determinants in A_1 and A_2 materials, and Moreno et al[2] have suggested that the existence of two types of A-determinant might form the basis for such structural differences. They proposed that A_1 materials carry both Type 1 and Type 2 A-determinants whereas only Type 2 A-determinants are present in A_2 materials.

```
        Fuc                            Fuc
        | α 1,2                         | α 1,2
GalNAc-α(1-3)Gal-β(1-3)GlcNAc   GalNAc-α(1-3)Gal-β(1-4)GlcNAc
```

Another view holds that the difference between A_1 and A_2 is quantitative[3]. The same A-determinants are thought to be present in materials of both subgroups, but more A-determinants are believed to be carried by A_1 materials. To distinguish between these two theories, A-active oligosaccharides have been isolated from glycoproteins obtained from blood group A_1 and A_2 persons.

EXPERIMENTAL AND RESULTS

One sample of a purified human ovarian cyst glycoprotein from a blood group A_1 individual and one from a blood group A_2 individual were treated with 0.05 $\underline{M}$ NaOH in 1 $\underline{M}$ NaBH$_4$. The oligosaccharides released were de-$\underline{N}$-acetylated by heating with 0.2$\underline{M}$ Ba(OH)$_2$ at 100° overnight and then hydrolysed with 0.25 $\underline{M}$ H$_2$SO$_4$ at 100° for 3 h. The products were fractionated on a column of AG 50 (H$^+$) resin and the fraction which was eluted with 0.75 $\underline{M}$ H$_2$SO$_4$ was separated preparatively on an amino acid analyser. The fractions obtained were re-$\underline{N}$-acetylated with [^{14}C] acetic anhydride and further purified by paper chromatography. Trisaccharides I, II and III were isolated from both the A_1 and A_2 glycoproteins, and those from each source were characterised separately using glycosidase digestion, Smith degradation and methylation. The results obtained indicated that the trisaccharides have the following structures :

I   GalNAc-α(1-3)Gal-β(1-3)GlcNAc
II   GalNAc-α(1-3)Gal-β(1-4)GlcNAc
III   GalNAc-α(1-3)Gal-β(1-3)$\underline{N}$-acetylgalactosaminitol

The agglutination of A_1 cells by human anti-A_1 and anti-A reagents were inhibited almost equally by all three trisaccharides.

Additional chromatographic evidence has been obtained for the presence of the three trisaccharide structures in a number of other ovarian cyst and some salivary glycoproteins. These materials were degraded as described above and the products were separated on a column of AG 50 (H$^+$) resin. The 0.75 $\underline{M}$ H$_2$SO$_4$ eluate was neutralised and NaBH$_4$ reduced and then examined on an amino acid analyser using two different buffer systems. Trisaccharides I, II and III, isolated and characterised as described above, were used as standards. With the first buffer system peaks corresponding to reduced, de-$\underline{N}$-acetyl-ated trisaccharides I, II and III were detected in the products of all the A_1, A_2, A_1B and A_2B glycoproteins examined. These peaks were absent in the products of a number of non- A-active glycoprot-eins used as controls. When the glycoprotein products were run in the second buffer system, those from the A_1, A_2, A_1B and A_2B materials all gave peaks consistent with the presence of reduced, de-$\underline{N}$-acetylated trisaccharides I and II. All the glycoprotein products, including those from the non- A-controls, gave a peak which coincided with the peak of the reduced and de-$\underline{N}$-acetylated trisaccharide III. Identification of this compound was therefore not possible with this buffer system.

DISCUSSION

Trisaccharides I, II and III have been isolated and fully characterised from one example each of an A_1 and A_2 glycoprotein and chromatographic evidence has indicated that these trisaccharide structures are constituents of a number of other A_1 and A_2 glycoproteins. Trisaccharide III has not previously been obtained from human glycoproteins although there is evidence that it is present on the erythrocytes[5], and the isolation of this compound indicates that there are at least three types of A-determinant present in mucous glycoproteins and not two as previously believed. As the same three trisaccharides were obtained from A_1 and A_2 materials, the findings reported here are not indicative of structural differences between the A-determinants of the A_1 and A_2 glycoproteins and clearly refute the hypothesis of Moreno et al[2] which predicts that the trisaccharide I should not be found in A_2 materials.

The findings reported here are also relevant to a theory recently proposed by Oriol[6] which suggests that blood group H-determinants are synthesised by two distinct α-2-fucosyltransferases which are coded for by the Se and H genes. The action of the Se transferase is thought to be restricted to the type 1 precursor chains and to form the H-determinants in secreted glycoproteins, whereas the H coded transferase is believed to form the Type 2 H-determinants found on erythrocytes and is not active on secreted glycoproteins. Thus, according to this theory, Type 2 H-determinants and the Type 2 A-determinants that are synthesised from them should not be found in secretions. The evidence reported here indicates that Type 2 A-determinants are present in both ovarian cyst and salivary glycoproteins.

REFERENCES

1.   W.M. Watkins in : " Glycoproteins ", A. Gotschalk, ed., Part B, p 830, Elsevier, Amsterdam (1972)
2.   C. Moreno, A. Lundblad and E.A. Kabat, J. Exp. Med. 134 : 439 (1971)

3.   W.M. Watkins and W.T.J. Morgan, Acta Genet. Statist. Med. 6 : 521 (1956/57)

4.   A.S.R. Donald, J. Chromatogr. 134 : 199 (1977)
5.   S. Takasaki, K. Yamashita and A. Kobata, J. Biol. Chem. 253 : 6086 (1978)
6.   R. Oriol, Blood Transfusion and Immunohaematology 23 : 517 (1980)

STRUCTURE AND FUNCTION OF MUCUS

Alexander Silberberg and Frank A. Meyer

Weizmann Institute of Science
Department of Polymer Research
Rehovot, Israel 76100

INTRODUCTION

A rather far reaching assumption is implied in writing the
above title. Is it indeed possible to treat "mucus" as a unique
secretion whose properties in health and disease can be discussed
in a single context, i.e. without reference to its origin, to its
detailed function or to the kind of disease which may affect its
proper action. Is mucus, even in principle, the same everywhere?

This is certainly not true in detail but sufficient a similarity exists, at the level of chemical composition, structure and
function to make it well worthwhile to stress these aspects
rather than the dissimilarities which we know do arise. Certainly
it has proved to be a most useful working hypothesis (1-4) to treat
mucus from various sources as belonging to the same class of materials. A very similar glycoprotein molecular building unit is involved, a similarity in structure exists and almost everywhere,
where mucus is secreted, it provides a similarly constituted
mechanically interesting system by means of which a problem of
gliding transfer of a foreign body, or bodies, relative to a part, or
the whole, of an organism or of an organism, relative to an outside
framework, is solved, modified and controlled. It also seems that
mucus layers very often act as barriers, protecting the underlying tissue from the effects of concentration gradients, either
dynamically or in equilibrium.

It is thus our plan in this discussion to treat mucus and all
the information that has been gathered about it from many sources
as if it indeed did relate to a unique material. In other words,

53

we shall treat the information as essentially supportive, i.e.
averageable, each bit of data contributing additively to our understanding of the system which we call mucus. Only later will we
analyze what does not fit in so well, discuss the why of it and try
to relate it to the functional aspects.

We shall divide the first part into three sections; discuss
first the existence of a basic glycoprotein building unit, then
consider the cross-links which hold it together in a network and
finally consider the mechanical and physico-chemical properties of
mucus which make it suitable to fulfil its physiological function.

THE BASIC GLYCOPROTEIN STRUCTURE UNIT

The essential features of the glycoprotein unit which provides a common link between all mucus secretions is shown in Fig.1.

A single protein backbone chain, some 800 amino acids long,
is involved. This, over some 63% of its length, is covered by
short carbohydrate side chains some 10 sugar moieties long on the
average. These side chains cover a single block of the protein
backbone in a very tight layer. There are some 200 side chains on
the average in this block. The remaining 300 amino acids constitute one,or possibly two,'naked' peptide regions at the ends of
the chain. The 'naked' peptides are subject to proteolytic digestion. The sugar covered portion is essentially protected. It has
generally been assumed that there is only one 'naked' peptide region (bottle-brush model), since there is some evidence that only
the C-terminal end is attacked (5-7). However, it is much easier
to account for some of the features of the overall system if one
can assume that two regions, one at each end, are involved. One of
these could, of course, be much shorter than the other (rolling
pin model).

The carbohydrate side chains are linked to the backbone via
O-glycosidic bonds and the amino acids involved are threonine and
serine. Indeed, one finds a large amount of these two amino acids
in the sugar bearing peptide, in fact there is a considerable
excess of these two amino acids over the 200 or so of them which
have been shown to get glycosylated.

As Fig.1 shows the sugar bearing section contains practically
no cysteine while some 16 out of 300 amino acids of the 'naked'
peptide region are cysteines. The 'naked' peptide region (or
regions) is thus generally considered to be involved with the cross-
link. The 'naked' peptide is also high in aspartic and other basic
and acidic amino acids. Perhaps no single unique peptide chain
common to all mucus glycoproteins is involved but a satisfactily
close resemblance between backbone protein chains from the different mucus glycoprotein is found. Enough of a resemblance, indeed,

MUCUS GLYCOPROTEIN

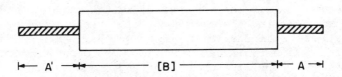

M.Wt : 530 000
A, A' : Naked Peptides (5% w/w of total; 35% w/w of protein)
[B] : Glycopeptide

Fig. 1.   The Mucus Glycoprotein Basic Unit

Two regions are recognized:

A,A':   'Naked' peptide regions, subject to proteolytic attack,
which contain almost all the cysteines in the chain and
are very high in charged amino acids, particularly
aspartic acid.

[B] :   A heavily glycosylated region, half of whose amino acids
are threonine and serine. It is covered by some 200 carbohydrate side chains attached O-glycosidically to either
serine or threonine. This part of the chain also contains
many prolines. It is thus almost certainly in a β-structure.

	A,A'*	[B]
Number of Amino Acids		
Total	|300|	|500|
Thr + Ser	37	|250|
Cys (1/2)	|16|	2
Asp	40	10
Number of Sugar Moieties		
Total	0	2000
No. of side chains	0	|200|
Protein Chain Length (or diameter)		
Collapsed (globular)	10 nm	–
α-chain	|45 nm|	75 nm
β-chain	105 nm	|175 nm|
Thickness		5-10 nm

Notes:* The calculations are to be interpreted additively
for A and A' together. Results of particular interest are
boxed.

to justify talking of an average composition (Fig. 2) and perhaps
even sequence.

Much greater variety seems to exist insofar as the carbohydrate composition is concerned. Fig. 3 summarizes a number of
different cases. The first sugar in the chain is always N-acetylgalactosamine and the chains are often terminated by fucose or by
a sialic acid with at least one branch per chain on the average.
It is a well established fact that considerable microheterogeneity
exists among the side chains in length and complexity (8,9).
Clearly mucus function does not depend upon the presence of a
highly specific cover composed of a set of unique carbohydrate side
chains. It is known that mucus can lose part of its sugar and yet
suffer no impairment in function. It might be argued in fact that
the main purpose of the dense fairly uniform hydrophilic coat of
sugar is to be just that, namely a dense fairly uniform hydrophilic
cover (8,9). The specific sugar sequences which do exist may have
been created by the mucus synthesizing cells not because these
sequences are needed, but because they happen to represent the way
the cells of the organism know how to build up carbohydrates. One
may even question this. A unique sequence plan which by microheterogeneity is left partially incomplete in some of the chains,
may not exist. Perhaps all the cell really knows and wants to do
is to add sugars in a certain preferred order statistically. Created
would be highly probable sequences resulting from a small number of
preferred synthesis plans (partially left incomplete) where the
sequence is selected by the availability and the reaction rates
of the various reaction components and by enzyme determined sequence
rules. Even such a process would create a sufficient number of
side chains, among the 200 or so in each sugar coated block, which
would have a desired specificity (or perhaps an irrelevant one).
To the extent indeed that specificity is needed it might be an
advantage to have these sites diluted along the chain and to have
a variety of different specificities co-existing. If this were
true, how could specific sequences have been isolated and characterized? The answer to this may be simple. Since not too many
variations are possible, the refinement procedure used could indeed
have selected out a fairly uniform, perhaps even perfectly homogeneous, sample. Even if not, however, and even if the composition
and sequence of the sample is not unique, but only highly probable,
the resulting analysis would give a sequence, the highly probable
one. This is then called "unique" not because it is so really, but
because it has somewhere been assumed that the cell is programmed
to produce a unique sequence.

O-glycosylation occurs in the Golgi apparatus of the cell
utilizing a battery of glycosyl transferases which specifically add
each sugar to the growing side chain (9). Since side chains are
closely spaced some will obviously be left stunted in cases where
the neighboring chains have grown too fast and the enzyme can no

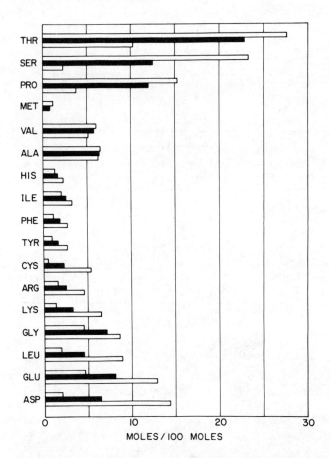

Fig. 2. Amino Acid Composition of Mucus Glycoprotein and its
          Components

    The results present an average from a number of cases and a
    variety of sources.

    Black bar: Overall composition.   Data taken from references
                (7,27,28,35,36,37,38).

    Upper open bar: Composition of the glycopeptide after
                     proteolytic enzyme digestion. Data taken
                     from references (7,28,36,38).

    Lower open bar: Composition of the 'naked' peptide, i.e.
                     the proteolytic enzyme digested part.
                     Calculated by difference.

Note: The order in which the amino acid appear is unconventional.
       On top of the list are the amino acids which predominate in
       the glycosylated peptide region. At the bottom those which
       predominate in the 'naked' peptide region.

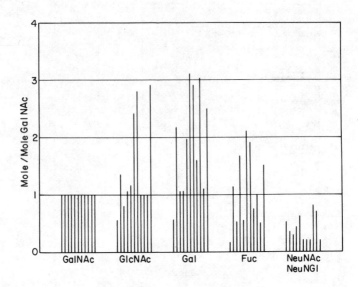

Fig. 3.   Carbohydrate Side Chain Composition of Mucus Glycoprotein

Compositions are normalized (GalNAc: N-acetyl galactosamine = 1) and refer, from left to right, in the case
of each sugar,to tracheal mucus, rabbit, dog, cat, goose
(39); to cervical mucus; cow (27); to gastric mucus,
human and pig (28); to small intestinal mucus, human
and pig (28,37); and to colonic mucus; human and pig (28).

GlcNac: N-acetyl-glucosamine
Gal   : galactose
Fuc   : fucose
Neu NAc: N-acetyl neuraminic acid
Neu NGl: N-glycollyl neuraminic acid.

longer reach a chain that has been left behind. We may in fact
assume that glycosylation proceeds until there are no enzyme sized
gaps left for the addition of sugars, at least until the protein
backbone is adequately covered. Termination with fucose or sialic
acid may be the signal to stop glycosylation of a particular branch
and depending upon whether this occurs prematurely or with delay,
side chains of different overall lengths and extent of branching

would result. Indeed the entire system might make better sense if,
as said, less the sequence than the amount and density of the sugar
moieties in the coat should turn out to be the important variables
and the cell were programmed mainly to synthesize a glycoprotein
of a certain stiffness, size and adequate solubility. The more
specific features required to build a supra-molecular structure
and thus produce the desired mechanical properties would be built
in primarily into the non-sugar covered region.

THE CROSS-LINK

     In the form that mucus is generally secreted the glycopro-
tein constitutes only a very small percentage of the total mass.
Most of the volume is fluid, physiological saline, with some additional diffusible macromolecules, mainly proteins, dissolved in
it. The secretion is viscoelastic and when put in contact with
pure physiological saline it does not go into solution, in fact
it does not even swell to any major extent. These are the characteristics of a gel, i.e. a system based on macromolecules in
solution where the macromolecules are all linked together (at
least one cross-link per chain) to give rise to an infinite network
or, at any rate, to very large, macroscopic aggregates. In such
systems a volume increase is either impossible or so slow to occur
as not to be detectable. Indeed the disentanglement of large
interpenetrated, sticky aggregates by thermal motion alone may be
difficult to achieve within a reasonable time span.

     This point is important since it is highly unlikely that
mucus, as synthesized by the cell, is totally, covalently cross-
linked. Mucus is stored in the cell in lipid membrane covered
granules. Only upon exocytosis (secretion) is the lipid cover of
the granules removed by being joined to the plasma membrane. Only
after secretion has taken place, therefore, are the contents of
the granules exposed to each other. The maximum size of the mucus
glycoproteins intracellularly is thus determined by the size of
the granules (200 to 1800 nm) (10) and the concentration of the
mucus glycoprotein inside the granule. This is unknown but,
however high it may be, the size of the glycoprotein molecule
secreted into the extracellular environment must be viewed as
limited. The possibility obviously exists that the covalent
bonding could arise subsequent to secretion. Unused reaction sites
(-SH and -C=O, for example) might be available.

     Certainly, it is important that immediately subsequent to
secretion the mucus be capable of flow, as it rises from crypts,
or through glandular craters or ascends the chimney-like gaps
between ciliated cells. Flow, or more precisely the ability to
flow, is a feature characteristic of sols. It requires that the
system involve only elements of finite extent which are mechanically independent of each other and thus can undergo infinite

relative displacement. On the other hand, whether permanently
cross-linked or not, the molecular units of the mucus glycoprotein
must, subsequently, overlap each other sufficiently to form the
linked macroscopic system that alone is able to resist swelling
and dissolution and which mechanically can function as mucus.

Table 1 lists some of the limitations which the above cited
facts put upon the size of the largest and smallest covalent unit
that can come in question. The molecular weight computed from the
size of the granule assumes that the entire contents constitute
one covalent molecule; the minimum size computed is worked out
from the concentration. It is based on a model which builds a
cubic monkey puzzle network from straight pieces which are taken
to constitute the molecules of minimum size to give the overall
concentration. In this model each end of each molecule is in
interaction with five other ends. The actual molecules could, of
course, be smaller and the chain of minimum length could involve
shorter mucus glycoprotein chains linked end-to-end.

Obviously, the ideas behind these numbers are extremely
simple. The results do show, however, that the basic molecular
unit of Fig. 1 fits well into the general picture. In Table 2
we cite some observations which make it very probable that these
estimates create a fair picture.

Table 1. Largest and Smallest Molecular Weight of Mucus
          Glycoprotein in Daltons

Glycopro-protein Concent-ration	Maximum Molecular Weight From Granule Size			Minimum Molecular Weight For Overlap
	200 nm dia.	600 nm dia.	1800 nm dia.	
1.0 mg/ml	10^6	30×10^6	900×10^6	280,000
50 " "	50×10^6	1500×10^6	45×10^9	40,000

In order to arrive at the results which are contained in
Table 2 mucus has to be solubilized. The following methods are
available:

(i)   By reducing disulfide bonds. This is a very fast process
      and in general such reagents as dithiothretiol tend to
      give total reduction and total dissolution.

(ii) By treatment with proteolytic enzymes. This attacks the
      'naked' peptide region. The sugar coated section is
      apparently sufficiently protected to be unaffected. Since

Table 2

Mucus Glycoproteins

Molecular Weight and Physico-Chemical Parameters

Source of Mucus	First Unit into Solution (by stirring)				Basic Unit (after both enzyme and/or disulfide bond reduction)			
	M.Wt. $\times 10^{-6}$	$s^{o}_{25,w}$ (S)	$D^{o}_{25,w} \times 10^{7}$ (cm^2/sec)	$[\eta]$ (ml/g)	M.Wt. $\times 10^{-6}$	$s^{o}_{25,w}$ (S)	$D^{o}_{25,w} \times 10^{7}$ (cm^2/sec)	$[\eta]$ (ml/g)
Cervix (cow) (14)	16.7 [a]	72	–	750	0.3 [c]	10.8	2.5	–
Bronchi (man) (14)	6.6 [b]	49	–	550	0.6 [a]	12.1	–	–
Trachea (dog) (31)	–	–	–	–	0.6	–	–	–
Stomach (pig) (7)	2.09	28.7	0.94	320	0.54	14.6–21.1	1.92–2.49	60–90
Stomach (man) (32)	1.8–2.2	24–33	0.9–1.0	–	–	–	–	–
Small Intestine (pig) (33)	1.8	–	–	500	0.24	–	–	–
Colon (pig) (34)	15	87	0.38	–	0.76	13.5	1.2	–

(a) by light scattering; (b) estimated from s^{o} and $[\eta]$; (c) from s^{o} and D^{o}. Light scattering gives a value of 0.49×10^{5}.

only some 5% of the total weight is lost, the digestion very often gives a unit which physico-chemically is very similar to the unit that comes out in (i).

(iii) By mild mechanical agitation. Usually a much larger unit than in (i) or (ii) is obtained. Since mechanical pull will separate macromolecules first at energetically weak points, the links that are separated in this process are dominantly not covalent, but thermally longer-lived, weaker bonds. A smaller number of strategic covalent bonds will, however, inevitably also be broken, particularly if higher shear rates are used. Dissolution may be aided by medium changes designed to break secondary bonds. Such medium changes do not, however, by themselves as a rule, produce dissolution.

(iv) By thermal activation. Dissolution of mucus occurs spon-taneously with time. Dissolution is relatively fast at 100°C but at 0°C the mucus is almost stable. The same activ-ation energy characterizes the entire accessible temperature range. The same basic unit which arises in (i) and (ii) seems to arise here as well, at least as a more stable intermediate.

The data in Table 2 suggests that, accepting the structure of Fig. 1, the sugar coated portion of the basic unit is essentially stable to all of the above four techniques. Hence, what is ana-lyzed in solution is this basic unit or oligomers of it.

We know that the various treatments affect the 'naked' peptide differently. Each degradation process must bring about its own kind of rupture, but if we accept that it is within the 'naked' peptide region that cross-link formation is concentrated the details of cross-link breakdown will be difficult to detect in a system where 95% by weight of the separating unit or units are not altered or where the entire 'naked' peptide region may have disappeared.

If the basic rod-like unit of Fig. 1 is designated by A[B] or A[B]A', where [B] is the sugar coated part and A and possibly a second portion A' at the other end are 'naked' peptide regions, we can lay down the following general principles of mucus build-up: We have seen that A is definitely involved in the cross-link and possesses the chemical where-with-all to form a covalent bond with another A or with an A'. Let us assume for the moment that [B] cannot interact with A, A' or another [B]. If A then were able only to react with A and there is no A', the process could build dimers. If A could intereact with several A's the process would produce stars. If there is a region A' as well as A and A and A' can react with each other (but only once) long chains would arise. Networks or large aggregates will only be formed if A or A' can react twice, or more, with the same A or the same A'. Only this

case could lead to branching and to infinite covalent network
formation.

Of course, if there is no A', and A could interact with [B]
and do so occasionally twice or more with the same unit, we could
again get long chains or network. Table 3 summarizes these cases.

Table 3. Cross-Link Formation by A Interacting with: A,A'
          ('Naked' Peptides) [B](Sugar Coated Peptide)

Case:	A	A'	[B]	Remarks
(a)	+,++,+++,···	–	–	Stars
(b)	+	+	–	Chains
(c)	+,++,+++,···	+,++,+++,···	–	Branches and networks
(d)	–	–	+	Chains
(e)	–	–	+,++,+++,···	Branches and networks
(f)	+,++,+++,···	+,++,+++,···	+,++,+++,···	Branches and networks

Let us now consider the following picture of the cross-linked
mucus which the above results seem to support:

Mucus is based on two kinds of cross-links, one involving
disulfide bonds the other entanglements. The very large number of
cysteines in the non-glycosylated region provides the means of
forming direct intermolecular links not only with one, but with
two, or several, other such regions if this is needed. Also feasible here is the formation of a very specific conformational fold
in the A region, using the disulfide bridges intramolecularly,
thereby creating a lectin-like surface capable of interacting
strongly and very specifically with the glycosylated [B]-part of
the basic unit. One may assume that despite microheterogeneity
the specific sugar sequence required would arise not only once but
several times among the 200 or so side chains involved. In fact,
if this turns out to be the nature of the cross-link it will be
seen to have been almost a requirement that only a few binding
sites arise over the [B] section of the unit. If this were not
the case, the entire glycoprotein fraction would tend to aggregate
and precipitate. Not only the direct disulfide link, but also
this lectin bond are 'covalently strong' and create 'permanent'
chains and aggregates.

The other kind of bond which has to be stipulated is entanglement. There is bound to be heavy entanglement if long, linear
multimer chains of the basic unit are formed by AxA' or Ax[B] bonds
even if the overall mucus concentration is only of the order of

1 mg/ml. Mechanically, as we know and shall discuss later, these
entanglements are almost equivalent to permanent cross-links. What
remains hard to explain is why, if only separate entities are
involved, the mucus when exposed to excess saline does not swell
and does not dissolve spontaneously by diffusion.

If as packaged in the granules, the mucus glycoproteins are
covalently assembled into linear chains of some 15-20 million
molecular weight, these chains would, when secreted, tend to
expand to micron sized spheres. Even at 1 mg/ml concentration,
however, parts of some 10 of these molecules would have to occupy
the same region of space and considerable interpenetration and
overlap is thus assured. Moreover, since the chains are very
"rough", relative motion in a direct contact would be severely
constrained.

Molecules of some 15-20 million m.wt. are the structures
being released by mild shear. The molecules as they exist in the
mucus could, some or all of them, perhaps be even larger. If these
large molecules are linear mucus glycoprotein chains they would
tend to be rather stiff as a comparison of some mass per unit
length data shows (Table 4). They are probably stiffer than
double stranded DNA or triple helical collagen. Very large random

Table 4. Mass Per Unit Length of
Some Biopolymers

System	Dalton/nm
Proteins (β-chains)	290
" (α-helix	670
Collagen (triple helix)	1000
DNA (double stranded)	1900
Mucus glycoprotein	2500
Collagen (microfibril)	5000

coils could thus result. Hence if they are crowded into the
granule due to lateral confinement, they would tend to spread
after secretion (9). As, therefore, the granular contents come
into contact with each other they would tend to merge due to chain
expansion and interpenetration. It should be pointed out,moreover,
that ring closures between the unreacted A of the first unit and
say the unreacted A' of the last is a distinct possibility. Part
of the granular content could thus involve ring structures which
are topologically entwined. Further unreacted A and A' units
would, upon secretion, tend to find partners giving rise to some

very large chains which would act to enhance covalent coherence
without necessarily giving rise to networks. As already pointed
out, networks would only form if AxA or AxA' interactions can
occur more than once on the same A or, in the case of Ax[B] bonds, if
the same [B] can react with more than one A. Should the reaction
be strictly such that A and A' react only once, only linear chains
or rings could result. It is thus possible to visualize a system
which, while not truly a three-dimensional covalently cross-linked
network, would have great difficulty swelling and disentangling
over a time span of practical interest.

For several years, the original observation of Meyer (11) that
mucus undergoes spontaneous thermal degradation created the impression that the cross-link is perhaps not covalent at all but might
involve some more mobile bond. It could be shown that the mono-
molecular degradation process proceeded with an activation energy
of 22.3 Kcal/mol. The same energy characterized the entire
accessible temperature range (0°C-100°C). The activation energy
and all the kinetics were unaffected by a wide variety of medium
changes. This ruled out all conceivable secondary bonds and rather
specialized models for the cross-link were proposed which involved
a disulfide bond dependent mechanical link (12). Such proposals
could be made consistent with most, but not all observations.
These models are no longer necessary. Further work has now led to
a simple explanation. Degradation very probably is due to peptide
bond hydrolysis, which by preference occurs next to aspartate, and
has a very similar activation energy (13). It will be noted (Fig.
2) that the 'naked' peptide region contains a particularly large
amount of aspartate. Hence this region should be particularly prone
to attack. The expected creation of new N-terminals involving
aspartate has indeed been found and very interestingly and con-
vincingly the density of the degradation product increases by an
amount consistent with the loss of the 'naked' peptide regions by
preference (14). Thermal degradation thus occurs by peptide bond
breakage preferably in the cross-link region. Hence, the creation
of breakdown products, which in their structural characteristics
would tend to leave the basic glycoprotein unit intact, is
explained. Eventually, of course, peptide bonds in the [B] region
will also be affected and indeed entities smaller than 500,000
m.wt. have now been established after long time degradation (14).
There is thus no further need to consider this process as proble-
matic. Its influence on mucus stability is understood and it is
not related directly to the cross-link. However, as the results
of Meyer (14) show, thermal degradation should be considered as a
process limiting the life time of the mucus. Over the entire
temperature range all other bonds which hold mucus together will
have to be longer lived and stronger. Hence, a covalent AxA' type
interaction, involving disulfide bonds, is to be preferred over
the possibility that a lectin-like Ax[B] interaction is the cross-
link.

FUNCTIONAL ASPECTS

Mucus is a product of certain secretory epithelia. Epithelia
of this kind line internal body cavities where tissue is in contact
with air (as in the lungs) or with non-tissue fluid or solid
material (as in the intestine). The mucus secreted has almost
always a protective role to play often in conjunction with cilia.
The underlying tissue is guarded against attack by bacteria or
viruses or protected from pollution by air borne particulate matter.
The mucus layer has to prevent these foreign bodies from penetra-
ting into the tissue and has to help to clear them, as well as
cell debris, from the surface. There are instances where the mucus
has to be a mechanical buffer and guard the underlying tissue
against mechanical damage. In other instances it has to protect
tissue from material losses or from molecular invasion due to steep
gradients in chemical or electrochemical potential between the
tissue and the particular body cavity ambiance.

Mucus restrains molecular or particulate penetration by blockage and entrapment. It is a barrier, diffusive, say to macromolecules or viruses, mechanical, say to bacteria. These and still
larger particles are held and engulfed by mucus to be finally
removed, i.e. cleared, together with the effusion itself by lateral
transport. For this purpose the epithelium also, generally, involves
cells, provided with a large complement of cilia, whose coordina-
ted beat provides the means of propulsion. The individual cilium
must be able to penetrate the mucus layer, transfer momentum to it
with a minimum of dissipative slip and then disengage in order to
perform a return stroke through the non-viscoelastic, periciliary
fluid layer which lies between the tips of the cilia and their
roots, i.e. between the bottom of the mucus sheet and the tops of
the cells (15).

In other instances, for example in the intestine, where the
force of propulsion derives from the surrounding musculature the
mucus keeps the contents of the cavity from scraping at the cellular lining. It is a lubricant.

These functions are mechanical in nature and thus intrinsi-
cally a consequence of the organization of the mucus glycoproteins into an entangled network. In addition, mucus, probably,
also has various biochemical roles to fulfill but these may not
necessarily be built-in features of the mucus glycoprotein. They
more likely are the function of proteins and other diffusible
molecules dissolved and possibly stored in the gel. For example,
mucus may contain immunoglobulins, hormones and a variety of
enzymes with defined, defensive or regulatory functions.

We shall concentrate mainly upon the mechanical aspects of
the system.

Mucus As a Gel

Gels depending upon the nature of the cross-links, can be classified as permanent or reversible. The gel substance, the mucus glycoprotein in our case, would be united into one large, "infinite" macromolecule if the mucus were a permanent gel. In that case, every atom of the gel substance could be reached from every other atom by going over covalent bonds only. A permanent gel thus has a definite reference configuration and overall shape and some part, at least, of any deformation to which it is subjected can be totally recovered. This kind of gel would possess an infinite relaxation time and a perfect memory for part of its deformation history. It is clear that mucus is not of this kind.

Mucus flows and, given enough time, totally relaxes. We know that in muco-ciliary clearance it is necessary for the cilium to penetrate into the mucus and interact directly with the mucus gel in the power stroke. Momentum transfer through a medium of low (water-like) viscosity would not be effective in energy transfer. On the other hand, if the medium, the periciliary fluid, were very viscous the ciliary beat (which takes place with movement of most of the ciliary surface relative to this medium) would be very wasteful of energy indeed.

Mucus must provide a compromise in its interaction with the cilia. As the cilium penetrates, end on, into the mucus layer, it experiences minimum resistance while the time taken over this process is long enough for any elastic deformation to have relaxed. When momentum transfer occurs in the power stroke the movement is side on. The cilium makes use of its aspect of maximum interaction and the speed of its movement is fast enough so that in the time taken over this process, force is transferred to the gel network without loss, i.e. without relaxation. Hence, there is an effective transfer of momentum. Disengagement of the cilium occurs more slowly. Here again, use is made of movement in the direction of the low resistance aspect of the rod-like slender body. Mucus thus has built into its network a sufficient number of relaxation mechanisms which will carry the force and transfer the momentum without loss during the fast power stroke. On the other hand, these mechanisms relax fast enough, over the period in which the cilium enters or leaves the mucus, so as minimally to hinder the engagement/disengagement process. While large parts of the mucus network stay together during the fast power stroke, they can rearrange and fall apart over the remainder of the cycle.

Experimentally we can check this out using the cilia beating on a detached mucus depleted frog's palate as a test device (2, 16,17). As we change the properties of the mucus, or of a system, which is rheologically equivalent to it, we alter transport velocity (16,17). One finds that only the storage, i.e. the elastic component of the complex dynamic shear modulus affects the velocity of transport (17). Changes in the loss modulus leave the results essentially unchanged (17). Most effective are systems which have a storage modulus G' at about 1 Hz of about 2 dyne/cm^2 (17-19). Systems which have either a lower or higher modulus are less effective since they would be inefficient in momentum transfer or in ciliary penetration, respectively (15). These results agree well with the results of Litt (18,19). King (20), tends to find a better correlation with the loss tangent, i.e. with the ratio of loss to storage modulus.

It is of considerable interest that the presence of sialic acid peripherally over the [B] block has no influence on the mechanical properties and on clearance (21). Mucus which has 80% of its sialic acid removed experiences no change in mechanical properties nor was the hydrodynamic behavior of the basic glycoprotein unit in saline affected (21). It would seem that no bonding of any significance goes via these acid groups and that the molecule is so stiff and Debye-Hückel shielding is so effective that no major conformational change accompanies the removal of the acid groups at physiological ionic strength.

It is of interest also to link the load bearing capacity of a system to the process of gelation. It is in fact one criterion which may be used to characterize the gel point. One compares the onset of solid-like properties with the ability of the system to support a weight of a macroscopically sized column of itself for a reasonable time; say a column of 1 cm for 1 sec. In the case of aqueous systems this requires a storage modulus of some 10^3-10^4 dynes/cm^2 at above 1 Hz. In the case of load bearing by mucus the particles are smaller and the mucus layer is much thinner, of the order of 10µ. A storage modulus of only 1 to 10 dyne/cm^2, at 1 Hz is thus required. This corresponds rather reasonably to the value of G' at the optimum. Such a material layer could support a load intensity of 0.01 to 0.1 mg/mm^2 assuming a 10% compression of the layer under the load. This is much lower a degree of loading than can be cleared effectively from epithelia such as the frog's palate (2). In our test procedure, we have used load densities of some 10 mg/mm^2 routinely and earlier results of Stewart (22) suggest that transport velocity is not affected till loads of 20 mg/mm^2 are exceeded. Under these high load densities the particle will penetrate the mucus layer, sink right through it, until it becomes supported by the ciliary bed, and then be pushed over the top of the cilia, through the intermediacy of the island mucus which surrounds it, as the latter is transported laterally over the epithelium. It is possible to estimate the load bearing capacity of the ciliary bed and 20 mg/mm^2 does indeed present no difficulty. On the other hand, 10-100µ particles, or smaller, would be well supported on top of the mucus.

Insofar as the storage modulus is concerned it is possible

to estimate its value on the basis of the Rouse-Zimm model (23). This model allows the computation both of the longest relaxation time and the value of the modulus from the concentration and molecular weight of the macromolecules, or their aggregates, in solution. A storage modulus of 1-10 dyne/cm^2 at a frequency of 1 Hz is predicted to arise when linear molecular entities of an average molecular weight of 15x10^6 and a chemical lifetime in excess of 10 sec are assumed to exist at a concentration of some 1-2 mg/ml.

This is very close to the situation which characterizes oestrus bovine cervical mucus. Entities of some 15x10^6 are indeed the molecular units identified in solution after mild stirring (11). Molecules of this size approximately also correspond to estimates based on laser light scattering from intact mucus gels. Such molecular entities would comprise from 10 to 100 basic units and, as said, only entanglements or other interaction which have a survival time in excess of 1 to 10 secs contribute to the storage modulus at 1 Hz. Such a system, as we know, interacts satisfactorily with a bed of beating cilia to produce lateral transference of the mucus layer.

It has been observed that mucus does not swell back to its original volume after it has been dehydrated. This is unusual since gels, originally in equilibrium with excess swelling medium, would return to this state after their concentration had been increased, say by evaporation. To account for the result with mucus, one has to assume that, as concentration changes, the degree of cross-linking is increased. This could indeed happen in the case of a system where, as we have just supposed, many cross-links are not covalent, but merely very long lived interactions, entanglements of a particularly involved kind. If M is the molecular weight of linear glycoprotein chain between cross-links and c is the overall concentration, an equation developed by Flory predicts that the product Mc^2 would have to be constant for such an effect to arise (24). This leads to a further prediction. Since the storage modulus G' should be proportional to c/M (23), i.e. to the number of effective cross-links per unit volume, G' would have to be proportional to c^3, if Mc^2= const. as c is varied. Studies with reconstituted human cervical mucus (18) and human ear mucus (19) as a function of glycoprotein concentration, are entirely consistent with a G' $\sim c^3$ relationship.

Little is known about the efficiency, of mucus as a lubricant. Studies of the apparent viscosity of frogs palatal mucus as a function of the rate of shear have shown that mucus tends to behave as a Newtonian fluid at rates of shear higher than about 1 sec^{-1} and that it possesses an apparent viscosity of about 10^3 poise at these rates of shear (25). An estimate of the rate of shear likely to arise, say, in the intestine, is indeed of the order of 1 sec^{-1}

just falling within the Newtonian range. We can calculate, therefore, that the wall shear stress will be of the order of 10^3 dynes/cm^2.

These stresses are not too far from those which would begin to detach cells adhering to artificial surfaces, but epithelial cells are much more securely anchored in place. Mucus, being viscoelastic, will, moreover, produce a normal stress which would act to oppose the lateral pressure exerted by the intestinal contents on the mucus layer.

In general, therefore, we can account for many of the mechanical features of mucus by treating mucus as an incipient gel, using the basic unit of Fig. 1 as the building block and applying some of the ideas, which we have outlined above, for the cross-link or links.

Permeability data indicates that the mucus glycoprotein is homogeneously distributed in the system (15,26). This leaves rather small openings only, even when mucus concentration is of the order 1 mg/ml, for the passage of other molecules through the mucus gel by diffusion. Gaps of the order of 10 nm are to be expected at 1 mg/ml and, say 3 nm at 50 mg/ml. As mucus goes from one extreme of this concentration range to the other, it would become impermeable to most biopolymers. A collapse of the mucus, and of the carbohydrate cover of the protein backbone, say, due to a lowered solubility at low pH, could create still less permeable gels which in the stomach would not only help to counteract the high pH gradient but also partially protect the mucus glycoproten from proteolysis by pepsin.

MUCUS FROM DIFFERENT EPITHELIA; THE QUESTION OF "UNIQUENESS" RE-EXAMINED

The scatter of the individual data points around the average amino acid composition of the mucus glycoproteins in Fig. 2 is really quite large. The scatter, however, is random and the direct compositional averages add up to 100 without normalization. Reasons for error are not hard to find. It turns out to be rather difficult to purify these glycoproteins, to purify them without degradation and to obtain the amino acids from a molecule so extensively associated with carbohydrate without loss. If a linker protein is present, moreover, some of it may still be associated with the glycoprotein basic unit, even after disulfide bond reduction. It is not at all easy to reduce every one of the disulfide bonds in every case (4). Hence, there are good explanations for the scatter and the true amino acid composition and sequence could indeed be common and even unique for all these secretions.

extent of glycosylation, too, seems to be common though there is
scatter (Fig. 3) is very much larger and specific differences in
composition are well documented. Gastric mucus contains much less
sialic acid than, say, cervical mucus (27,28). The total amount
of sugar also varies and in some instances ester sulfate groups
are present. Though chains of different lengths are isolated, it is
believed that (for any particular mucin at least) a specific
sequence is to be found (29). Considerable doubts have been expressed, however, as to whether any message contained in these
sequences is relevant (8,9). Perhaps no significance at all
attaches to them and only the "mix" and the amount of sugar on the
glycoprotein is important.

If the amino acid sequence is unique then the A and A'
regions are also specified and will be common to all basic units.
Differences in bonding would arise only through the intervention
of a linker protein, or by assuming that the bond is an Ax[B]
lectin-like interaction and that the variability in the sugar coat
will influence the number of bonds which are being made. Pearson
and Allen (30) have indeed separated a linker protein and proved
that four basic glycoprotein units are associated with it.

This is in the case of gastric mucin which functions at quite
a different glycoprotein concentration than, say, cervical mucus.
While the concentration of the former is of the order of 30 mg/ml,
bovine cervical mucus at oestrus contains about 1 mg/ml. It is
thus quite conceivable that a different "packaging" arrangement is
required and that the linker protein is present in gastric mucus
for this reason.

Even if mucus functions in much the same way, there are very
different requirements to be satisfied quantitatively on any
given epithelium. Overall, the same principle may be employed to
build up the supra-molecular structures and the main differences
in rheological properties may depend mainly upon concentration.
If indeed the differences in functional response are related merely
to the amount of mucus glycoproteins present in unit volume, a
unique variation of properties, if not a unique set of properties,
may then link different mucus secretions into one description. It
is already known that the cyclical changes of cervical mucus are
mainly mucin concentration dependent (8). If this is indeed one
manifestation of a wider principle we may even be more justified
than, heretofore, to discuss mucus structure and function in terms
of a common approach.

SUMMARY

Discussing the available evidence a fairly strong case can
be made for the existence of a basic glycoprotein unit, characterized by what may be a common protein backbone (Fig. 1). The

This is far less likely for the carbohydrate portion. The
considerably more variability in the amount and composition of
the carbohydrate coat and species and organ differences may arise
because of this fact. Very large aggregates are built up from
the basic unit using cross-links of disulfide bonds either inter-
molecularly, i.e. directly, or intramolecularly, i.e. indirectly
via a possible lectin-like structure which forms its bond with
some of the carbohydrate side chains. Structures of the order of
10-100 million molecular weight are to be expected which, being
heavily entangled, give rise to the special rheological character
of the mucus. In most instances mucus behaves rheologically like
a gel. The concentration of glycoprotein in the mucus may be
the most important parameter which determines the special rheological features required in a special functional context. A unified point of view, when discussing mucus structure and function,
was taken. On the evidence available, it seems well justified
to continue to do so.

REFERENCES

1. R.A. Gibbons and G.P. Roberts, Some Aspects of the Structure
    of Macromolecular Constituents of Epithelial Mucus, Ann.N.Y.
    Acad.Sci. 106:218-232 (1963).
2. J. Sadé, N. Eliezer, A. Silberberg, and A.C. Nevo, The Role of
    Mucus in the Transport by Cilia, Am.Rev.Resp.Dis. 102:48-52
    (1970).
3. F.A. Meyer, N. Eliezer, A. Silberberg, J. Vered, N. Sharon,
    and J. Sadé, An Approach to the Biochemical Basis for the
    Transport Function of Epithelial Mucus, Bull.Physio-path.resp.
    9:250-272 (1973).
4. F.A. Meyer, Comparison of Structural Glycoproteins from Mucus
    of Different Sources, Biochim.Biophys.Acta. 493:272-282 (1977).
5. M. Lhermitte, G. Lamblin, J.J. Lafitte, J. Rousseau, P. Degand,
    and P. Roussel, Properties of human neutral bronchial mucins
    after modification of the peptide or the carbohydrate
    moieties, Biochimé, 58:367-372 (1976).
6. K.S.P. Bhushana Rao and P.L. Masson, Study of the Primary
    Structure of the Peptide Core of the Bovine Estrus Cervical
    Mucin, J.Biol.Chem. 252:7788-7795 (1977).
7. M. Scawen and A. Allen, The Action of Proteolytic Enzymes on
    the Glycoprotein from Pig Gastric Mucin, Biochem.J.
    163:363-368 (1977).
8. R.A. Gibbons, Mucus of the Mammalian Genital Tract, Brit.Med.
    Bull. 34:34-38 (1978).
9. C.F. Phelps, Biosynthesis of Mucus Glycoprotein. Brit.Med.
    Bull. 34:43-48 (1978).
10. R. Jones and L. Reid, Secretory Cells and their Glycoproteins
    in Health and Disease, Brit.Med.Bull. 34:9-16 (1978).
11. F.A. Meyer, Mucus Structure: Relation to Biological Transport

Function, Biorheology 13:49-58 (1976).

12. F.A. Meyer and A. Silberberg, Structure and Function of Mucus, in "Respiratory Tract Mucus," Ciba Foundation Symposium 54 (new series) Elsevier, Amsterdam (1978); pp.203-218.

13. A. Courts, The N-Terminal Amino Acid Residues of Gelation. 2. Thermal degradation, Biochem.J. 58:77-79 (1954).

14. F.A. Meyer (to be published).

15. F.A. Meyer and A. Silberberg, The Rheology and Molecular Organization of Mucus, Biorheology 17:163-168 (1980).

16. M. King, A. Gilboa, F.A. Meyer and A. Silberberg, On the Transport of Mucus and its Rheological Simulants in Ciliated Systems, Am.Rev.Resp.Dis. 110:740-745 (1974).

17. R.A. Gelman and F.A. Meyer, Mucociliary Transference Rate and Mucus Viscoelasticity. Dependence on Dynamic Storage and Loss Modulus, Am.Rev.Resp.Dis. 120:553-557 (1979).

18. C.K. Shih, M. Litt, M.A. Khan and D.P. Wolf, Effect of Non-dialyzable Solids Concentration and Viscoelasticity on Ciliary Transport of Tracheal Mucus, Am.Rev.Resp.Dis. 115:989-995 (1977).

19. A.L. McCall, W.P. Potsic, C.K. Shih and M. Litt, Physico-chemical Properties of Human Middle Ear Effusions (Mucus) and their Relation to Ciliary Transport, The Larynscope 88:729-738 (1978).

20. M. King, Interrelation between Mechanical Properties of Mucus and Mucociliary Transport: Effect of Pharmacologic Inter-actions, Biorheology 16:57-68 (1979).

21. F.A. Meyer, M. King and R.A. Gelman, On the Role of Sialic Acid in the Rheological Properties of Mucus, Biochim. Biophys.Acta. 392:223-232 (1975).

22. W.C. Stewart, Weight-Carrying Capacity and Excitability of Excised Ciliated Epithelium, Amer.J. of Physiol. 152:1-5 (1948).

23. J.D. Ferry, "Viscoelastic Properties of Polymers," 2nd ed. Wiley, New York (1970); p.204.

24. T.L. Hill, "Introduction to Statistical Thermodynamics," Addison-Wesley,Reading (1960); pp.410-414.

25. A. Gilboa and A. Silberberg, Characterization of Epithelial Mucus and its Function in Clearance by Ciliary Propulsion, in "Air Pollution and the Lung," E.E. Aharonson, A. Ben-David, M.A. Klingberg (eds.), Wiley, New York (1975); pp.49-63.

26. T. van Vliet, N. Weiss and A. Silberberg (to be published).

27. F.A. Meyer, J. Vered and N. Sharon, Studies of Glycoproteins from Mucociliary Secretions, in "Mucus in Health and Disease" M. Elstein and D.V. Parke (eds.), Plenum Press, New York (1977); pp. 239-249.

28. A. Allen, Structure of Gastrointestinal Mucus Glycoproteins and the Viscous and Gel-Forming Properties of Mucus, Brit.Med.Bull. 34:28-33 (1978).

29. M.D.G. Oates, A.C. Rosbottom and J. Schrager, The Composition of Human Gastric Mucin, Mod.Probl.Paediat. 19:11-21 (1977).

30. J.P. Pearson and A. Allen, A Protein, 70,000 Molecular Weight, is Joined by Disulphide Bridges to Pig Gastric Mucus Glyco-proteins, Trans.Biochem.Sol. 8:388-389 (1980).

31. G.P. Sachdev O.F. Fox, G. Wen, T. Schroeder, R.C. Elkins, and R. Carubelli, Isolation and Characterization of Glyco-proteins from Canine Tracheal Mucus, Biochim.Biophys.Acta. 536:184-196 (1978).

32. J. Pearson, A. Allen and C. Venables, Gastric Mucus: Isolation and Polymeric Structure of the Undegraded Glycoprotein : Its Breakdown by Pepsin, Gastroenterology 78:709-715 (1980).

33. A. Allen, M. Mantle and J.P. Pearson, The Polymeric Structure of Mucus Glycoproteins, in "Glycoconjugates," R. Schauer et al. (eds.), Georg Thieme, Stuttgart (1979); pp. 42-43.

34. T. Marshall and A. Allen, The Isolation and Characterization of the High-Molecular-Weight Glycoprotein from Pig Colonic Mucus, Biochem.J. 173:569-578 (1978).

35. D.M. Carlson, Chemistry and Biosynthesis of Mucin Glycopro-teins, in "Mucus in Health and Disease," M. Elstein and D.V. Parke (eds.), Plenum Press, New York (1977); pp.251-273.

36. K.S.P. Bushana-Rao and P.L. Masson, A Tentative Model for the Structure of Bovine Oestrus Cervical Mucin, in "Mucus in Health and Disease," M. Elstein and D.V. Parke (eds.), Plenum Press, New York (1977); pp. 275-282.

37. J.F. Forstner, I. Jabbal, R. Qures , D.I.C. Kells, and G.G. Forstner, The Role of Disulphide Bonds in Human Intestinal Mucin, Biochem.J. 181:725-732 (1979).

38. G.P. Roberts, Structural Studies on the Glycoproteins from Bovine Cervical Mucus, Biochem.J. 173:941-947 (1978).

39. P.W. Kent, Chemical Aspects of Tracheal Glycoproteins, in "Respiratory Tract Mucus," Ciba Foundation Symposium 54 (new series), Elsevier, Amsterdam (1978); pp. 155-174.

MODIFICATION OF THE RHEOLOGICAL PROPERTIES OF MUCUS BY DRUGS

Christopher Marriott

Department of Pharmacy, Brighton Polytechnic
Moulsecoomb, Brighton BN2 4GJ, U.K.

The rheological properties of mucus secretions from various sites in the body are related to their particular physiological function. Although in normal circumstances the body produces secretions with optimal rheological properties to fulfil such functions, during disease the mucus produced may be too thick or too thin and in both cases may behave sub-optimally. In such situations it becomes attractive to administer chemical agents or drugs, either locally or systemically, which modify the rheological properties in an attempt to correct function. Furthermore, in certain cases it may be desirable to artificially manipulate the consistency of the mucus in order to produce a specific effect.

It is also important to be aware if a drug which is to be administered to man has any effect on mucus structure since such activity may give rise to side effects.

The significance of drug mucus interactions makes it essential that proper methods of measurement are used in the evaluation of any effect. This paper will briefly review both the methods of measurement and those agents which have been shown to affect mucus structure.

Methods of Rheological Evaluation

The methods which have been used to evaluate the rheological or flow properties of mucus are shown in Table 1.

The viscometric techniques all suffer the major disadvantage in that they are only capable of measuring the viscous properties

Table 1.   The types of rheological test that have
           been used for mucus

Capillary viscometers   - U-tube
                               - suspended level
Falling sphere viscometer
Efflux viscometer
Rotational viscometer   - concentric cylinder
                               - cone-plate
Creep compliance test
Oscillatory test

of mucus and do not provide any measure of elasticity. Also, the
capillary, falling sphere and efflux methods have the added
disadvantage in that the mucus must either be diluted or
homogenized before the instrument can be filled and even then the
exclusion of air bubbles is virtually impossible. The fact that
the native secretion has to be altered in order to measure its
properties is a severe drawback.

In contrast, rotational viscometers can be loaded, with
proper care, without any significant alteration in the gel structure.
However, the subsequent test is in itself destructive because of
the high rates of shear that are developed during measurement.
Often the type of hysteresis loop which is produced when shear rate
is plotted as a function of shear stress (Figure 1) is taken to
indicate that the mucus is thixotropic (i.e. the gel structure
has been reversibly reduced during shearing and will recover on
standing). This is not the case and repeat determination on the
same sample does not produce a similar curve to the original;
curve B (Figure 1) is more usual. Thus, the method of measurement
destroys the very structure it is intended to measure. Consequently
values of viscosity, which can be obtained from the reciprocal of
tangents drawn to the curve in Figure 1, are only of the order
10^1 - 10^3 P for sputum.

In common with all gels, mucus is viscoelastic which means
that it exhibits both viscous and elastic properties simultaneously.
Consequently, any measurement technique must be capable of measuring
both elements of such behaviour. The situation is further complicated
by the delicate nature of the mucus gel which will only behave in
a truly viscoelastic manner at very small strains (displacements).
Two different types of test exist, the static or creep compliance
test and the dynamic or oscillatory test. In the former a small
constant force or stress is applied to the sample and the resultant
displacement or strain is measured. The shape of the displacement
curve which is produced is the same for all viscoelastic liquids

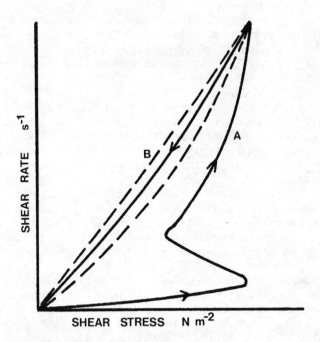

Fig 1   Typical rheogram for a mucus gel

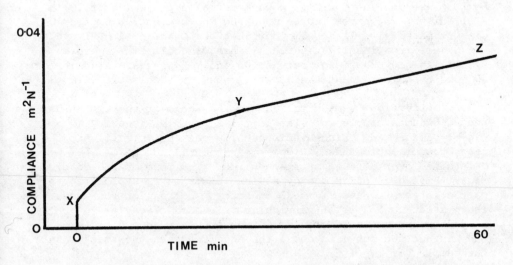

Fig 2   Creep compliance curve produced for a mucus gel

and it is only the magnitude which varies. If the strain is
divided by the stress then, since the latter is constant, the shape
of the curve will not be affected but the units of the ordinate
will be converted into those of compliance (M^2N^{-1}) which is the
reciprocal of elasticity. A typical creep compliance curve for
tracheobronchial mucus is shown in Figure 2.

The region o - x corresponds to an initial elastic response
and the reciprocal is usually given the symbol J_0 . The linear
region y - z can be associated with purely viscous behaviour and
since the strain (and rate of strain) are very low then the
viscosity (η_0), which is calculated from the reciprocal of the
slope of this region, is typically 10^5 - 10^6 P. The curved section
which links the elastic and viscous portions is usually referred
to as the retarded elastic region (x - y) and typifies viscoelastic
behaviour in that the mucus flows under the influence of elastic
retardation. It therefore can only be quantified in terms of both a
viscosity and an elasticity.

Since the results are obtained in the form of a compliance
then experiments with a thicker sample would produce a curve which
was closer to the abscissa in Figure 2 whilst in the case of a
thinner sample it would move towards the ordinate.

In the dynamic test the sample is exposed to a sinusoidally
oscillating stress or strain and the respective displacement or
stress in the sample, which should also vary sinusoidally, is
monitored. For a perfectly elastic material the two sine waves will
be in phase whereas for a viscous material a phase lag of 90^0
will exist. The phase lag between the two cycles for a viscoelastic material will be somewhere between O and 90^0. This value
can be used to classify materials since viscoelastic liquids
should exhibit phase lags between 45 and 90^0 whereas values below
45^0 would indicate that the material was a viscoelastic solid.

If provision is made to measure the ratio of the amplitudes
then a dynamic modulus of elasticity (G') and a viscosity (η')
can be calculated. Since it is usually quite simple to vary the
rate of oscillation then it does mean that G' and η' can be
determined over a range of frequencies. The form of the double
logarithmic plot of frequency against G' or η' is typical for a
particular system and Figure 3 indicates such a plot for mucus.
The plateau region is of particular relevance since its length
is indicative of the degree of entanglement in the gel. Also,
the further the curve is above the abscissa then the thicker is
the sample.

A correlation exists between the static and the dynamic test
and a creep compliance experiment is equivalent to an oscillatory
experiment at low frequency.

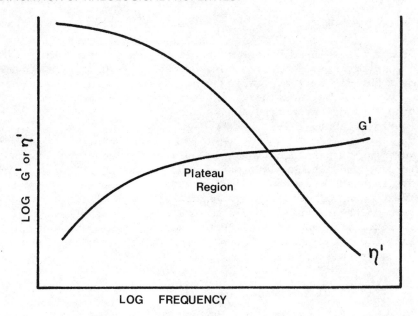

Fig 3   Double logarithmic plot of frequency
against G' and η' for a mucus gel

The use of the excised frog palate to provide a measure of
mucociliary transport is in effect a rheological measurement since
it has been shown to be extremely sensitive to changes in elasticity
(Shih et al, 1977). If the ciliated upper palate of the frog is
removed and kept in a saturated environment at a suitable
temperature then it will continue to secrete and clear mucus.
Eventually mucus production ceases although the cilia continue
to beat, and the depleted palate can then be used to evaluate the
clearance rate of exogenous mucus samples. Also, before depletion,
the endogenous mucus clearance can be used to evaluate the effect
of topically and parenterally administered drugs.

This technique has been widely used in recent years to provide
information on the significance of rheological measurements made
with creep and oscillatory methods.

Agents Affecting Mucus Gel Structure

These can be most conveniently divided into those which thin
and those which thicken mucus gels.

Mucus Thinning Agents

It has become traditional to refer to those agents which
reduce the viscosity and elasticity of mucus as mucolytics. Such
a term is particularly misleading since few of the active compounds
induce lysis of the mucus gel. Alternative terms that have been
suggested include mucotropic (Davis and Deverell, 1977) and
mucokinetic (Ziment, 1978).

Numerous preparations which are claimed to be expectorants
contain compounds which might modify mucus clearance from the lung.
However, in most cases the ingredients of such preparations will
not affect mucus directly. This paper will be limited to those
materials which are known to exert either a direct, after local
administration, or an indirect effect, after oral or parenteral
administration, on mucus structure. The important groups of compounds
are considered below.

Iodides

Iodides exert both a direct and an indirect effect on mucus
structure. Although a bronchomucotropic effect is observed in
vivo (Ziment, 1978) direct application of potassium iodide to mucus
gels produces a reduction in the rheological properties (Marriott
and Richards, 1974; Martin et al, 1980). It is probable that the
effect is due either to an induced change in the configuration of
the glycoprotein molecule or that it is lyotropic in nature. Both
inorganic and organic iodides are used therapeutically.

Cysteine Derivatives

The active cysteine derivatives exert their activity by breaking
disulphide bonds which exist in mucus. L-cysteine, N-acetylcysteine
cysteamine and methyl cysteine have all been shown to be directly
effective (Lieberman, 1968; Davis and Deverell, 1977; Martin et al,
1980). In contrast the blocked thiol compounds such as S-carboxy-
methylcysteine and S-methylcysteine do not affect mucus structure
directly (Martin et al, 1980). However, certain blocked thiol
compounds have been shown to release the free sulphydryl in vivo
and to subsequently exert an effect (Tabachnik et al, 1980).
Although N-acetylcysteine has been used by the topical route for
many years (Chodosh, 1980) it has recently been observed that an
effect is produced following oral administration (Verstraeten, 1980).

Sulphydryl Bond Breakers

Other compounds have the capacity to disrupt disulphide bonds
and a wide range have been evaluated. Examples of active compounds
include dithiothreitol, 2-mercaptoethanesulphonate, glutathione,
D- and DL-penicillamine, N-acetyl-penicillamine and sodium meta-

bisulphite (Lieberman, 1968; Hirsch et al, 1969; McNiff et al, 1975; Khan et al, 1976; Davis and Deverell, 1977; Marriott et al, 1981b). All these compounds exhibit high activity and this is particularly true with dithiothreitol. However, nebulization of this compound causes intense irritation (Lightowler and Lightowler, 1971) and it is suggested that compounds of this type may be too effective for clinical use.

Enzymes

An extensive range of enzymes have been demonstrated to exhibit a mucus thinning effect. The most active would appear to be deoxyribonuclease although trypsin, chymotrypsin, elastase, ficin, papain and protease have all been shown to be active (Lieberman, 1968; Marriott et al, 1981a). The contact time is critical and it is apparent that the reduction in elasticity is not as great as that in viscosity (Marriott et al, 1981a).

Miscellaneous Agents

These include materials as diverse as simple elctrolytes and surfactants. An illustrative selection would include caesium bromide and chloride, sodium chloride, sodium and potassium thiocyanate, guanidine, urea, Triton X-100, sodium dodecyl sulphate and cetyltrimethylammonium bromide. The activities are wide ranging and although few are used clinically (Ziment, 1978) a number are used to solubilise mucus gels for experimental work (Khan et al, 1976).

Mucus Thickening Agents

These have attracted much less attention than the compounds mentioned in the previous section. Indeed it is only in recent years that authors have bothered to report anything other than a mucus thinning effect.

Di- and trivalent cations have been demonstrated to thicken mucus gels and the effect is concentration dependent (Forstner and Forstner, 1975; Marriott et al, 1979, Crowther and Marriott, 1979). It appears that the changes produced are due to conformational alterations of the glycoprotein in solution. It has also been observed that cation levels in human cervical mucus alter with the stage of the cycle (Kosasky et al, 1973) and this may be associated with the fertility controlling function.

The tetracycline group of antibiotics all thicken mucus with the exception of pyrrolidinomethyltetracycline (Marriott and Kellaway, 1975), a highly water soluble analogue. The activity is dependent upon the molecular structure, oxytetracycline being

the most active. The effect can be explained at least in part
by the fact that tetracycline molecules bind to glycoproteins
(Brown, 1980).

Mucus gels can also be thickened by the addition of other
macromolecules. Bovine serum albumin, IgG, IgA and DNA (Marriott
et al, 1981a) are all potent in this respect and the thickening
is presumably of significance in lung disease once infection occurs.

Boric acid and borates crosslink macromolecules containing
hydroxyl groups and therefore thicken mucus gels (Davis and Deverell,
1977). The action depends on the borate ion and is therefore more
efficient and rapid under alkaline conditions. The bifunctional
agent glutaraldehyde forms intermolecular links between amino
groups and Gleman and Meyer (1979) have shown that if effectively
crosslinks bovine cervical mucus. Azo dyes such as congo red
crosslink mucus in a similar manner to borate ions (Davis and
Deverell, 1977).

Progesterone delivered locally from an intra-uterine device
exerts contraceptive activity and this compound has also been
shown to interact with mucus gels to produce a thickening effect
(Martin et al, 1981). Finally pentobarbitone anaesthesia has
been shown to thicken mucus secretions in dogs (King et al, 1979).

Agents which modify mucus gel structure appear to abound.
However, the lack of successful clinical application would suggest
that the problem is one of delivery rather than activity.

REFERENCES

Brown, D.T., 1980, Ph.D. Thesis, University of Nottingham.
Chodosh, S., 1980, Acetylcysteine in chronic bronchitis,
    Eur.J.Respir.Dis., 61(S111):90.
Crowther, R.S. and Marriott, C., 1979, Cation induced changes in
    the biophysical properties of mucins, J.Pharm.Pharmac., 31:89P.
Davis, S.S. and Deverell, L.C., 1977, Rheological factors in
    mucociliary clearance, Mod.Probl.Paediat., 19:207.
Forstner, J.F. and Forstner, G.G., 1975, Calcium binding to
    intestinal goblet cell mucin, Biochim.Biophys.Acta, 386:283.
Gelman, R.A. and Meyer, F.A., 1979, Mucociliary transference rate
    and mucus viscoelasticity dependence on dynamic storage and
    loss modulus, Am.Rev.resp.Dis., 120:553.

Hirsch, S.R., Zastrow, J.E. and Kory, R.C., 1969, Sputum liquefying
    agents: a comparative in vitro evaluation, J.Lab.Clin.Med.,
    74:346.
Khan, M.A., Wolf, D.P. and Litt, M., 1976, Effect of mucolytic
    agents on the rheological properties of tracheal mucus,
    Biochim.Biophys.Acta, 444:369.
King, M., Engel, L.A. and Macklem, P.T., 1979, Effect of
    pentobarbital anesthesia on rheology and transport of
    canine tracheal mucus, J.Appl.Physiol., 46:504.
Kosasky, H.J., Kopito, E.L., Somers, H.S. and Shwachman, H.,
    1973, Changes in water and electrolytes in human cervical
    mucus during treatment with chlormadinone acetate,
    Fertil.Steril., 24:507.
Lieberman, J., 1968, Measurement of sputum viscosity in a cone-
    plate viscometer, Amer.Rev.resp.Dis., 97:662.
Lightowler, J.E. and Lightowler, N.M., 1971, Comparative mucolytic
    studies on dithiothreitol, N-acetylcysteine and L-cysteine
    on human respiratory mucus in vitro and their effects on
    the role of flow of mucus in the exposed trachea of the rat
    on topical application, Arch.int.Pharmacodyn., 189:53.
Martin, R., Litt, M. and Marriott, C., 1980, The effect of
    mucolytic agents on the rheologic and transport properties
    of canine tracheal mucus, Amer.Rev.resp.Dis., 121:495.
Martin, G., Marriott, C. and Kellaway, I.W., 1981, The interaction
    of progesterone with mucus glycoproteins, Pharm.Acta Helv.,
    56:5.
Marriott, C., Beeson, M.F. and Brown, D.T., 1981a, Biopolymer
    induced changes in mucus viscoelasticity, this conference.
Marriott, C., Brown, D.T. and Beeson, M.F., 1981b, Evaluation of
    mucolytic activity using a purified mucus glycoprotein gel,
    this conference.
Marriott, C. and Kellaway, I.W., 1975, The effect of tetracyclines
    on the viscoelastic properties of bronchial mucus,
    Biorheology, 12:391.
Marriott, C. and Richards, J.H., 1974, The effects of storage and
    of potassium iodide, urea, N-acetylcysteine and Triton X-100
    on the viscosity of bronchial mucus, Brit.J.Dis.Chest, 68:171.
Marriott, C., Shih, C.K. and Litt, M., 1979, Changes in the gel
    properties of tracheal mucus induced by divalent cations,
    Biorheology, 16:331.
McNiff, E., Clemente, E. and Fung, H.L., 1975, In vitro comparison
    of the mucolytic activity of sodium metabisulfite,
    N-acetylcysteine and dithiothreitol, Drug Dev.Comm., 1:507.
Shih, C.K., Litt, M., Khan, M.A. and Wolf, D.P., 1977, Effect of
    non-dialysable solids concentration and viscoelasticity on
    ciliary transport of tracheal mucus., Am.Rev.resp.Dis., 115:989.
Tabachnik, N.F., Peterson, C.M. and Cerami, A., 1980, Studies on
    the reduction of sputum viscosity in cystic fibrosis using
    an orally absorbed protected thiol, J.Pharmacol.Exp.Ther.,
    214:246.

Verstraeten, J -M., 1980, Mucolytic treatment in chronic airways
     obstruction. Double-blind clinical trial with acetylcysteine,
     bromhexine and placebo., Eur.J.Respir.Dis., 61(S111):77.
Ziment, I., 1978, Respiratory Pharmacology and Therapeutics,
     W.B. Saunders, Philadelphia.

A RHEOLOGICAL STUDY OF MUCUS - ANTIBIOTIC INTERACTIONS

David T. Brown, Christopher Marriott* and
Malcolm F. Beeson†

Department of Pharmacy, University of Nottingham
Nottingham NG7 2RD, U.K., *Department of Pharmacy
Brighton Polytechnic, Brighton BN2 4GJ, U.K. and
†Beecham Pharmaceuticals Research Division, Great
Burgh, Epsom KT18 5XQ, U.K.

Antibiotic therapy is commonly used both during acute
exacerbations of bronchitis and prophylactically. Unless the
antibiotic is instilled directly into the lung, then the transfer
of the drug across the blood-bronchial barrier is a critical step.
The maintenance of a homogeneous bactericidal or bacteriostatic
concentration in the mucus is dependent upon the ability of the
drug to diffuse through the mucus gel. The thickened hypersecretory mucus which is produced during infection may present
a barrier to diffusion and this in turn may be affected by the
antibiotic itself. Earlier work (Marriott and Kellaway, 1975)
has indicated that tetracyclines exert a thickening effect on
sputum. In this study we have examined the effect of a wide range
of antibiotics on the rheological and mucociliary transport
properties of a purified glycoprotein gel (Marriott et al, 1979).

U-tube viscometry was used to determine the limiting viscosity
number (LVN) and Huggins constant (K'): results are shown in
Table 1.

The only changes in K' which are significantly different
from the control (p > 0.01) are those for tetracycline and
oxytetracycline, whilst in the case of LVN, the value for
doxycycline was also increased. An increase in LVN represents
an increase in hydrodynamic volume and it has been suggested
(Eirich and Riseman, 1949) that a decrease in K' indicates that
a polymer has been transferred to a better solvent, the molecule
becoming more extended, kinked, looped and 'softened' allowing

85

Table 1. The Effect of Antibiotics on the Limiting
Viscosity Number and Huggins Constant

Agent	Concentration (% w/v)	LVN ml g^{-1}	K'
Control	0.1	228.5	0.641
Penicillin G	0.1	229.2	0.538
Ampicillin	0.1	231.2	0.529
Cloxacillin	0.1	227.0	0.643
Doxycycline	0.1	238.2	0.546
Tetracycline	0.1	266.3	0.251
Oxytetracycline	0.1	271.3	0.212
Erythromycin	0.1	226.5	0.579
Gentamycin	0.1	229.6	0.472
Cephalexin	0.1	228.2	0.598

better penetration of solvent. The decreased value of K' can
therefore be considered to indicate an increase in the degree of
internal deformability of the glycoprotein which also exhibits
lower resistance to solvent penetration. Thus it appears that
the tetracycline solutions are superior solvents for the glycoproteins and suggests that an interaction occurs resulting in
molecular expansion.

    The results of creep compliance testing and mucociliary
clearance measurements are shown in Table 2.

    Once more the only antibiotics which are seen to affect mucus
gel structure are the tetracyclines which produce thickening in
the rank order doxycycline < tetracycline < oxytetracycline. More
importantly, the thickening resulted in a decrease in transport
rate on the frog palate and in the case of 1% w/v tetracycline
the decrease was as high as 75%. Such a change could potentially
affect clearance from the lung. Although changes in the pH of
the mucus gel could explain the decrease in clearance rate due
to an inhibitory effect on ciliary activity, the high buffer
capacity of the glycoprotein gels should counteract this
possibility (Marriott and Kellaway, 1975). The order of effectiveness is inversely related to the oil/water partition coefficient which is known to correlate directly with the degree
of tetracycline binding to serum albumin (Kellaway and Marriott,
1978), a process which has been shown to be essentially hydrophobic. Hence, these results indicate that the less hydrophobic

Table 2. The Effect of Antibiotics on Viscoelastic Parameters and Mucociliary Clearance

Agent	Concentration % w/v	Viscosity Poise	Elasticity Nm^{-2}	% of Control Clearance Rate
Control		720	1.39	100
Oxytetracycline	0.05	943	1.82	73
	0.10	1979	2.23	52
	0.50	7408	3.73	47
	1.00	9396	5.10	25
Control		706	1.58	100
Tetracycline	0.05	823	1.87	85
	0.10	1265	2.12	71
	0.50	4631	2.51	62
	0.91	6490	4.43	48
Control		662	1.46	100
Doxycycline	0.05	807	1.54	91
	0.10	981	1.63	82
	0.50	3422	1.60	78
	1.00	4057	2.12	63
	n = 3	p > 0.01 (0.5%)		
Control		844	4.73	
Cloxacillin	1.00	868	3.82	103
Ampicillin	1.00	733	4.27	94
Penicillin G	1.00	921	3.90	100
Tetracycline	0.91	35458	15.41	48
Control		2125	2.97	–
Cephalexin	1.00	2542	2.92	–
Gentamycin	1.00	2643	3.17	–
Control		4242	4.72	–
Cephradine	1.00	3842	3.74	–
Erythromycin	1.00	4161	3.63	–
	n = 3	No significant difference		

the tetracycline, the greater the extent of interaction with mucus glycoproteins. The interaction must therefore be predominantly hydrophilic or ionic in nature. Since the least active tetracycline was found to be doxycycline which is one of the tetracyclines which does not chelate calcium ions then it is tempting to speculate that such ions may be involved in the thickening interaction. Since the level of calcium in the mucus gel was only 0.3 mM l^{-1} then it is unlikely that such an explanation is feasible.

The expansion of the glycoprotein demonstrated by U-tube viscometry explains the increase in gel strength since a higher degree of entanglement would result. The degree of protein binding to the mucus glycoproteins is low and, since the penicillins used in this work exhibit a range of affinities for proteins, it is concluded that binding of tetracycline to the glycoprotein cannot solely explain the observed effects and that the changes are most likely related to alterations in molecular configuration or water structure.

REFERENCES

Eirich, F. and Riseman, J., 1949, Some remarks on the first
     interaction coefficient of the viscosity-concentration
     equation, J.Polym.Sci., 4:417.
Kellaway, I.W. and Marriott, C., 1978, Influence of drug hydro-
     phobicity on the binding of tetracyclines to albumin,
     Canad.J.Pharm.Sci., 30:90.
Marriott, C. and Kellaway, I.W., 1975, The effect of tetracyclines
     on the viscoelastic properties of bronchial mucus,
     Biorheology, 12:391.

BIOPOLYMER INDUCED CHANGES IN MUCUS VISCOELASTICITY

Christopher Marriott, Malcolm F. Beeson* and
David T. Brown†

Department of Pharmacy, Brighton Polytechnic
Brighton BN2 4GJ, U.K., *Beecham Pharmaceuticals
Research Division, Great Burgh, Epsom KT18 5XQ,
U.K. and †Department of Pharmacy, University of
Nottingham, Nottingham NG7 2RD, U.K.

The occurrence of macromolecules, other than glycoproteins, in the sputum of patients suffering from chronic obstructive airways disease is well documented (Ziment, 1978). The difficulties associated with the collection of 'normal' tracheo-bronchial mucus from healthy individuals renders it difficult to determine the effect that the contaminants which appear during disease exert. For example Creeth et al (1977) have demonstrated that it is impossible to remove free protein by isopynic ultracentrifugation in CsBr and a further separation in CsCl is necessary. This may suggest that a certain amount of free protein is associated with the glycoprotein even in the normal tracheobronchial tree. The transudation of serum proteins during infection together with the extracellular and breakdown products of bacteria are usually considered to be associated with an increase in the consistency of the sputum produced. However, conflicting reports have been published concerning the effect of DNA on the rheological properties of sputum (Puchelle et al, 1973; Bornstein et al, 1978). Therefore, it does appear important to establish the contribution of other constituents of sputum to the viscoelasticity; in this work we have used a mucus gel purified from human sputum for this purpose (Marriott et al, 1979). The substances tested were bovine serum albumin (BSA), human secretory IgA and IgG, calf thymus deoxyribonucleic acid (DNA), dipalmitoyl lecithin (DPL) and a lipid fraction (LF) separated from human sputum. In addition, the effect of calcium ions was determined. Table 1 shows the effect of these materials on the viscoelastic properties and mucociliary transport rates of the mucus gels.

Table 1.   The Effect of Sputum Contaminants on the
           Viscosity, Elasticity and Transport Rate
           (MTR) of Mucus

Agent	Concentration	MTR	Viscosity Poise	Elasticity Nm^{-2}
Control			2469	3.05
LF			1962	3.55
DPL			3899	5.26
Control Ca^{2+}			981	2.49
	5 mM		1012	1.85
	10 mM		1211	2.04
	15 mM		1511	2.43
	20 mM		1622	3.00
	30 mM		2631	3.27
	n = 3	$p > 0.01$ (30 mM)		
Control		7.66	1269	2.18
BSA	2%	7.22	1994	2.95
	4%	5.78	3063	3.13
	5%	3.62	3403	3.47
	10%	2.76	5766	4.51
	20%	2.96	22399	6.05
Control		7.33	613	2.64
IgG	2%	6.71	843	3.80
	4%	5.09	1050	6.27
	6%	1.95	3249	6.86
Control		7.66	1044	1.69
IgA	2%	6.26	2003	2.80
	4%	4.34	4901	5.40
	6%	1.73	58945	16.47
Control		7.33	2561	2.78
DNA	0.05%	3.09	2654	3.33
	0.1%	2.42	4541	4.08
	0.5%	1.93	5058	5.52
	1.0%	1.42	6098	11.32
	2.0%	0.72	10818	19.23
	4.0%	0.17	183428	84.74
Control		7.00	491	2.11
BSA	5.0%	3.80	918	3.82
IgG	5.0%	2.39	1354	7.39
IgA	5.0%	1.88	2161	10.42
DNA	5.0%	0	92082	33.90
	n = 3	$p > 0.01$ (5%)		

The lipid fraction produced no significant change in the
viscoelastic parameters and although there is an apparent change
in the case of dipalmitoyl lecithin, because of the wide spread
in the results which were induced by this material, the changes
are not significant at the 95% level. This would support the
findings of Martin et al (1978) who found that purified egg
phosphatidylcholine (lecithin) had no effect on mucus gel structure
whereas lysophosphatidylcholine was strongly mucolytic. In a
similar manner, the changes induced by calcium ions are only
significant ($p > 0.01$) in the instance of viscosity and not the
elasticity of the mucus gel. The results of other workers
(Forstner and Forstner, 1976; Marriott et al, 1979) have differed
according to the type of material used and concentration, but it
is clear that calcium ions induce conformational changes rather
than producing cross-links.

All the biopolymers which were added thickened the mucus gels
and the order of effectiveness is directly related to the molecular
weight of the added species. The viscosities of solutions of the
same concentration as added to the mucus gels were determined and,
with the exception of DNA, the highest value obtained was approximately 2 cP. In the case of DNA, solutions of greater than 1%
were viscoelastic and creep analysis indicated that for a 5% gel
the viscosity was 4439 P. The resultant viscoelasticity of the
mixtures is much greater than the result of the summation or
multiplication of the viscosities of the separate gel and protein
preparations; with DNA, however, this is only true for the
additive viscosities. The increase in viscoelasticity is obviously
of consequence since marked changes in the transport of the
thickened gels by a ciliated epithelium were observed (Table 1).
The gels containing more than 2% DNA exhibited transport rates
which were less than 10% of those observed for the control sample
and in the case of the 5% sample no movement was observed. This
supports the hypothesis that highly viscoelastic materials (even
mucus) cannot be transported by a ciliated epithelium.

When the viscosity is plotted as a function of biopolymer
concentration then a marked increase is observed at a concentration
of 2% for DNA and 4% for IgA. It is supposed that some type of
molecular 'filling' occurs at these concentrations. Gel chromatography experiments, together with the fact that the viscoelasticity
of the mixtures was reduced by the addition of neutral electrolytes
such as sodium chloride, have indicated that the interaction between the macromolecules is non-covalent in nature and that it most
likely involves hydrophobic, hydrophilic and ionic interactions.

All the polymers studied are considered to be capable of ful-
filling a structural role in the mucus gel which may involve a
significant change in the spatial arrangement of the glycoproteins.
It could well be that the changes which are observed in mucus

viscoelasticity in vivo are related to differences in protein/
glycoprotein ratio. For example, in the lung serum transudation
and bacterial infection are a source of the substances tested in
this work and will almost certainly account for the increase and
variation in the rheological properties of the sputum produced:
lung clearance of this material is notoriously difficult. The
fall in the protein/mucin ratio which has been observed in peri-
ovulatory cervical mucus could also explain the concomitant
decrease in consistency and the resultant ability to transfer
spermatozoa: luteal phase mucus contains more protein and is
hostile to spermatozoa (Van Kooij et al, 1979). Similar protein-
glycoprotein interations could account for the thickening of
mucus which occurs in kerato-conjunctivitis sica, vernal catarrh
and meconium ileus.

REFERENCES

Bornstein, A.A., Chen, T.-M. and Dulfano, M.J., 1978, Disulphide
    bonds and sputum viscoelasticity, Biorheology, 15:261.
Creeth, J.M., Bhaskar, K.R., Horton, J.R., Das, I., Lopez-Vidriero,
    M. and Reid, L., 1977, The separation and characterization
    of bronchial glycoproteins by density gradient methods,
    Biochem.J., 167:557.
Forstner, J.F. and Forstner, G.G., 1976, Effects of calcium on
    intestinal mucin: implications for cystic fibrosis,
    Pediat.Res., 10:609.
Marriott, C., Barrett-Bee, K. and Brown, D.T., 1979, A comparative
    evaluation of mucus from different sources, in: "Glycocon-
    jugates", R. Schauer et al, eds., Georg Thieme, Stuttgart.
Martin, G.P., Marriott, C. and Kellaway, I.W., 1978, Direct
    effect of bile salts and phospholipids on the physical
    properties of mucus, Gut, 19:103.
Puchelle, E., Zahm, J.M. and Havez, R., 1973, Relationship
    between the biochemical constituents and rheological
    properties of sputum, Bull.Physio-path.Resp., 9:237.
van Kooij, R.J., Roelofs, H.M.J. and Kramer, M.F., 1979,
    Composition of human cervical mucus and synthetic activity
    of endocervical cells during the menstrual cycle, in
    "Glycoconjugates", R. Schauer et al, eds., Georg Thieme,
    Stuttgart.
Ziment, I., 1978 "Respiratory Pharmacology and Therapeutics",
    W.B. Saunders, Philadelphia, p.42.

EVALUATION OF MUCOLYTIC ACTIVITY USING A PURIFIED MUCUS

GLYCOPROTEIN GEL

Christopher Marriott, David T. Brown* and Malcolm
F. Beeson†

Department of Pharmacy, Brighton Polytechnic
Brighton BN2 4GJ, U.K., *Department of Pharmacy
University of Nottingham, Nottingham NG7 2RD, U.K.
†and Beecham Pharmaceuticals Research Division
Great Burgh, Epsom KT18 5XQ, U.K.

The prime characteristic of chronic obstructive airways
disease is hypersecretion of bronchial mucus which is invariably
too thick to be cleared by the lung mucociliary system. In this
situation the objective of therapy must be to reduce the viscoelastic properties of the abnormal secretion to the optimum
value for removal by lung clearance mechanisms. The group of
drugs which are designed for this purpose have been known as
mucolytics and, while many compounds have been shown to be effective in vitro, in clinical practice few can be considered successful (Ziment, 1978). Comparison of reported in vitro experiments
is difficult because of the diversity and quality of the mucus
systems used. The recent development of a model mucus preparation
apparently consisting entirely of purified human bronchial mucus
glycoproteins (Marriott et al, 1979) which are capable of gelation
at low concentrations, should provide a suitable basis for the
comparative evaluation of mucolytic agents.

The viscoelastic properties of the mucus gels were measured
by creep viscometry and results are summarised in terms of the
residual shear viscosity, η_0, and the elasticity, E_0, for the
compounds in Table 1.

In the case of sodium dodecyl sulphate the resultant 'gel'
was so fluid that an elasticity could not be determined. Some
of the observed effect may have been due to a reduction in inter-
facial tension resulting in sample slippage in the rheometer. In
addition, dithiothreitol, N-acetylcysteine, d-penicillamine and

93

urea were all shown to be mucolytic in that they produced a marked
decrease in both viscosity and elasticity.

Table 1. The Effect of Mucolytic Agents on the Viscoelastic
          Properties of Mucus Gels

Agent	Viscosity, η_0 (Poise)	Elasticity, E_0 (Nm^{-2})
Control	4360	2.31
Dithiothreitol 1%	23	0.18
N-acetylcysteine 1%	402	0.18
d-penicillamine 1%	43	0.42
Urea 1 M	174	0.70
KCNS 0.1 M	526	0.88
KI 0.1 M	771	0.79
CsBr 0.1 M	1052	1.21
CsCl 0.1 M	912	1.66
Protease 0.1%	753	1.18
Papain 0.1%	1278	2.06
Disodium tetraborate 1%	4500	13.26
Sodium dodecyl sulphate 0.1%	0.12	-

(Each value is the mean of at least three determinations).

   The limiting viscosity numbers derived by U-tube viscometry
measurements after extrapolation to infinite dilution are shown
in Table 2.

   Although sodium dodecyl sulphate was potently mucolytic to
mucus gels it is apparent that the glycoprotein molecule is
unaffected. In contrast, N-acetylcysteine obviously reduces
either the molecular size or shape of the glycoprotein.

Table 2. The Effect of Mucolytic Agents on the Limiting
          Viscosity Number (LVN) of Mucus Glycoproteins

Agent	Concentration	LVN ml g^{-1}
Control	-	228.5
Sodium dodecyl sulphate	0.030 M	217.8
Potassium thiocyanate	0.022 M	215.0
N-acetylcysteine	0.030 M	120.0

   The pronounced effects of the agents which disrupt disulphide
bonds underlines the importance of this type of bond in both the
intra- and inter-molecular stabilisation of the mucus gel. However,
the fact that agents which only affect non-covalent forces produce

approximately a 75% reduction in the gel structure indicates that secondary bonds also make a significant contribution. Indeed we have observed that sputum can be completely dispersed on agitation in 0.22M potassium thiocyanate. The relatively poor effect observed with the proteolytic enzymes may well be due to the fact that the 'naked' peptide region is shrouded by adjacent sugar side chains.

The results show good agreement with those of Martin et al (1979) using canine tracheal mucus and would indicate that the purified mucus gel is a realistic and senisitive model for mucolytic evaluation. The results with d-penicillamine are of particular interest since the high mucolytic activity could explain the gastrointestinal side effects, including the reactivation of peptic ulcer, which have been observed when this compound is administered orally.

REFERENCES

Marriott, C., Barrett-Bee, K. and Brown, D.T., 1979, A comparative evaluation of mucus from different sources, in: "Glycoconjugates", R. Schauer et al, eds., Georg Thieme, Stuttgart.
Martin, R., Litt, M. and Marriott, C., 1980, The effect of mucolytic agents on the rheologic and transport properties of canine tracheal mucus, Amer.Rev.resp.Dis., 121:495.
Ziment, I., 1978, Respiratory Pharmacology and Therapeutics, W.B. Saunders, Philadelphia.

RHEOLOGICAL STUDIES ON NATIVE PIG GASTRIC MUCUS GEL

Alan E. Bell[*], Adrian Allen[*],
Edwin Morris and David A. Rees
*Department of Physiology, University of Newcastle
upon Tyne NE1 7RU, U.K. and Unilever Research
Colworth House, Sharnbrook, Bedford MK44 1LQ, U.K.

Gastric mucus is secreted as a discrete gel phase which is not readily soluble in its aqueous environment. Previous studies (Allen, 1977) have demonstrated that the gel matrix is formed by non-covalent interactions between glycoprotein molecules of 2×10^6 molecular weight. In the present work we have used mechanical spectroscopy to study the structure and properties of native gel taken directly from the surface of the gastric mucosa and the gel reconstituted from the isolated glycoprotein purified by gel filtration on Sepharose 4B.

Both native and reconstituted gels contain approximately the same concentration of glycoprotein (55 mg mg ml^{-1}, 18 determinations). Storage (G') and loss (G") moduli which respectively characterise the solid-like and liquid-like behaviour of the gel, were measured over a frequency range of 10^{-2}-10^2 rad s^{-1}. The profiles obtained for both gels were similar and typical of a weakly cross linked network with a G' somewhat greater than G" throughout the whole of the frequency range investigated (Fig. 1). In contrast to perm-anently cross-linked polysaccharide gels, e.g. agar, where G' and G" are almost totally independent of frequency and solid like character predominates (G'/G" 10-100) the mechanical properties of mucus show appreciable frequency dependence (G'/G" 3, i.e. signif-icant liquid like character). In these respects the behaviour of mucus gels resembles that of viscous biological fluids such as the vitreous humour of the eye and synovial fluid of joints where net-work formation is by simple physical entanglement of hyaluronate chains. Transient interactions of this type of low or zero binding energy would be consistent with the ability of mucus gels to recover from mechanical damage (i.e. anneal if sectioned) and to flow over the mucosal surface.

Table 1   Mechanical spectroscopy studies on native pig gastric
           mucus: effect of pH, temperature and bile.

Treatment of sample	log G' (10^{-1} rad sec)		Reduction in G'
	Control	Sample	(%)
4h. pH1, 4°C	2.32	2.30	5
4h. pH2, 37°C	2.53	1.88	78
6h. pH2, 37°C	2.51	1.49	91
6h. pH6, 37°C	2.51	2.43	17
6h. pH8, 37°C	2.51	2.56	–
24h. pH5.5, 37°C*	2.32	1.0	95
18h. pH2 bile	2.15	2.02	26
18h. pH6 bile	2.57	2.42	29
18h. pH8 bile	2.44	2.30	27

All gel samples were dialysed against solvent as stated and
assayed at 25°C.
        *Data taken from Figure 1.

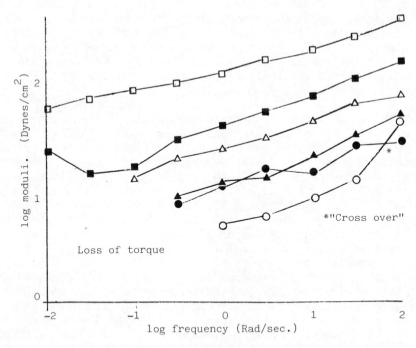

Fig. I.   CHANGE OF MODULI WITH TIME OVER A FIXED FREQUENCY RANGE
           Incubated at 37°C for times zero (□,■), 8 hours (△,▲) and
           24 hours (O,●). Open symbols storage modulus G', closed,
           loss modulus, G".

However, the properties of the native mucus gel differ in two important respects from those of networks formed by simple physical entanglement of a linear polymer such as hyaluronic acid (Morris et al., 1980). Firstly solid like character remains dominant at frequencies as low as 10^{-2} rad sec^{-1} indicating a much longer average time scale of intermolecular association. Secondly on exposure to an aqueous environment the native mucus gel swells to a finite limit but does not readily dissolve, in contrast to entangled systems where infinite dilution is possible. It would appear therefore that there may be some mechanism of interchain association within the mucus gel in addition to physical entanglement which is of relatively long duration and is strong enough to withstand osmotic pressure and solubilisation, but not sufficiently strong to resist gel spreading or mechanical disruption.

Reduction or proteolysis of mucus results in a collapse of gel structure to form a viscous solution. When mechanical spectroscopy is used to follow this process an overall fall in the value of G' and G" are seen with G' falling more rapidly until a cross over occurs and G" (liquid like behaviour) predominates. Such a collapse in gel structure is seen when native mucus gel is incubated at 37°C at pH 5.5 for 24 hours (Fig. 1). This autolysis of the native mucus was investigated further by varying time, temperature and pH (Table 1). Incubation time of 6 hours at pH 6 & 8, 37°C produced only a limited change in G' (< 20%). Incubation at pH 2, 37°C resulted in a much greater decrease in G' with complete collapse of the gel after 6 hours. That this collapse was not due to the acid environment is shown by the relative small decrease (< 20%) in G' on incubation of the gel at pH 1 for 4 hours at 4°C. We therefore conclude that autolysis of the gel at 37°C is enzymic in nature presumably due to endogenous pepsin associated with mucus gel. This was confirmed by gel filtration studies on Sepharose 2B which showed that the lower molecular weight, pepsin-degraded, glycoprotein subunit was formed during mucus autolysis.

The effect of bile on gel structure was investigated by leaving the gel in direct contact with or dialysing against pig bile for 18 hours at 4°C. While in all cases bile produced a limited fall in G' over 18 hours of between 15-35% (Table 1) in no experiment was there any evidence for the drastic reduction in G' associated with the loss of the gel structure as is seen on autolysis (Fig. 1). These studies show that both H$^+$ and bile, agents implicated in gastric mucosal damage do not, in contrast to pepsin, lead to collapse of the gastric mucus gel structure.

Allen, A. 1977, In Mucus in Health & Disease, edited by
    M. Elstein & D.V. Parke, pp. 275-299 Plenum Press,
    New York.

Morris, E.R., Rees, D.A. & Welsh, E.J., 1980, J. Mol. Biol.137.

MUCUS HYDRATION: A DONNAN EQUILIBRIUM CONTROLLED PROCESS

Patrick Y. Tam and Pedro Verdugo

University of Washington, Center for Bioengineering

WD-12, Seattle, Washington 98105 USA

Cervical mucus of certain mammalian species changes from a thick rubbery gel, impenetrable by sperm (at the luteal phase of the sexual cycle) to a watery gel, easily penetrable by sperm (around the time of ovulation). These variations in the physical properties of mucus serve an important function in fertility regulation.

Mucus has a small solids content (from 5-10%) whose main constituents are long chain macromolecular glycoproteins (1,2). It has been postulated that observed changes in mucus physical properties are due to variations in the degree of covalent cross-linking of these glycoproteins. However, recent experiments in mucus dispersion (3) and observations by laser scattering spectroscopy (4) suggest that the mucus macromolecular structure is probably an entangled rather than a cross-linked network. This new molecular model can explain these changes in the physical properties of mucus as due to variation in its degrees of hydration (4). It has been demonstrated that the molecular matrix of gels formed by ionic polymers can function as a semi-permeable membrane (5). An attractive model of mucus hydration is that the mucus macromolecular network might indeed behave as a semi-permeable membrane that prevents the polyionic mucins from diffusing to the outside of the gel. Thereby, mucus might undergo swelling according to a Donnan equilibrium (6) in which entrapped polyionic species could generate the osmotic drive, and the mucosa could regulate the hydration by controlling the transepithelial movement of H^+, water and electrolytes. The present experiments were designed to investigate if cow cervical mucus can be swollen as a polyionic gel. Results indicate that luteal mucus can indeed be swollen to become an estrous-like gel and that this hydration follows the behavior of a Donnan equilibrium, i.e., it is strictly dependent upon, and takes place at, physiologic ranges of pH and salt concentration in the swelling medium.

Swelling was measured, as the relative volume expansion, using
an osmometer-like chamber. The swelling medium was a water salt
solution containing either 0, 20 or 200 mM NaCl at neutral pH; or
else 150 mM NaCl at pH 8.0, 7.6, 6.99 or 6.49. The pH was adjusted
using a phosphate buffer. Figure 1 shows the typical transient
swelling response of luteal and estrous mucus. Estrous mucus
demonstrated very small amount of swelling. Luteal mucus allowed
a 13% equilibrium swelling. Figure 2 and 3 show the steady state
relationship between swelling and salt concentration and swelling
and pH respectively. Note that in both cases estrous mucus showed
no significant amount of swelling while the swelling of luteal mucus
followed a linear relationship to log [NaCl] and to pH of the solvent.

It has been shown that mucus is discharged as densely packed
granules (7). These granules are then hydrated in some unknown
way (8). The results presented here show that mucus can indeed be
hydrated according to a Donnan equilibrium process. Although the
present experiments were limited to investigating the swelling of
cervical mucus at different salt concentrations and pH, with only
minor constraints the principle that mucus hydration follows a
Donnan equilibrium remains valid for other types of mucus and other
ionic influences. These findings, together with the observation
that the transport of water and electrolytes in the mucus-secretory
epithelia is under physiologic control (8), suggest that mucus

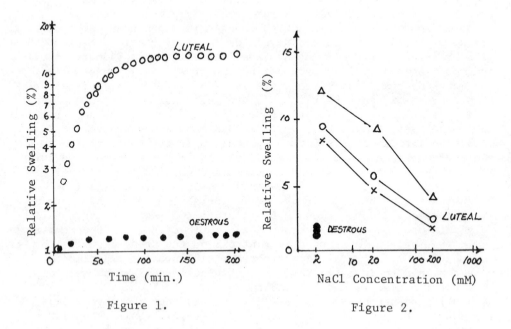

Figure 1.   Figure 2.

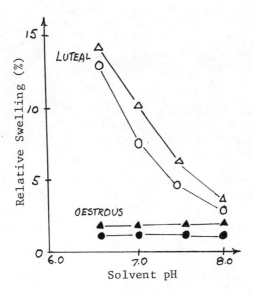

Figure 3.

hydration could be regulated by the transepithelial movement of H^+ water and electrolytes. The proposed model for regulation of mucus hydration is consistent with the limited evidence available and can be used to guide further experimentation.

REFERENCES

1. Gibbons, R.A.   Biochem J. 73, 217, 1959
2. Odeblad, E. Acta Obstet Gynec Scand 47, Suppl. 1, 1968
3. Lee, W.I., Verdugo, P., Blandau, R.J. and Gaddum-Rosse, R.
   Gynec. Invest. 8, 254, 1977
4. Meyer, F.A. Biorheology 13, 49, 1976
5. Katchalsky, A., Lifson, S. and Einsenberg, H.J. Polymer Sci
   7, 571, 1951
6. Donnan, F.J. and Guggenheim, E.A.Z. Physik Chem A162, 346, 1932
7. Nicosia, S.V., Wolf, D.P. and Streibel, E. Fertil Steril 29,
   237, 1978
8. Nadel, J.A., Davis, B. and Phipps, R.J. Ann Rev Physiol 41, 369,
   1979

DISSOLUTION OF TETRACYCLINE HYDROCHLORIDE IN

MUCIN SOLUTIONS

Pat Kearney and Christopher Marriott

Department of Pharmacy, Brighton Polytechnic
Brighton BN2 4GJ, U.K.

It has been shown that the diffusion rates and indirectly, the dissolution rates of simple electrolytes and drug molecules are not simply a function of solution viscosity as given by the Stokes-Einstein equation, $D = \dfrac{kT}{6\pi r\eta}$ (Farng and Nelson, 1977; Florence et al, 1973).

The concepts of 'effective' and 'microscopic' viscosity were initiated in order to explain the apparent discrepancies between measured bulk viscosity and both diffusion and dissolution rate. This is of obvious importance when considering drug dissolution from formulations containing polymer adjuvants where the dissolution environment is dictated by the nature of the polymer. In addition there is the natural environment of the stomach to consider. For a fasting individual taking a formulation with water and disregarding enzymic secretions, the main component of the gastric contents would be a dilute mucin solution of low pH. In order to assess the effects of this bio-polymer environment, comparative dissolutions of a representative drug, tetracycline hydrochloride, were carried out in $0.1\,M$ HCl (pH 1.1) containing either polethylene glycol 1500 (PEG), polyvinyl-pyrrolidone 700,000 (PVP), mucin or sucrose.

The mucin used was a crude pig gastric type (Sigma Chemical Co., Poole, U.K.) which was purified by gel filtration on a Sepharose 4B column and concentrated in an ultrafiltration system (Amicon). The dissolution apparatus was mounted in a water bath at $37^\circ C$ and consisted of a 100 ml beaker with a removable nylon base into which was mounted a disc of compressed drug with one surface exposed. The sample size was 40 mls and was stirred with a two blade paddle at 20 rpm. Drug concentration was measured spectrophotometrically at 354 nm and the dissolution rate assessed

from the initial slope of absorbance versus time curves. The
dissolution rate in 0.1 M HCl with no additive was taken as the
baseline rate. Test discs were compressed from 200 mg of crystalline
tetracycline hydrochloride (Sigma Chemical Co., Poole, U.K.) at 7.1
tonnes per cm^2 using an infra-red KBr die, 1.3 cm diameter.

In Fig. 1 the reduction in baseline dissolution rate as a
function of additive concentration is shown and the order of
effectiveness initially follows the order mucin > PVP >> PEG >
sucrose with the suggestion that PVP and mucin reach a maximum
reduction of around 40 - 50%, whereas sucrose and PEG exhibit a
continued rise in rate reduction with concentration to approximately
70%. Fig. 2 shows the relative viscosity of the solutions as a
function of additive concentration and again the effectiveness
decreases in the order mucin > PVP >> PEG > sucrose but the curves
are exponential with no tendency to reach a maximum. As dissolution
rate does indeed decrease with increased medium viscosity, then the
order of effectiveness shown in Fig. 1 would follow from that in
Fig. 2. However, if diffusion and hence dissolution rate were
governed by viscosity according to the Stokes-Einstein equation, a
plot of dissolution rate reduction versus ηrel for each additive
would produce identical curves. It is clear from Fig. 3 however,
that this is not so as the tendency of the additives to reduce
dissolution rate at a given viscosity is different. The order of

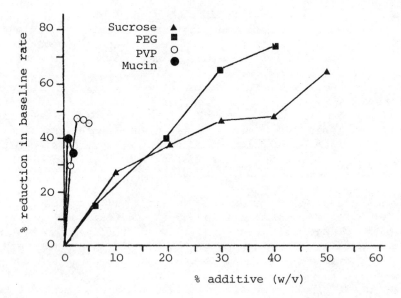

Fig. 1 The effect of additive on baseline dissolution rate

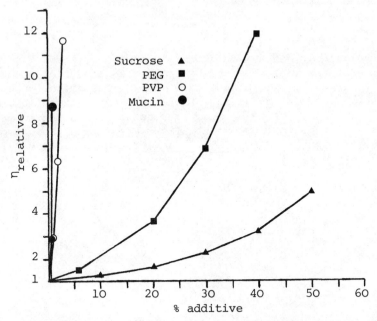

Fig. 2   The effect of additive on the viscosity of the medium

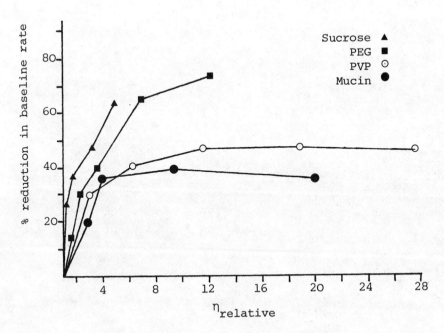

Fig. 3   The effect of viscosity on baseline dissolution rate

effectiveness is mucin > PVP > PEG > sucrose which is indicative
of a progressive increase in 'effective' viscosity from mucin to
sucrose.

Consequently, in any in vitro dissolution system, simple bulk
viscosity of the medium cannot be treated as a usable parameter
since it has a complex relationship with solute diffusion coefficient.
Attempts at formulating an in vitro simulated gastric fluid using
simple thickening agents such as sucrose or indeed the lower
polymer additives would substantially overestimate the dissolution
retardation experienced in vivo. The retardation observed in mucin
solutions however, is still too large to be ignored.

It is suggested therefore, that a model gastric fluid using
dilute mucin as the medium or at least a solution of high polymer
such as PVP 700,000, would provide a more realistic assessment
of the in vivo environment for dissolution studies.

REFERENCES

Farng, K.F., Nelson, K.G., 1977, J.Pharm.Sci., 66:1611
Florence, A.T., Elworthy, P.H. and Rahman, A., 1973,
     J.Pharm.Pharmac., 25:779

THE USE OF HIGH RESOLUTION CARBON-13 NMR IN THE STUDY OF MUCUS

K. Barrett-Bee, G. Bedford and P. Loftus

ICI Pharmaceuticals Division, Alderley Park

Macclesfield, Cheshire SK10 4TG, U.K.

The glycoproteins of mucus secretions are large polymers with molecular weights of several million daltons. The molecules are believed to have a peptide backbone with oligosaccharide chains which conform to a bottlebrush structure. Physical investigations on mucus glycoproteins have been hampered by the presence of co-secreted proteins. The presence of these proteins means that purification of the glycoprotein is generally needed. Several methods of purification have been used which have yielded preparations which can be studied by chemical and physical methods e.g. density gradient centrifugation (1) and gel exclusion chromatography (2). The work described here utilises carbon-13 NMR on native mucus and glycoprotein obtained from both centrifugation and gel chromatography.

Experimental

Native mucus was obtained from a dog equipped with a tracheal pouch (3), this material was a visco-elastic gel and CsBr equilibrium density gradient ultracentrifugation showed that it contained non-mucus proteins. A sample of mucus was fractionated according to the method of Marriott et al (2) using sepharose 2B and the excluded peak examined, and also separated by means of CsBr ultracentrifugation (1). NMR spectra were recorded using a JEOL FX-100Q spectrometer at a frequency of 25MHz for carbon-13.

The spectrum obtained from a sample of whole dog mucus consists of a number of broadly defined spectral regions (Amino acid and N-acylated sugar carbonyls 178-170 ; amino acid aromatics 132-128 ; sugar anomerics 105-95 ; sugars 80-60 ; amino acid α-carbons 60-50 ; amino acid aliphatics 45-15). In addition to

these broad signals there are also sharp resonances at 16.6 corresponding to the methyl signal from L-fucose, 23.2 and 23.6   due to
the methyl signals of N-acylated sugars such as N-acetyl-D-Glucos-
amine and at 40.8 , thought to arise from sialic acid. Treatment
of the sample with 5% v/v β-mercapto ethanol dramatically improved
the resolution. It is possible to identify resonances corresponding
to many individual amino acids. It seems that the protein component
is particularly rich in hydrophobic side chains and that these, due
to the sharpness of their signals, exist in a highly conformation-
ally mobile environment. Such a level of mobility would not be
expected for the backbone residues of a peptidoglycan and so the
signals observed are almost certainly derived from extraneous
material trapped within the mucus matrix.

The corresponding spectrum was obtained from a fractionated
sample of pig gastric mucus. The most striking feature was the
almost complete disappearance of the protein signals, clearly
indicating that they arose from the presence of extraneous material
as proposed above.   Treatment with β-mercapto-ethanol, lead to a
dramatic improvement in resolution (Fig. 1). The carbonyl and
methyl signals from the N-acetyl sugars are clearly seen at 175.0,
23.6 and 23.1 . The anomeric region is now resolved into six
distinct peaks corresponding to the individual sugars and linkage
patterns present. Of particular interest is the lowest field signal
at 105.9 which corresponds very closely with the expected β1-4
galactose-glucose linkage which has a value of 106.1 in lactose(5).
The central sugar region contains several well resolved peaks but
the extensive overlap of sugar chemical shifts means that it is not
possible to make an unambiguous assignment. The small signals at
65-50 correspond to a mixture of C_6 methylene carbons (cf. β1-4
glucose-glucose values of 60.5 and 59.7 and β1-4 galactose-glucose
of 63.9 and 63.1). Finally, the signal at 16.5 corresponds to the
methyl signal in L-fucose which could also be detected in the whole
mucus sample above.

It is interesting that in figure 1 , the 40ppm signal is missing,
i.e. that corresponding to sialic acid. This result agrees with the
analytical data where very little sialic acid was measured. The
analytical results, both for purity (as detected by CsBr ultracentrifugation) and chemical analysis (sugars, amino acids etc.) agree
closely with the NMR results.

Calculations based on peak areas of standard sugars indicate
that the majority of the signal for fucose and sialic acid
resonances are detected in the spectrum whereas a smaller proportion of the sugars such as glucosamine and galactosamine which are
expected to be bound deeply into the structure(s) are observed.

These types of experiments suggest that high resolution NMR may
play a role in investigations of the structure of mucus glyco-

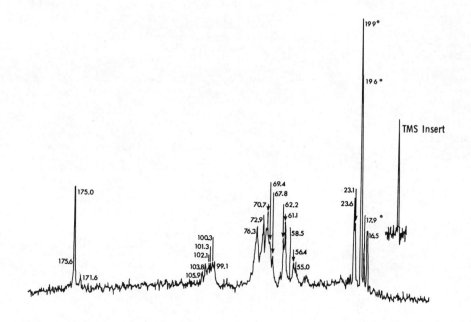

Figure 1

     Purified pig mucus glycoprotein treated with 2-mercaptoethanol.
Asterisked signals are from 2-mercaptoethanol, resonance values are
relative to TMS, for explanation see text.

proteins particularly in relation to the binding of ions and
other molecules. Even though the molecular weight is very
large, there is clearly much flexibility within the sugar
chains which allows the NMR signals to be detected.

References

(1) Creeth. Characterization of secretory glycoproteins
by ultracentrifuge methods in "Glycoconjugates" 94. ed. Schauer,
George Thieme Stuttgart (1979).
(2) Marriott, Barrett-Bee, Brown. A comparative evaluation of
mucus from different sources. Ibid 528 (1979).
(3) Wardell, Chakrin, Payne. Am.Rev.Resp.Dis. 101, 741 (1970).
(4) Voelter, Bilik, Breitmaier. Collect.Czech.Chem.Comm. 38
2054 (1973).

GASTERO-INTESTINAL MUCUS

THE STRUCTURE AND PHYSIOLOGY OF GASTROINTESTINAL MUCUS

Adrian Allen, Alan Bell, Michele Mantle & Jeffrey P. Pearson

Department of Physiology
The University, Newcastle upon Tyne NE1 7RU, U.K.

The function of mucus

The primary function of gastrointestinal mucus is considered to be protection of the surface mucosal cells (Hollander, 1954; Florey, 1955). Mucus forms a gel which, throughout the gut, protects the mucosal surfaces from the vigorous shear forces that attend digestion (Fig. 1). The mucus gel provides a slimy lubricant for the passage of solid material through the gut while some of the gel layer remains firmly stuck to the mucosa to protect it from the next round of mechanical abuse. Mucus has particular physical properties which allow it to flow and if sectioned, anneal. Such properties, which show mucus to be a weaker gel than a rigid gel such as agar, facilitate the spread of the mucus over the mucosal surface. However, mucus will not dissolve with infinite dilution and is quite distinct from a viscous liquid.

In the stomach and the first part of the duodenum mucus is implicated as part of the protective mechanism against the corrosive components of the gastric juice namely HCl and pepsin (Hollander, 1954; Allen & Garner, 1980). Heatley (1959), postulated that within the mucus gel there was an alkaline secretion produced by the mucosa that neutralised the HCl before it reached the mucosal cells. The function of the mucus was to provide an immobile phase which was permeable to ions but prevents the relatively small amount of alkaline secretion from mixing with the bulk of the gastric juice. In this way it was not swamped by the excess of acid in the lumen (Fig. 1). Direct evidence for such a protective mechanism has come from the demonstration of bicarbonate secretion by the gastric and duodenal mucosa in amphibians and mammals (Flemstrom, 1977; Garner & Flemstrom, 1978). Further, several studies show that

115

mucus and bicarbonate secretion are stimulated and inhibited by a
variety of ulcer healing and ulcerogenic drugs respectively.
(Glass & Slomiany, 1977; Parke, 1978; Allen & Garner, 1980; Allen,
1981a).   A consequence of surface neutralisation of acid is that a
pH gradient should exist across the mucus gel.   Such a pH gradient
has been shown across the mucus adherent to isolated rabbit gastric
mucosa from a pH of 2.31 at the luminal surface to a pH of 7.26 at
the mucosal surface of the mucus (Williams & Turnberg, 1981).   The
potential for the mucus gel to act as a mixing barrier is evident
from the structure of the gastric mucus gel which contains relatively high concentrations (about 50 mg/ml) of gastric mucus glycoprotein (Allen et al. 1976; Bell et al. 1980).   However such a mucus
gel is still 95% water and freely permeable to ions, e.g. H^+ and
small molecules, e.g. salicylate (Heatley, 1959; Davenport, 1967;
Williams & Turnberg, 1980).   From what is known at present of
gastric mucus gel structure (discussed below) it would appear that
H^+ can diffuse freely throughout the immobilised phase (unstirred
layer of solution) contained within the gel matrix.   The thickness
of such a layer of mucus is unlikely to exceed 1mm and the diffusion
of H^+ through such an unstirred layer would be very rapid
(Davenport, 1967; Bickle & Kauffman, 1981).   If, as suggested here,
mucus acts primarily as an immobilised phase (mixing barrier) and not
as a diffusion barrier, an approximate calculation suggests that
mucus and measured rates of gastric mucosal HCO_3^- secretion could

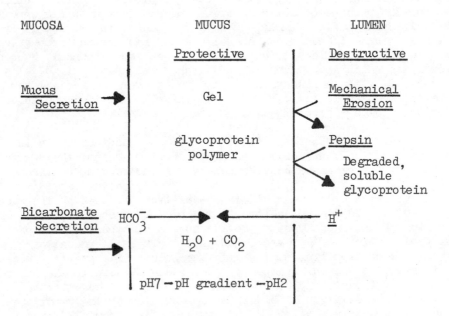

Fig. 1.   Mucus and gastric mucosal protection

provide an effective first line of defence against acid if the pH
of the lumen is somewhere above pH2 (Allen & Garner, 1980).
Although the intraluminal pH of the stomach is probably above pH2
for much of the time the pH will fall below this and under such
conditions a second line of defence against acid would appear
necessary presumably, at the level of the mucosal cells themselves,
as originally, presumably Hollander, (1954). The measured rates
of mucosal bicarbonate secretion in the duodenum are higher than
those for the gastric mucosa (Flemstrom, 1980 and Flemstrom
et al., 1981) and here surface alkaline neutralisation within the
duodenal mucus gel might provide comprehensive protection of the
mucosa against any fall in the pH of the luminal juice in the first
part of the duodenum.

The other corrosive factors of gastric juice are pepsins and
these in common with other proteolytic enzymes will dissolve the
surface mucus gel and release degraded mucus glycoprotein into the
lumen (Scawen & Allen, 1977; Allen, 1981b). There is a dynamic
balance at the mucosal surface between formation of the surface
mucus gel and its erosion by proteolysis and mechanical shear.
However experiments in which we have investigated the permeability
of mucus to larger sized molecules show that it is not permeable to
pepsin. In these experiments molecules as large as vitamin B_{12}
(molecular weight 1,346) could pass through a layer of mucus 1mm
thick but, myoglobin (molecular weight 17,000) did not permeate this
layer over a period of 48 hours (Allen, 1981b). Pepsin, like myoglobin, is a globular protein but twice its size (molecular weight
35,000) therefore it too will not diffuse into the mucus gel. It
follows, that although pepsin will continually erode the luminal
surface of the mucus gel layer, while the gel layer is maintained by
secretion, pepsin cannot attack the surface mucosal cells. Mucus
does not cover the cells of the gastric glands which secrete acid
and pepsin, but it is present at the neck of the glands and could
well block their entrance to the lumen. If this is so then the
question arises as to how newly secreted acid and pepsin gain access
to the gastric lumen. One possible mechanism is that the flow of
gastric juice could push the newly secreted pepsin and acid through
the mucus layer which might then reseal after secretion of the gastric
juices had finished. The cells of the gastric glands, in contrast
to the surface epithelial cells of the rest of the mucosa, must be
quite resistant to their own secretions.

Isolation of the mucus glycoproteins

To understand the structure of the mucus gel it is necessary to
characterise the component glycoproteins in a native state where
they retain the viscous and gel-forming properties of the original
mucus secretion. In this and subsequent sections we describe some
of our work on the glycoproteins from pig gastric, pig intestinal and
human gastric mucus. It is essential to start with mucus gel

scraped off the surface of the mucosa since luminal mucus can
contain large amounts of proteolytically degraded glycoprotein
(Scawen & Allen, 1977; Allen et al. 1980).   Proteolysis by a
variety of enzymes or reduction of disulphide bridges with agents
such as 0.2M mercaptoethanol readily solubilises the gel but
results in a degraded mucus glycoprotein which has lost the viscous
and gel forming properties of the mucus gel (Allen & Snary, 1972).
Native pig gastric mucus glycoprotein that retains the viscous and
gel forming properties of the mucus can be obtained by dialysing
mucus scrapings for 24 hours against 0.2M NaCl (Snary & Allen, 1971).
By gently stirring mucus gel for 24 hours between 30-70% of the
glycoprotein is solubilised depending on the preparation (A. Bell,
unpublished results).   Most, if not all, the glycoprotein from pig
gastric and small intestinal mucus gel can be completely solubilised by shear e.g. homogenisation (1-2 minutes) to yield a preparation of native glycoprotein which is undegraded and has the viscous
and gel forming properties of the mucus (Allen et al. 1976; Mantle
& Allen, 1981).

Mucus gel scraped from the mucosal surface is heavily contaminated with cell debris etc., e.g. pig gastric mucus gel scrapings
are about 60% by weight mucus glycoprotein and the rest is protein
with a little nucleic acid.   The mucus glycoprotein can be considerably purified by gel-filtration on Sepharose 4B which separates
the large molecular weight native glycoprotein (excluded) from the
majority of the lower molecular weight protein.   However such
glycoprotein preparations still contain some lower molecular weight,
non-covalently bound, extraneous protein (20% by weight of the
total protein present in the glycoprotein preparation) which is
strongly bound to the glycoprotein and excluded with it on gel-
filtration in 0.2M NaCl (Starkey et al. 1974).   The only satisfactory method for completely purifying gastrointestinal mucus
glycoproteins free of non-covalently bound protein is by equilibrium centrifugation in a density gradient of CsCl (Creeth &
Denborough, 1970; Starkey et al. 1974).   By a combination of gel
filtration on Sepharose 4B followed by equilibrium centrifugation,
pig gastric mucus can be obtained free of non-covalently bound
protein.   Mucus scraped from the pig small intestine is very
heavily contaminated with an analysis of 15%,80% and 5% by weight
of glycoprotein, protein and nucleic acid respectively.   Here a
three stage preparative procedure was found necessary to purify the
mucus glycoprotein free from detectable amounts of protein and
nucleic acid; this consisted of two consecutive equilibrium cent-
rifugations in CsCl density gradients followed by a final fractionation on Sepharose 4B (Mantle & Allen, 1981).   By these preparative
methods the pure native gastric (molecular weight 2×10^6) and
small intestinal (molecular weight 1.7×10^6) pig mucus glycoproteins can be shown to account for at least 90 & 60% by weight
respectively of the total glycoprotein in the original mucus
scrapings.

It is important that the purified mucus glycoproteins are free from all detectable traces of non-covalently bound protein. The peptide component of these gastrointestinal glycoproteins is between 13-18% by weight of the molecule (Table 1). Therefore even a 5% by weight contamination with free protein, which is not an unreasonable amount in a partially purified glycoprotein preparation, will make any analysis of the peptide of the glycoprotein impracticable. The presence of non-covalently bound protein has been shown in both gastric mucus glycoprotein (Allen, 1977) and small intestinal mucus glycoprotein (Mantle & Allen, 1981) to reduce the viscosity and gel forming properties of the glycoprotein. For example, the progressive removal of the remaining 30% by weight non-covalently bound protein from the small intestinal glycoprotein resulted in a rise in the intrinsic viscosity from 0.14 to 0.50 ml/mg and only when free of non-covalently bound protein will the glycoprotein form a gel at concentrations (10-12 mg/ml) comparable to that of glycoprotein in the native mucus gel (Mantle & Allen, 1981). From such evidence it would appear that the glycoprotein alone can account for the gel forming properties of these mucus secretions. This contrasts with work by others where it has been proposed that firmly, yet non-covalently, bound protein could have an integral role in the enhancement of rheological properties of the mucus glycoprotein (Creeth, 1978; List et al. 1979).

The most sensitive method for demonstrating the absence of non-covalently bound extraneous protein in purified pig gastric and small intestinal mucus glycoproteins has been by polyacrylamide gel electrophoresis in 1% SDS (Allen et al. 1980; Pearson et al. 1981). Prior to electrophoresis the glycoproteins are boiled in 1% SDS for 2 minutes and a high loading of 250-500 μg of glycoprotein is placed on the gel. After electrophoresis no protein bands were visible by staining with Coomassie blue, apart from the band at the origin where the glycoprotein had not entered the gel. Other evidence that the glycoprotein is pure comes from the absence of detectable protein following analytical density gradient centrifugation in Cs_2SO_4 or CsCl containing 4M guanidinium chloride as well as constant individual sugar and amino acid compositions for several different preparations of the respective glycoproteins (Scawen & Allen, 1977; Mantle & Allen, 1981).

Structure of gastro-intestinal mucus glycoproteins

The compositions of the purified glycoproteins isolated from the mucus secretions of pig stomach, small intestine and colon are shown in Table 1 and are similar to glycoproteins from other mucus secretions (Carlson, 1977; Horowitz & Pigman, 1977). In overall sugar composition the glycoproteins from gastric and colonic mucus are very similar except for a greater amount of negatively charged sialic acid residues on the colonic glycoprotein. Small intestinal mucus glycoprotein differs markedly in composition from the

Table 1. The composition of pig gastro-intestinal glycoproteins

Constituents	Composition % dry weight glycoprotein		
	Gastric[+]	Small intestinal[*]	Colonic[o]
Carbohydrate	82.9	77.5	83.4
galNAc	11.0	25.5	10.6
gluNAc	29.5	14.1	30.5
gal	25.4	13.2	21.4
fuc	14.6	14.8	10.7
sialic acid	2.5	19.9	10.2
ester sulphate	3.2	2.8	3.4
protein	13.2	19.6	13.7
A & H blood group activity	+	+	+

Data calculated from Allen, 1977[+], Mantle & Allen, 1981[*] and
  Marshall & Allen, 1978[o].

other two glycoproteins in that it contains a much higher proportion of N-acetylgalactosamine to the other sugars as well as more
negatively charged sialic acid and ester sulphate residues.   The
complex structure of the sugar chains of pig gastric mucus glycoprotein, up to 19 sugar residues in length has been elucidated by
Slomiany & Meyer (1972).   Pig submaxillary mucus glycoprotein has a
relatively simple structure of up to 5 sugar residues per chain
(Carlson, 1968).   The structures of the sugar chains of pig colonic
and small intestinal mucus glycoproteins have not been elucidated.
If it is assumed that N-acetylgalactosamine only occurs at the ends
of the sugar chains as in other mucus glycoproteins then from the
high molar ratio of N-acetylgalactosamine to the other sugars the
carbohydrate chains of the small intestinal mucus glycoproteins must
be relatively short, with an average of not more than 6-8 sugar
residues per chain (Mantle & Allen, 1981).   It is clear that the
carbohydrate chains of mucus glycoproteins from the different regions
of the pig gastrointestinal tract differ considerably in their
structural complexity, chain length and negative charge.   Comparison of mucus glycoproteins from a number of sources show that their
carbohydrate chains also have common structural features for example,
the carbohydrate chains are linked to the peptide by an O-glycosidic

linkage from N-acetylgalactosamine to serine and threonine and in man (A,B,H) and pig (A,H) possess blood group activity (Carlson, 1977).

Our own work on purified pig gastrointestinal mucus glycoproteins has concentrated on their polymeric structure of glycoprotein subunits joined together by disulphide bridges.   These disulphide bridges are located in regions of the peptide core of the glycoprotein that are free from detectable carbohydrate and accessible to proteolytic enzymes.   Both reduction and proteolysis cleave the glycoproteins into lower molecular weight glycoprotein subunits with the loss of their viscous and gel forming properties (Allen & Snary, 1972, Allen, 1978, Allen et al. 1980).   Native pig gastric mucus glycoprotein has a molecular weight of 2×10^6 and is a polymer, on average, of four glycoprotein subunits (molecular weight 5×10^5) joined by disulphide bridges.   It is a single covalent entity since it is not dissociated in boiling 1% sodium dodecyl sulphate, 4M guanidinium chloride, 3.5M CsCl or 2M NaCl (Scawen & Allen, 1977; Pearson et al. 1981).   On reduction of its disulphide bridges with 0.2M mercaptoethanol the pig gastric mucus glycoprotein is split into glycoprotein subunits of a molecular weight 5×10^5.   A protein of 70,000 molecular weight is also released on reduction of the native glycoprotein in amounts equivalent to a molar ratio of one protein, 70,000 molecular weight, per glycoprotein, 2×10^6 molecular weight (Allen et al. 1980 ; Pearson et al. 1981).   Proteolysis also cleaves the native glycoprotein into subunits of molecular weight 5×10^5 and with the loss of 40% of the total peptide but no detectable loss of carbohydrate (Table II).   SDS gel electrophoresis of the papain digested glycoprotein shows that the 70,000 protein represents a major portion of the protein removed on proteolytic digestion.   The data obtained are compatible with the model shown in Fig. 2 where the four glycoprotein subunits are joined together by the central 70,000 molecular weight protein.   This model is supported by the detailed amino acid analysis of the components (Pearson & Allen, 1981; this volume) which shows in particular: (1) the 70,000 protein and the non-glycosylated region of the glycoprotein both rich in cysteine residues and therefore have the potential for disulphide bridge formation: (2) the progressive removal on reduction and then proteolysis of all the other amino acids apart from serine, threonine and proline (Table II).   These three amino acids which characterise the glycosylated peptide regions comprise 43 & 70% by weight of the native and papain digested glycoproteins respectively.   Further proteolysis or reduction of the proteolytically digested glycoprotein subunits does not decrease their size below 5×10^5 molecular weight.   This is in keeping with the absence of detectable cysteine in the proteolytically digested glycoprotein subunit and its resistance to further proteolytic digestion suggests that the remaining peptide core is comprehensively protected by its sheath of carbohydrate chains. A less likely model where the four glycoprotein subunits are joined

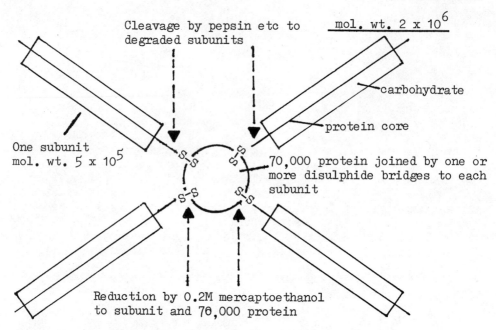

Cleavage by pepsin etc to degraded subunits

mol. wt. 2×10^{6}

carbohydrate

protein core

One subunit mol. wt. 5×10^{5}

70,000 protein joined by one or more disulphide bridges to each subunit

Reduction by 0.2M mercaptoethanol to subunit and 70,000 protein

Fig.2 A diagrammatic representation of the structure proposed for pig gastric mucus glycoprotein.

to each other by disulphide bridges and the 70,000 molecular weight protein is attached separately is also possible from the existing data.

Pig small intestinal mucus glycoprotein has a molecular weight of 1.7×10^{6}, close to that of pig gastric mucus glycoprotein and again it is not dissociated by non-covalent bond breaking solvents (Mantle et al. 1981). It has a polymeric structure that is split on proteolysis into glycoprotein subunits (molecular weight 4.5×10^{5}) about one quarter the molecular weight of the native glycoprotein. This is accompanied by a reduction of 29% by weight in the peptide component without detectable loss of carbohydrate (Table II). However in contrast to the gastric mucus glycoprotein, the proteolytically digested small intestinal glycoprotein is dissociated on reduction with 0.2M mercaptoethanol into a further 2 subunits (molecular weight 2.7×10^{5}). A similar size glycoprotein subunit (molecular weight 2.4×10^{5}) is obtained by direct reduction with 0.2M mercaptoethanol of the native glycoprotein and the latter evidently consists of an average of 7 or 8 glycoprotein subunits of equal size joined by disulphide bridges. On reduction

Table 2. Reduction and proteolysis of pig gastric and small intestinal mucus glycoproteins

	Gastric				Small intestinal glycoprotein	
	Native	Reduced	70,000 protein fraction	Papain digested	Native	Pronase digested
Protein content (% by wt. glycoprotein)	13.3	9.6		6.4	18.0	12.5
Thr. + ser + prol (% by wt. protein)	40.4	53.5	14.3	77.2	52.4	68
Cys. (% by wt. protein)	4.3	3.3	3.5	absent	4.3	3.2
Molecular weight (S,D).	2×10^6	5.2×10^5	70,000 (SDS gel)	5.3×10^5	1.72×10^6 Reduced in 0.2M mercaptoethanol $\downarrow$ 2.7×10^5	4.5×10^5 $\downarrow$ 2.4×10^5
Intrinsic viscosity mlg^{-1}	320	60			500 Reduced glycoprotein $\longrightarrow$	123 Reduced glycoprotein 79

of the native small intestinal glycoprotein a 90,000 molecular
weight protein is released on a 1:1 molar basis by weight and this
protein is destroyed by proteolysis. The polymeric structure of
pig small intestinal glycoprotein is consequently more complicated
than that of pig gastric mucus glycoprotein. A further difference
is the difficulty in exhaustively reducing the small intestinal
glycoprotein which requires 0.2M mercaptoethanol for 48 hours for
full dissociation compared with 0.01M mercaptoethanol for 20 hours
which will dissociate the gastric glycoprotein. The macromolecular
structure of the pig colonic mucus glycoprotein is not as well
characterised as those of the pig gastric and small intestinal glycoproteins. The native glycoprotein, as isolated from the water
soluble mucus, has a very large molecular weight of 15×10^6
(Marshall & Allen, 1978) and this makes it difficult to study the
glycoprotein in detail by analytical ultracentrifugation. In
particular it is not clear how much of the large size of this
colonic glycoprotein is the result of non-covalent association by
smaller glycoprotein units. However, it is broken down into
various sizes of glycoprotein units by reduction with 0.2M mercaptoethanol (molecular weight 6.0×10^6), proteolysis with Pronase
(molecular weight 1.5×10^6) and further by proteolysis followed by
mercaptoethanol (molecular weight 0.76×10^6). Proteolysis of pig
colonic mucus glycoprotein results in the loss of 29% by weight of
the peptide without any detectable loss of carbohydrate and conser-
vation of three amino acids, threonine, serine and proline.

The evidence for the polymeric structure in these mucus glycoproteins, discussed above, comes from analytical ultracentrifugation
backed up by gel-filtration studies. The molecular weight values
in the case of pig gastric and small intestinal mucus glycoproteins
and their subunits are consistent within ± 10% for several estimat-
ions. There is good agreement between molecular weights calculated by two different methods firstly, a combination of the sedimentation and diffusion coefficients and secondly, by high speed
equilibrium centrifugation. In all cases the glycoproteins
sediment as unimodal, although polydisperse, peaks, showing they
have a homogenous size distribution (Gibbon, 1972). The changes
in molecular size on reduction and proteolysis of the glycoproteins
can be demonstrated by gel-filtration using long columns of Sepharose
2B which clearly separates the excluded polymeric glycoprotein from
the included subunits (Pearson et al. 1980; Mantle et al. 1981).
Glycoprotein preparations with a heterogeneous size distribution and
containing molecules of intermediate size between the native polymeric glycoprotein and its subunits can be detected by analytical
ultracentrifugation or gel-filtration (Pearson et al. 1979; Mantle
et al. 1981). By these criteria preparations of pig gastrointestinal mucus glycoproteins when isolated directly from the surface
mucus gel are obtained undegraded, in the polymeric form, in contrast, to glycoproteins from the luminal juice which are at least
partially degraded by proteolysis.

The native glycoprotein of human gastric mucus gel has been isolated from mucus scraped from antral gastric mucosa, removed post-operatively and purified by equilibrium centrifugation in a CsCl density gradient (Pearson et al. 1980). This glycoprotein is the same size as that from pig (molecular weight 2×10^6) and with apparently the same polymeric structure in that it is cleaved into subunits (molecular weight about 5×10^5) by reduction with 0.2M mercaptoethanol or by proteolysis with pepsin and other proteolytic enzymes. A 70,000 molecular weight protein is released on reduction of the human gastric mucus glycoprotein, as shown by SDS gel electrophoresis, but the details are not yet as clear as those described for pig gastric mucus glycoprotein. This is because the human gastric glycoprotein can be obtained only in limited amounts and has not been purified by the full procedure used for pig mucus glycoproteins (Pearson, et al. 1981)

If the model proposed for pig gastric mucus above is correct then the 70,000 and 90,000 molecular weight proteins have a central structural role in the polymeric structure of these gastrointestinal mucus glycoproteins. This may have important implications for the structure of glycoproteins from other mucus secretions that depend on disulphide bridges for their structure and have regions of protein susceptible to proteolysis. Such glycoproteins have been demonstrated in mucus from the respiratory tract (Roberts, 1978), the cervical tract (Gibbons, 1978) the ovarian cyst (Dunstone & Morgan, 1965; Donald, 1973), as well as a variety of sources in the gastrointestinal tract (Allen, 1981a). An example are the detailed studies on bronchial mucus of Creeth et al. (1977) who observed the release of variable amounts of protein and a substantial decrease in the molecular weight on thiol reduction of the purified mucus glycoprotein. It should be emphasised that the 70,000 and 90,000 molecular weight proteins we have isolated are part of the covalent peptide structures of the native pig gastric and small intestinal glycoproteins (molecular weights 2×10^6 and 1.7×10^6 respectively). It is these native glycoproteins which are the covalent entities that interact non-covalently to form the mucus gels and the 70,000 and 90,000 molecular weight protein components do not function as separate non-covalent linking proteins between the glycoprotein molecules interacting to form the gel.

The structure of the mucus gel

The gel matrix of pig gastric mucus is formed by non-covalent interactions between the glycoprotein molecules of molecular weight 2×10^6 (Allen et al. 1976). Analytical ultracentrifugation and viscosity studies show that in dilute solutions the gastric mucus glycoprotein is a highly expanded molecule occupying a large solution volume. Calculation of the effective hydrodynamic volume from the frictional coefficients indicates that about a concentration of 20 mg/ml the glycoprotein molecules will be spread over

the whole of the volume of the solution and the probability of intermolecular interaction becomes appreciable.   It is above this concentration of 20 mg/ml that the viscosity of the glycoprotein solution
is seen experimentally to rise asymtotically, due to increasing
intermolecular interactions.   These intermolecular interactions
increase until, about 50 mg/ml, the glycoprotein solution assumes
the characteristics of the native mucus gel taken directly from the
gastric mucosal surface.   The concentration of glycoprotein in the
native gastric mucus gel is 55 mg/ml and when this gel is compared
by mechanical spectroscopy with that reconstituted from the isolated
purified gastric glycoprotein their moduli versus frequency scans
have similar profiles (Bell et al. 1980 & 1981, this volume).   This
profile is that of a weak gel where the magnitude of the value for
the storage modulus (G', solid character) of the gel is markedly
lower than that of a rigid gel e.g. agar but is consistently greater
than the value for the loss modulus (G", liquid character) over a
frequency range of $10^{-2} - 10^2$ rad S^{-1}.   The resulting flow and
annealing properties of the mucus gel mean that the non-covalent
interactions of the gel matrix must be transient and of low energy,
strong enough to resist the forces of solvation yet can make and
break over a finite time scale or under mild shear.   The molecular
nature of these non-covalent interactions is not known but, from the
above, a high concentration of glycoprotein (molecular weight
2×10^6) with considerable interdigitation of molecular domains is
necessary before gel-formation occurs.   Agar where the non-covalent
interactions are of an essentially permanent nature will form a gel
at concentrations of polysaccharide over an order of magnitude lower
than the mucus glycoprotein.   The one difference observed between
the native mucus gel and the gel reconstituted from the purified
gastric mucus glycoprotein is that the latter will dissolve in excess
solvent although if left at $4^{\circ}C$ over a period of days it becomes
'aged' and will not dissolve.   It would appear from this that the
native gel as a result of the biosynthetic process has a more
ordered structure which is optimal for the maximum number of non-
covalent interactions to form the gel matrix.   In the reconstituted
gel this ordered structure is only attained over a period of time
and therefore the non-covalent interactions are weaker and the gel
less resistant to solubilisation.

A comparison of the properties of pig gastric and small intestinal mucus glycoproteins is interesting in view of their being about
the same size yet with their wide differences in length of carbohydrate side chain, the amount of negative charge, the different
amino acid analysis and the number and size of subunits in
their polymeric structure as described above.   However despite
these differences the pattern of gel-formation by the small intestinal mucus glycoprotein appears the same as that described for
gastric mucus glycoprotein (Mantle & Allen, 1981).   As the concentration of the isolated pig small intestinal mucus glycoprotein
(molecular weight 1.7×10^6) is increased above 1 mg/ml so the

viscosity rises asymptotically until at about 10 mg/ml the solution
takes on the properties of a mucus gel (Allen et al. 1980). This
is about the same concentration of glycoprotein that is present in
the native gel taken directly from the small intestine. The only
difference detected between the physical properties of the pig
gastric and small intestinal mucus glycoproteins is that the latter
has a higher intrinsic viscosity (500 compared to 300 mlg^{-1}) with a
larger calculated hydrodynamic volume and these properties are
reflected in the lower concentration of glycoprotein in the small
intestinal mucus gel one fifth of that in the gastric mucus gel.

   One structural feature of the gastric mucus glycoprotein that
is essential for its gel-forming and viscous properties is the
polymeric structure (Snary et al. 1970; Allen & Snary, 1972). Both
reduction and proteolysis solubilise the gel and associated with this
is the breakdown of the native glycoprotein (molecular weight
2 x 10^6) into the degraded glycoprotein subunits (molecular weight
5 x 10^5). These glycoprotein subunits have much lower intrinsic
and specific viscosities than the native glycoprotein (Table II)
and at the concentration of native glycoprotein in the gel (about
50 mg/ml) have entirely liquid properties as measured by mechanical
spectroscopy (Bell et al. 1980). The importance of the polymeric
structure of the glycoprotein in the formation of pig small intestinal mucus gel is apparent from the intrinsic viscosities (Table II)
and the low specific viscosities of the reduced (molecular weight
2.6 x 10^5) and proteolytically digested (molecular weight 4.5 x 10^5)
glycoproteins: less than one tenth of the native glycoprotein at
the same concentration of 8 mg/ml (Allen et al. 1980). The existence of a similar polymeric structure in glycoproteins from other
gastrointestinal, respiratory and cervical mucus secretions, together with the solubilisation of their gels by proteolysis or
reduction, suggests that such polymeric structures may be the basis
for gel-formation in a large number of mucus secretions.

   The role of the carbohydrate chains in gel formation by the pig
gastric and small intestinal mucus glycoproteins is not yet known.
These carbohydrate chains constitute over 75% by weight of the glycoproteins and will confer on the molecules considerable hydrophylicity.
However the two glycoproteins have markedly different compositions
and average chain lengths for their carbohydrate chains although
they both form similar mucus gels. This together with the wide
variety of carbohydrate chain structures in the glycoproteins from
other mucus secretions (Gottschalk, 1972; Horowitz & Pigman, 1977)
poses the question as to the significance in gel-formation of precise
changes in the structure of the carbohydrate chains in these mucus
glycoproteins.

   Both gastric and small intestinal mucus glycoproteins are
negatively charged but with the latter containing more sialic acid
(7-8 times more by weight). Changes in the ionic strength from

above 10mM up to 2.5M NaCl do not affect the viscosity of the isolated glycoproteins.   When the ionic strength is decreased below
10mM the intrinsic viscosity of solutions of both the gastric and
small intestinal glycoproteins increases about tenfold and the
solutions assume gel like characteristics at much lower concentrations, below 10 mg/ml for the gastric glycoprotein.   The increased
viscosity at these low, aphysiological, salt concentrations can be
explained by the absence of the shielding by counterions of the
negatively charged residues, namely sialic acid, sulphate and the
acidic amino acids.   In the absence of shielding by cations the
negatively charged residues will repel each other with expansion of
the molecule in solution.   This molecular expansion is reflected in
increased viscosity and enhanced intermolecular interactions of the
glycoprotein molecules.

Mucus in vivo

        The degree of protection given by mucus in vivo will depend on
the structure and depth of the gel layer covering the mucosal surface (Fig. 1).   The structure of the gel will depend on a variety of
factors including: the covalent structure of the constituent glycoproteins; the concentration of the secreted glycoprotein which will
determine whether the threshold concentration for gel-formation is
exceeded; and the strength of the non-covalent interactions between
the glycoprotein molecules that form the gel matrix which will
determine its resistance to solubilisation and shear.

        From the studies described above one specific feature that
clearly affects the gel-forming ability of the mucus secretion is
the polymeric structure of the constituent glycoproteins.   By
using methods developed from the studies on pig gastric mucus,
changes in the polymeric structure of the purified glycoprotein in
antral gastric mucus from ulcer patients have been shown (see
Younan et al. 1981, this volume).   The native undegraded mucus
glycoprotein from pig gastric mucus is completely excluded by gel-
filtration on Sepharose 2B, while the exhaustively proteolytically
digested or fully reduced subunit is well included.   Purified
glycoprotein from human antral gastric mucus gel, on analysis by
gel-filtration contains an excluded peak of native glycoprotein
together with material that is included on the column and therefore
of lower size than the native material but, all of which elutes
either before or in a position of the subunit peak.   The percent-
ages of this lower molecular weight material in glycoprotein isolated from the human mucus gel and analysed under the same conditions
were 33.4%, 50.2% and 65.1% from apparently normal, duodenal ulcer
and gastric ulcer patients respectively.   The glycoprotein preparations under study have been first purified by equilibrium centrifugation in a CsCl density gradient and should be free from serum
glycoproteins (lower density fractions) and proteoglycans (higher
density fractions).   Further evidence that all the detectable

carbohydrate containing material present is mucus glycoprotein is
the absence of any PAS positive material of lower size than the
proteolytically digested mucus glycoprotein subunit following
exhaustive proteolysis (Pearson et al. 1980). Both serum glycoproteins and proteoglycans on proteolysis would give lower molecular weight material which would be totally included by Sepharose
2B. It is not yet known whether this decrease in the amount of
native polymeric glycoprotein in the mucus gel in ulcer patients is
a reflection of changes in biosynthesis and/or degradation of the
glycoprotein. It is interesting that pig gastric mucus glycoprotein isolated from the surface mucus gel contains no detectable lower
molecular weight, included, glycoprotein however, on partial proteolysis such included glycoprotein rapidly appears and is paralleled by a sharp decrease in the viscosity of the solution (Pearson
et al. 1979). This shows that any decrease in polymeric glycoprotein, as detected by gel-filtration on 2B, is associated with a
loss of the viscous and gel-forming properties of the glycoprotein.
Extrapolation of this to the results from ulcer patients would mean
that the decreases observed in the content of polymeric glycoproteins in the gel will result in a weaker structure.

   The greater the depth of the gel layer on the mucosal surface
the more it will resist total removal by food etc, or penetration by
pepsin and the better will be the unstirred layer within which
surface neutralisation can occur. The depth of the gel layer will
depend on the rate of secretion of the mucus balanced by its rate of
erosion by shear and pepsin. Therefore in studying changes in mucus
in vivo it is equally important to consider mucus erosion as well as
mucus secretion (Allen, 1978, 1981a). Previously several workers
have equated changes in luminal mucus glycoprotein content with
changes in mucus secretion. This is not necessarily correct since
a rise in luminal mucus glycoprotein, for example, could be due to
increased erosion by shear and pepsin without any change in mucus
secretion and a thinner gel would result.

   That peptic erosion of the surface mucus gel is important in
vivo can be shown by measuring the relative amounts of native and
lower molecular weight mucus glycoprotein in gastric washouts from
patients undergoing secretory studies (Allen et al. 1980b).
Following insulin stimulation there is a significant threefold rise
in the luminal mucus glycoprotein but, 78% of this glycoprotein was
included on Sepharose 2B and therefore degraded (Younan et al. 1981,
this volume). That this rise in luminal glycoprotein was in a
large part due to pepsin erosion of the gel is supported by the
parallel rise in pepsin activity on insulin stimulation and the
absence of both the rise in glycoprotein and pepsin activity following vagotomy.

   The polymeric structure of the glycoprotein and its degradation by pepsin is one aspect of mucus that can change in vivo and

130

that can now be investigated. An understanding of its significance in mucus physiology has come directly from structural studies
on the isolated mucus glycoproteins in vitro particularly in the
animal model, the pig. Other important questions to be answered
by structural studies on the isolated mucus and which are yet to be
resolved are: the exact nature of the non-covalent interactions
between the glycoprotein molecules which form the gel matrix and
the influence on mucus gel properties of the carbohydrate chains
with their wide variety of structures and negative charge.

References

Allen, A., 1977, Structure and function of gastric mucus, in:
    Mucus in Health and Disease, M. Elstein & D.V. Parke, ed.,
    Plenum Press, New York and London, pp. 275-299.
Allen, A., 1978, The structure of gastrointestinal mucus glyco-
    proteins and the viscous and gel-forming properties of mucus,
    Brit. Med. Bull., 34:28-33.
Allen, A., 1981a, Structure and function of gastrointestinal mucus,
    in: Physiology of the Gastrointestinal Tract, L.R. Johnson, ed.,
    Raven Press, New York, Chap. 22.
Allen, A., 1981b, The structure and function of gastrointestinal
    mucus, in: Gastrointestinal Mucosal Protection, J. Harmon Wilkins
    and Wilkins, ed., Baltimore, Chap. 25.
Allen, A., and Garner, A., 1980, Gastric mucus and bicarbonate
    secretion and their possible role in mucosal protection,
    Gut, 21:249-262.
Allen, A., Mantle, M., and Pearson, J.P., 1980, in: Perspectives in
    Cystic Fibrosis, Proceedings of Eighth International Congress on
    Cystic Fibrosis, J. Sturgess, ed., Imperial Press Ltd.,
    Mississauga, Ontario, pp.102-112.
Allen, A., Pain, R.H., and Robson, T., 1976, Model for the structure
    of the gastric mucus gel, Nature, 264:88-89.
Allen, A., Pearson, J.P., Venables, C.W., and Younan, F., 1980b,
    Measurement of the degradation of gastric mucus gel by peptic
    erosion in vivo, J. Physiol., 306: 40P.
Allen, A., and Snary, D., 1972, The structure and function of gastric
    mucus, Gut, 13:666-672.
Bell, A.E., Allen, A., Morris, E., and Rees, D.A., 1980, The
    structure of native gastric mucus gel, Biochem. Soc. Trans.,
    8:716.
Bell, A.E., Allen, A., Morris, E., and Rees, D.A., 1981, Rheological
    studies on native pig gastric mucus gel, this volume.
Bickel, M., and Kauffman, G.L., 1981, Gastric gel mucus thickness:
    effect of distention, 16,16-dimethyl prostaglandin E2 and carb-
    enoxolone, Gastroenterology, 80:770-775.
Carlson, D.M., 1968, Structures and immunochemical properties of
    oligosaccharides isolated from pig submaxillary mucus,
    J. Biol. Chem., 243:616-626.

Carlson, D.M., 1977, Chemistry and biosynthesis of mucin glyco-proteins, in: Mucus in Health and Disease, M. Elstein and D.V. Parke, ed., Plenum Press, New York and London, p.251-273.

Creeth, J.M., 1978, Constituents of mucus and their separation. Brit. Med. Bull., 34:17-24.

Creeth, J.M., and Denborough, M.A., 1970, The use of equilibrium density methods for the preparation and characterisation of blood group specific glycoproteins, Biochem. J., 117:879-891.

Creeth, J.M., Bhaskar, K.R., Horton, J.R., Das, I., Lopez-Vidriero, M.T., and Reid, L., 1977, The separation and characterisation of bronchial glycoproteins by density gradient methods, Biochem. J., 167:557-569.

Davenport, H.W., 1967, Physiological structure of the gastric mucosa, in: Handbook of Physiology, Alimentary Canal, Vol. II, C.F. Code, ed., American Physiological Society. Washington D.C., pp.759-779.

Donald, A.S.R., 1973, The products of pronase digestion of purified blood group-specific glycoproteins, Biochem. Biophys. Acta, 317:420-436.

Dunstone, J.R., and Morgan, W.T., 1965, Further observations on the glycoproteins in human ovarian cyst fluids, Biochim. Biophys. Acta., 101:300-314.

Flemstrom, G., 1977, Active alkinization by amphibian gastric fundic mucosa in vitro, Am. J. Physiol., 233:E1-E12.

Flemstrom, G., Garner, A., Nylander, O., Hurst, B., and Haylings, J.R., 1981, Intraluminal acid, prostaglandin E_2 and glucagon, stimulate duodenal epithelial HCO_3^- transport in the cat and guinea pig, Acta Physiol. Scand., in press.

Florey, H., 1955, Mucin and the protection of the body, Proc. R. Soc., Lond., B, 143:144-158.

Garner, A., and Flemstrom, G., 1978, Gastric HCO_3^- secretion in the guinea pig, Am. J. Physiol., 234:E535-E541.

Gibbons, R.A., 1972, Physico-chemical methods for determination of purity, molecular size and shape of glycoproteins, in: Glyco-proteins, A. Gottschalk, ed., Elsevier, Amsterdam, p.31-109.

Gibbons, R.A., 1978, Mucus of the mammalian genital tract, Brit. Med. Bull., 34:34-38.

Glass, G.B.J., and Slomiany, B.L., 1977, Derangements in gastro-intestinal injury and disease, in: Mucus in Health and Disease, M. Elstein and D.V. Parke, ed., Plenum Press, New York, pp.311-347.

Gottschalk, A., 1972, ed., Glycoproteins: their composition, structure and function, revised 2nd ed., Elsevier, Amsterdam and London.

Heatley, N.G., 1959, Mucosubstance as a barrier to diffusion, Gastroenterology, 37:313-318.

Hollander, F., 1954, The two-component mucus barrier, Archives of Internal Medicine, 93:107-120.

Horowitz, M.I., and Pigman, W., 1977, ed., The Glycoconjugates, Vol. I., Academic Press Inc.

132

List, S.J., Findlay, B.P., Forstner, G.G. and Forstner, J.F., 1978, Enhancement of the viscosity of mucin by serum albumin, Biochem. J., 175:565-571.

Mantle, M., and Allen, A., 1981, Isolation and characterisation of the native glycoprotein from pig small intestinal mucus, Biochem. J., 195:267-275.

Mantle, M., Mantle, D., and Allen, A., 1981, Polymeric structure of pig small-intestinal mucus glycoprotein, Biochem. J., 195:277-285.

Marshall, T., and Allen, A., 1978, Isolation and characterisation of the high molecular weight glycoprotein from pig colonic mucus, Biochem. J., 173:569-578.

Parke, D.V., 1978, Pharmacology of mucus, Brit. Med. Bull., Brit. Med. Bull., 34:89-94.

Pearson, J.P., Allen, A., and Venables, C.W., 1979, Proteolytic digestion of human and pig gastric mucus glycoproteins, in: Glycoconjugates, R. Shauer, ed., Georg Thieme, Stuttgart, p.136.

Pearson, J.P., and Allen, A., 1981, Reduction by mercaptoethanol and proteolysis of the non-glycosylated peptide region of pig gastric mucus glycoprotein, this volume.

Pearson, J.P., Allen, A., and Venables, C.W., 1979, Proteolytic digestion of human and pig gastric mucus glycoproteins, in: Glycoconjugates, R. Schauer, ed., Georg Thieme, Stuttgart, p.136.

Pearson, J.P., Allen, A., and Venables, C.W., 1980, Gastric mucus: isolation and polymeric structure of the undegraded glycoprotein: its breakdown by pepsin, Gastroenterology, 78:709-715.

Pearson, J.P., Allen, A., and Parry, S., 1981, A 70,000 molecular weight protein isolated from purified pig gastric mucus glycoprotein by reduction of disulphide bridges and its implication in the polymeric structure, Biochem. J., in press.

Roberts, G.P., 1978, Chemical aspects of respiratory mucus, Brit. Med. Bull., 34:39-41.

Scawen, M., and Allen, A., 1977, The action of proteolytic enzymes on the glycoprotein from pig gastric mucus, Biochem. J., 163:363-368.

Slomiany, B.L., and Meyer, K., 1972, Isolation and structural studies of sulphated glycoproteins of hog gastric mucosa, J. Biol. Chem., 247:5062-5070.

Snary, D., and Allen, A., 1971, Studies on gastric mucoproteins. The isolation and characterisation of the mucoprotein of the water soluble mucus from pig cardiac gastric mucosa, Biochem. J., 123:845-853.

Snary, D., Allen, A., and Pain, R.H., 1970, Structural studies on gastric mucoproteins. Lowering of molecular weight after reduction with 2-mercaptoethanol, Biochem. Biophys. Res. Commun., 40:844-851.

Starkey, B.J., Snary, D., and Allen, A., 1974, Characterisation of gastric mucoproteins isolated by equilibrium density-gradient centrifugation in caesium chloride, Biochem. J.,141:633-639.

Williams, S.E. and Turnberg, L.A., 1980. Retardation of acid diffusion by pig gastric mucus: a potential role in mucosal protection, Gastroenterology, 79:299-304.

Williams, S.E., and Turnberg, L.A., 1981, Demonstration of a pH gradient across mucus adherent to rabbit gastric mucosa: evidence for a 'mucus-bicarbonate' barrier, Gut, 22:94-96.

Younan, F., Pearson, J.P., Allen, A., and Venables, C.W., 1981, Gastric mucus degradation in vivo in peptic ulcer patients and the effects of vagotomy, this volume.

STUDIES UPON THE SECRETION OF GASTRIC MUCUS FROM NORMAL SUBJECTS

John Clamp and Paul Brown

Department of Medicine
Bristol Royal Infirmary
Bristol BS2 8HW

INTRODUCTION

A number of pharmacological substances are known to affect the rate of gastric secretions. However, little is known about the action of such substances upon the production of mucus, that is upon the glycoprotein content of such secretions. Accordingly the effects of insulin and pentagastrin have been studied upon the volume and mucus glycoprotein content of gastric secretions from normal human subjects.

METHODS

Gastric secretions were aspirated from normal subjects who had fasted for at least 12 hr. The initial "fasting" aspirate was discarded and the "basal" secretions collected for 1 hr. An intra-venous injection of insulin was then given and the secretions collected for 2 hr. Finally pentagastrin was administered subcut-aneously and the secretions collected for a further hour. Each aspirate was measured and after neutralisation with sodium hydrogen carbonate was exhaustively dialyzed and subjected to gel-permeation chromatography on a column of Sepharose 2B. Three fractions were obtained, namely "high", "intermediate" and "low" molecular weight. The high molecular weight fraction consisted largely of mucus glycoprotein.

The carbohydrate content of each fraction was determined by gas-liquid chromatography.

RESULTS AND DISCUSSION

   The volume, content of non-dialyzable solids and
carbohydrate content are shown in Table 1. It can be seen that
there were no significant differences between the results from
secretors and non-secretors although in the monosaccharide
analyses, non-secretors had lower values for fucose and
N-acetylgalactosamine. In both groups insulin and pentagastrin
caused a marked increase in the volume of gastric secretions, but
there were no significant differences in the content of non-
dialyzable solids under the various conditions of stimulation.

   In the secretions from the unstimulated or resting stomach
(basal), the non-dialyzable solids consisted almost entirely of
mucus glycoprotein material, whereas after stimulation only about
50-60% of the material consisted of mucus glycoprotein.

   Thus both insulin and pentagastrin cause a reduction in the
output of mucus glycoproteins but an increase in the output of
other non-dialyzable material.

Table 1.   The Effect of Insulin and Pentagastrin on the Volume,
           Non-Dialyzable Solid and Carbohydrate Content of Gastric
           Secretions from Normal Subjects.

	Basal	Insulin Induced	Pentagastrin Induced
A-Secretor			
Volume (ml/hr)	70^+_-61	114^+_-19	158^+_-54
Non-dialyzable solids (mg/hr)	246^+_-223	233^+_-120	210^+_-105
Carbohydrate content of non-dialyzable solids	40%	23%	24%
Carbohydrate content of high molecular weight fraction	40%	51%	44%
Non-Secretor			
Volume (ml/hr)	72^+_-20	138^+_-34	232^+_-68
Non-dialyzable solids (mg/hr)	209^+_-33	247^+_-24	278^+_-42
Carbohydrate content of non-dialyzable solids	28%	17%	17%
Carbohydrate content of high molecular weight fraction	31%	38%	44%

THE NEUTRAL CARBOHYDRATE CHAINS OF GASTRIC AND DUODENAL

MUCOSUBSTANCES

Leslie Hough and Ramon L. Sidebotham

Department of Chemistry, Queen Elizabeth College
University of London, London W8 7AH

A complete stomach and duodenum were obtained at autopsy from an
individual with no history of gastroduodenal disease, and divided into
body-fundus, antrum, upper duodenum (approximating to the first and
second segments), and lower duodenum (third and fourth segments).
The gastroduodenal mucosa was histologically normal. The mucosa was
dissected from each specimen, and preparations of mucuous glycopolypeptides were obtained by a procedure of pronase digestion and
column chromatography.

The amino acid analysis of the mucous glycopolypeptides indicated
that all, or nearly all, the carbohydrate chains were $\underline{O}$-glycosidically
linked to serine or threonine residues in the peptide core.   In
addition, a comparison of the numbers of carbohydrate units, with the
combined numbers of serine and threonine units, in each of the mucous
glycopolypeptide preparations indicated (a) that the average (minimum)
length of the carbohydrate chains of the Brunner's glands
mucosubstance was~ 25% less than in the mucosubstances from the
(normal) gastric mucosa, or (b) that the carbohydrate chains in the
mucosubstances of the gastric and Brunner's glands, if of a comparable
length, were~ 25% fewer in number in the latter mucosubstance.

To examine these propositions, each of the mucous glycopolypeptide
preparations was treated with alkaline borohydride reagent[1] to detach
the carbohydrate chains (as reduced oligosaccharides), which were
fractionated on a column of Bio-Gel P-4.   It was reasoned that if the
carbohydrate chains were on average ~ 25% shorter in the Brunner's
glands mucosubstance, this would reflect itself in a detectably
higher concentration of lower molecular size components in the
oligosaccharide mixture.

137

The oligosaccharides released from the (normal) body-fundic and antral mucosubstances formed, as anticipated, identical (super-imposable) fractionation patterns, which were characterised by the presence of five major peaks. Unexpectedly, the oligosaccharides released from the mucosubstance of the Brunner's glands formed a fractionation pattern that was markedly different, and was characterised by the presence of three broad peaks.

To account for these findings, we suggest that the arrangement of the structural units within the (neutral) carbohydrate chains of the mucosubstances produced in the Brunner's glands and in the (normal) gastric mucosa must in some way be different. It was not apparent, when the fractionation patterns were compared, that the carbohydrate chains were, on average, shorter in the mucosubstance of the Brunner's glands. This point, however, is being further investigated.

A similar analysis was undertaken of mucous glycopolypeptides that were extracted from the body-fundus, antrum and duodenum of a second individual, who exhibited gross and histological evidence of an advanced gastritis. The fractionation pattern of the oligosaccharides released from the Brunner's glands mucosubstance was identical (superimposable) to that observed in the previous experiment. The fractionation patterns formed by the oligo-saccharides from the (gastritic) body-fundic and antral muco+substances were similar to each other (superimposable), but closely resembled the pattern that was obtained from the Brunner's glands mucosubstance not, as anticipated, the pattern that had characterised the (normal) body-fundic and antral mucosubstances. Differences comparable to these have also been demonstrated within a larger group of mucous glycopolypeptides extracted from normal and gastritic antral mucosa.

It is generally recognised that there are two groups of cells that produce mucosubstances in the histologically normal (disease free) human gastric mucosa: the columnar cells that line the mucosal surface and pits, and the mucous cells that occur within the glands of the pylorus, fundus and body mucosa[2]. The former group of cells appear to produce the greater part of the gastric mucosubstance[3]. Moreover, it has been observed that these cells are lost (to be replaced by less mature cells, with characteristics of cells found in the mucosal glands) during the gastritic process[4]. We conclude, therefore, that our results reflect that the (neutral) carbohydrate chains of the mucosubstances produced by the columnar cells of the epithelial surface, and the mucous cells within the gastric glands have dissimilar structures whereas the (neutral) carbohydrate chains associated with muco-substances of the Brunner's and gastric glands have comparable structures.

ACKNOWLEDGEMENT

        The authors thank the Medical Research Council for financial
support.

REFERENCES

1.   J. W. Mayo and D. M. Carlson, Carbohyd. Res. 15: 300 (1970)

2.   Histology (7th edition) A. W. Ham, ed.   J. B. Lippincott,
      Philadelphia and Toronto (1974).

3.   R. Lev, Lab. Invest. 14: 2080 (1966)

4.   G. B. J. Glass and C. S. Pitchumoni, Hum. Path. 6: 219 (1975).

THE COMPOSITION OF A MACROMOLECULAR EXTRACT OF COMBINED

HUMAN BODY AND FUNDIC MUCOSA

Leslie Hough and Ramon L. Sidebotham
Department of Chemistry, Queen Elizabeth College
University of London, London W8 7AH

A mixture of macromolecules was isolated from combined human
body and fundic mucosa (from which lipids had previously been removed
with acetone, and Folch reagent) by a procedure of pronase
digestion, and chromatography on Bio-Gel P-30.   In order to resolve
the various components, the mixture was subjected to preparative zone
electrophoresis in a block of vinyl chloride copolymer (Pevikon).[1]
The mixture was separated into twelve fractions, which were
designated P2 to P13.   A number of the Pevikon fractions were
combined and refractionated on Bio-Gel Agarose 0.5m (exclusion limit
500,000 Daltons for globular proteins), to give nine subfractions
which were designated A1 to A9.   The various separated components
were analysed to determine the contents of amino acids, carbohydrates, sulphate ester groups, and ABO(H) blood group
specificities.   The results of these analyses are summarised in
Table 1.   The largest single constituent of the extract (fraction
A1) proved to be a neutral glycopolypeptide of M.W.$>$ 500,000, that
exhibits A, H blood group activity, and in which the proportions
of amino acid and carbohydrate units are comparable to those
expected of a mucous glycoprotein.[2]   The proportions of amino
acid and carbohydrate units in the acidic glycopolypeptides

141

TABLE I   THE ANALYSIS OF A MACROMOLECULAR EXTRACT OF COMBINED HUMAN BODY AND FUNDIC MUCOSA

FRACTION	P2 - P4		P5	P6 - P7		P8	P9		P10	P11 - P13		
	A1	A2		A3	A4		A5	A6		A7	A8	A9
% DRY WEIGHT	31	11	5	2	9	6	2	7	6	6.5	6	6.5
COMPOSITION (%)												
CARBOHYDRATE ^{a}	77	48	39	29	30	14	23	<1	25	-	8	-
PEPTIDE ^{b}	19	28	42	56	54	62	10	63	32	4	6	76
SULPHATE ESTER ^{c}	0.3	0.6	0.4	ND	1	1	3	3	2	1	1	2
DNA ^{c}	-	-	-	-	-	-	<1	<1	41	89	53	12
RNA ^{d}	-	-		-	-	-	-	<1	<1	5	29	<1
UNACCOUNTED	4	23	18	15	15	23	63	31	-	1	3	9

$\underline{a}$ By Gas Chromatographic analysis (excludes contributions from DNA and RNA.) $\underline{b}$ Calculated from an amino acid analysis. $\underline{c}$ Standard used: DNA from Salmon Testes. $\underline{d}$ Standard used: RNA from Torula yeast.

of fractions A2, P5 and A4 vary in a way which suggests that they consist of (sulphated) mucous glycopolypeptides in combination with glycopolypeptides possessing structural elements (sialic acid, mannose and aspartic acid) usually associated with glycoproteins of plasma origin.[3]   Fractions A3, P8, A5 and P10 contain residues of glucuronic acid (indicative of a glycosaminoglycan), are partially degraded by testicular hyaluronidase, and have migrations on cellulose acetate electrophoresis comparable to hyaluronic acid (A3), and chondroitin sulphates (P8, A5 and P10).   The largest component within fraction P10, and the principal components of fractions A7 and A8 are nucleic acids.   Fractions A6 and A9 are unusual, in that the main constituents are polypeptides of $MW < 500,000$, in which aspartic acid, glutamic acid, glycine and proline are the main structural units.   These findings lead us to conclude that the latter fractions may contain fragments, or intact molecules, of pepsin.[4]

ACKNOWLEDGEMENTS

The authors thank the Medical Research Council for financial support.

REFERENCES

1 H. J. Muller-Eberhard, Scand. J. Clin. and Lab. Invest. 12:33(1960)

2 J. Schrager and M. D. G. Oates, Biochim. Biophys. Acta, 372:183(1974)

3 J. R. Clamp in Plasma Proteins, 2nd Ed. F. W. Putnam (ed.), Acad. Press, New York pp 163-211 (1975).

4 J. Tang, J. Mills, L. Chiang and L. de Chiang, Ann. N.Y. Acad. Sci. 140:688 (1967).

DIFFERENCES IN MUCUS GLYCOPROTEINS OF SMALL INTESTINE FROM

SUBJECTS WITH AND WITHOUT CYSTIC FIBROSIS

A.W. Wesley, A.R. Qureshi, G.G. Forstner, J.F. Forstner

The Hospital for Sick Children
Toronto, Ontario M5G 1X8
Canada

Studies on the composition of mucus from subjects with Cystic Fibrosis (CF) have previously shown increased fucose and decreased sialic acid (Dische et al, 1959; Clamp and Gough, 1979). To expand these studies we isolated human intestinal goblet cell mucus as previously described (I. Jabbal et al, 1976) with further purification by nuclease enzyme digestion and Sepharose 2B column chromatography. Mucins from 4 subjects with CF aged 9-23 years and 6 non CF controls aged 1 day to 90 years were compared with respect to amino acid, carbohydrate content and CsCl density gradient ultracentrifugation.

No differences between CF and non CF samples were noted in the amino acid profile, or in the degree of saturation of serine and threonine with N-acetyl-galactosamine. However, CF mucins contained on average, more total carbohydrate than controls (22.9 ± 3.39 vs 14.36 ± 4.20 n moles/ug protein $p < 0.01$). When individual sugars were analysed there was significantly more fucose (5.94 ± 2.30 vs 2.57 ± 1.18 $p < 0.01$), galactose (8.16 ± 1.05 vs 4.32 ± 1.93 $p < 0.005$) and N-acetyl-glucosamine (4.46 ± 0.63 vs 2.54 ± 1.32 $p < .02$) but there was no significant difference in sialic acid or N-acetyl-galactosamine.

Individual molar ratios of carbohydrate to N-acetyl-galactosamine tended to be higher in the CF patients with the exception of sialic acid, but only fucose/N-acetyl-galactosamine reached statistical significance (2.35 ± 0.84 vs 1.21 ± 0.83 $p < 0.05$). The fucose to sialic acid ratio was also significantly different (3.85 ± 2.18 vs 1.57 ± 1.24 $p < 0.05$). These differences were not due to differences in ABH or Lewis blood group specificity. There was a strong positive correlation between fucose per protein and

total carbohydrate per protein (r = 0.85), whereas there was no
similar correlation for sialic acid, the other sugar known to ter-
minate oligosaccharide chains. The buoyant density of the major
mucin species separated by CsCl density gradient ultracentrifugation
ranged from 1.39 to 1.49. In all four CF specimens, but only one
non CF specimen, almost all of the mucin exhibited a buoyant density
greater than 1.44, while in the other non CF specimens a substantial
proportion was of lower density.

These results show that significant differences exist in the
carbohydrate content and density of intestinal mucin from CF and
control patients. CF mucins are denser and more highly glycosylated
with a higher content of fucose, galactose and N-acetyl-glucosamine,
but no change in sialic acid. Since increased fucose correlated
with increased total carbohydrate, in CF mucins a more highly
branched oligosaccharide structure is likely. Increased glycosylation and branching would be expected to enhance the tendency
of CF mucins to gel and obstruct in vivo.

References

Clamp, J.R., and Gough, M., 1979, Study of the oligosaccharide units
     from mucus glycoproteins of meconium from normal infants and
     from cases of cystic fibrosis with meconium ileus, Clin.
     Sci., 57:445.
Dische, Z., di Sant'Agnese, P., Pallavichini, C., Youlos, J., 1959,
     Compositions of mucoprotein fractions from duodenal fluid
     of patients with cystic fibrosis of the pancreas and from
     controls, Pediatrics, 24:74.
Jabbal, I., Kells, D.I.C., Forstner, G., Forstner, J., 1976,
     Human intestinal goblet cell mucin, Can. J. Biochem., 54:707.

GASTRIC MUCOSUBSTANCES AND BENIGN GASTRIC ULCERATION

Shelagh M. Morrissey, J.G. Mehta and D. Hollanders*

Department of Physiology, Queen Elizabeth College
University of London, London, W8 7AH and
* Department of Medicine, Withington Hospital, Manchester

INTRODUCTION

Variations in the composition of mucosubstances have been demonstrated in the gastric contents of patients with chronic gastritis, pernicious anaemia and gastric carcinoma[1]. There have however been few histochemical studies of the epithelial mucins observed in the benign gastric ulcers[1]. Neutral mucosubstances have been identified by their reaction with periodic acid-Schiff's reagent to form magenta chromophores[2]. Haynes and Hartridge[3] and Steedman[4] have shown acidic mucosubstances to interact with Alcian Blue at different levels of pH to form blue chromophores. The method employed in this study is based on electronic densitometric measurements of chromophores. The energy absorbed by each chromophore is dependent upon two factors, the spectral characteristics of the chromophores and the concentration of the stained material.

MATERIALS AND METHODS

Endoscopic biopsies used in this study were obtained from gastric antrum and mid-lesser curve. Patients with dyspepsia but endoscopically normal stomach were used as controls. Cases with histologically proven benign, lesser curve gastric ulcer were biopsied 1 cm. from the observed edge or from the scar if healed and also from the antrum, both before and immediately after one month's treatment with tri-potassium dicitratobismuthate (Denol). Biopsy specimens were fixed in 10% formaldehyde solution. 4 μm paraffin wax sections were dewaxed and stained with Alcian Blue and Periodic Acid-Schiff's reagent respectively. The acid mucosubstances were differentiated according to the protocol reported[5,6]. A Vicker's M 86 Scanning and Integrating Microdensitometer was used to measure chromophores represented by each mucosubstance[7].   The chromophores of acidic mucosubstances were scanned at 610 nm and those of neutral mucosubstances at 550 nm.   The integrated density values were calculated as reported earlier[7].

RESULTS

Table 1.   Shows the mean integrated density readins of Total Neutral and Acidic Mucins in the lesser curve gastric biopsy described above.

Lesser Curve Biopsy

		Controls (n=5)	Gastric Ulcer Pretreated (n=10)	Post-treated (n=6)
	Total Neutral Mucins	74.5±1.5 SEM	87.1±1.7 SEM	97.2±3.4 SEM
A	Total Acid Mucins	23.1±0.9 SEM	62.8±1.8 SEM	41.8±1.5 SEM
B	Sialidase Resistant	16.4±0.8 SEM	43.9±1.2 SEM	29.2±1.9 SEM
C	Sulphomucins	11.6±0.6 SEM	30.8±0.9 SEM	21.0±0.8 SEM
D	Sulphomucins	7.3±0.3 SEM	19.7±0.5 SEM	12.9±0.4 SEM

Table 2.   Shows the mean integrated density readings of Total Neutral and Acidic mucins in the Antral biopsies taken from lesser curve ulcer patients.

Antral Biopsy

		Controls (n=14)	Pretreated (n=10)	Post-treated (n=6)
	Total Neutral Mucins	66.1±2.5 SEM	77.3±2.1 SEM	100.5±3.3 SEM
A	Total Acid Mucins	23.2±3.6 SEM	36.1±1.4 SEM	24.9±1.1 SEM
B	Sialidase Resistant	17.0±1.4 SEM	25.4±1.0 SEM	17.6±0.9 SEM
C	Sulphomucin	11.2±0.8 SEM	18.7±0.7 SEM	12.8±0.6 SEM
D	Sulphomucin	7.6±0.6 SEM	11.3±0.4 SEM	7.6±0.3 SEM

CONCLUSION

Neutral mucins increased over controls in the presence of active gastric ulcer in all samples, Table 1 and Table 2. Neutral mucins were further elevated once the ulcers were treated and had healed and did not return to the control level ($p < 0.001$).

The total acidic mucosubstances show a highly significant increase over controls in the presence of active ulcers ($p < 0.001$). There was a less significant increase in the antral tissue (Table 2). After treatment the total acidic mucins showed a return towards the control values, this is more evident in the antral mucosa. The changes in treated gastric ulcer may be due to either a) ulcer healing, b) Denol treatment or c) both factors. Our data does not allow us to distinguish between these possibilities.

J.M. was supported by Cystic Fibrosis Trust, Great Britain. We are grateful to Professor P. Gahan of the Biology Department, Queen Elizabeth College, for the use of an M 86 Vicker's Microdensitometer.

REFERENCES

1. R. Lev, The histochemistry of mucus-producing cells in normal and diseased gastrointestinal mucosa. Lab. Invest. 14:2080 (1965).
2. J.F.A. McManus, Histological demonstration of mucin after periodic acid. · Nature, London 158:202 (1946).
3. F. Haynes and H. Hartridge, "Histology for Medical Students", Oxford Medical Publications, London (1930).
4. H.F. Steedman, Alcian Blue 8GX: A new stain for mucin. W.J. Micr. Sci. 91:477 (1950).
5. C. McCarthy and L. Reid, Acid mucopolysaccharide in the bronchial tree in the mouse and rat. Q. Jl. Exp. Physiol. 49:81 (1963).
6. R. Jones and L. Reid, The effect of pH on Alcian Blue staining of epithelial acid glycoproteins. 1. Sialomucins and sulphomucins (singly or in simple combinations). Histochem. J. 5:9 (1973).
7. L. Hough, J.G. Mehta, S.M. Morrissey and R.L. Sidebotham. The differentiation and quantitation of mucosubstances in human gastric antrum. J. Physiol. 291:9P (1979).

REDUCTION BY MERCAPTOETHANOL AND PROTEOLYSIS OF THE NON-GLYCOSYLATED

PEPTIDE REGION OF PIG GASTRIC MUCUS GLYCOPROTEIN

Jeffrey P. Pearson and Adrian Allen

Department of Physiology
The University
Newcastle upon Tyne NE1 7RU, U.K.

The native pig gastric mucus glycoprotein was purified free
from non-covalently bound protein by a combination of gel-filtration
on Sepharose 4B and equilibrium centrifugation in a CsCl gradient
(Starkey et al., 1974; Pearson et al., 1981).   On reduction with
0.2M mercaptoethanol the native glycoprotein (2 x 10^6 molecular
weight) is split into an average of 4 equal sized subunits (5 x 10^5
molecular weight) and a 70,000 molecular weight protein fraction.
The 70,000 molecular weight protein fraction which contains little
or no carbohydrate can be separated from the reduced glycoprotein
subunits by equilibrium centrifugation in a CsCl gradient (Allen
et al., 1980; Pearson & Allen, 1980).   Here we report further
investigations on the structure of the non-glycosylated region of
glycoprotein defined as that part of the peptide core which is free
of carbohydrate and is removed when the molecule is split by proteolysis into 4 subunits of molecular weight 5 x 10^5 (Scawen & Allen,
1977).

The amino acid analysis of the isolated glycoprotein subunit
obtained by reduction with 0.2M mercaptoethanol shows a loss of
amino acids equivalent to the removal of the 70,000 molecular weight
protein fraction (Table 1).   Proteolysis of this reduced glycoprotein subunit removed further peptide to give an amino acid content
similar to that of the glycoprotein obtained by direct proteolysis
of the native glycoprotein.   Thus the non-glycosylated peptide
region can be split into two components: the 70,000 molecular
weight protein fraction and a further portion of peptide which is
not removed by reduction but is susceptible to proteolysis, these
two components comprise 3.7% and 2.9% of the glycoprotein respectively.   Both regions of this non-glycosylated peptide have a
complete spectrum of amino acids and are rich in cysteine (Table 1).

Table 1   Amino acid analysis of preparations of pig gastric mucus glycoprotein

Amino acid	Amino acid content (μmol/mg)		Glycoprotein Subunit	
	Native Glycoprotein	Reduced 70,000 mol.wt. protein fraction	Reduced	Reduced and papain digested
His	0.024	0.111	0.013	0.006
Lys	0.035	0.225	0.016	0.006
Arg	0.049	0.237	0.030	0.006
Asp	0.072	0.615	0.033	0.015
Thr	0.190	0.467	0.170	0.185
Ser	0.201	0.616	0.186	0.240
Glu	0.103	0.877	0.059	0.049
Pro	0.175	0.432	0.155	0.125
Gly	0.082	0.700	0.056	0.046
Ala	0.075	0.563	0.055	0.037
Cys*	0.056	0.253	0.031	absent
Val	0.094	0.600	0.063	0.037
Met	0.007	0.052	TRACE	absent
Ile	0.040	0.229	0.027	0.019
Leu	0.048	0.490	0.024	0.009
Tyr	0.026	0.265	0.009	absent
Phe	0.026	0.254	0.014	0.005

The protein content of the native glycoprotein, the reduced glycoprotein and the reduced papain digested glycoprotein were 13.3, 9.6 and 6.7% by wt. respectively.

In complete contrast the glycosylated peptide, that is the glycoprotein subunit of about 5×10^5 molecular weight from proteolysis,
is very high in threonine, serine and proline in keeping with no
detectable loss of carbohydrate on proteolysis (Scawen & Allen,
1977).

The native glycoprotein is split into lower molecular weight
glycoprotein subunits and with the concommitant release of 70,000
protein by concentrations of mercaptoethanol as low as 10mM for 24
hours.   Controlled proteolysis of the native glycoprotein with
papain caused a reduction in viscosity, over 30 min at $37^{\circ}C$ and this
was paralleled by an associated production of lower molecular weight
glycoprotein and the release of 70,000 molecular weight protein.
All the data obtained are compatible with a model for the gastric
glycoprotein where each of the four glycoprotein subunits contains a
region of non-glycosylated protein which is joined centrally to the
70,000 molecular weight protein.

REFERENCES

Allen, A., Mantle, M. and Pearson, J.P., 1980, The structure
    and properties of mucus glycoproteins. In: Perspectives
    in Cystic Fibrosis, edited by J. Sturgess, pp. 102-112.
    Canadian Cystic Fibrosis Foundation, Toronto.
Pearson, J.P., Allen, A. and Venables, C.W., 1980, Gastric
    mucus: isolation and polymeric structure of the undegrad-
    ed glycoprotein: its breakdown by pepsin. Gastroenterol-
    ogy, 78: 709-715.
Pearson, J.P., Allen, A. and Parry, S., 1981, A protein
    70,000 molecular weight, isolated from purified pig
    gastric mucus glycoprotein by reduction of disulphide
    bridges: implication in polymeric structure. Biochem.
    J. in press.
Scawen, M. and Allen, A., 1977, The action of proteolytic
    enzymes on the glycoprotein from pig gastric mucus.
    Biochem. J. 163: 363-368.
Starkey, B.J., Snary, D. and Allen, A., 1974, Characteris-
    ation of gastric mucoproteins isolated by equilibrium
    density gradient centrifugation in CsCl. Biochem. J.,
    141: 633-639.

AN INSOLUBLE MUCIN COMPLEX FROM RAT SMALL INTESTINE

Ingemar Carlstedt, Heléne Karlsson, Frank Sundler[*] and
Lars-Åke Fransson

Department of Physiological Chemistry 2 and [*]Department
of Histology, University of Lund
P.O. Box 750, S-220 07 Lund 7, Sweden

INTRODUCTION

The lining of the gastro-intestinal tract is protected by a
visco-elastic mucus layer. To understand the functions of this layer
its components must be studied. We have prepared mucin molecules from
rat small intestine using a procedure that avoids proteolysis, reduction of disulphide bonds and high shear.

EXPERIMENTAL AND RESULTS

In preliminary experiments mucosal scrapings were extracted with
guanidine hydrochloride (GuHCl), urea, neutral salts and detergents
over broad concentration ranges. Neither of these agents nor compatible combinations thereof were successful. In contrast, exposure of
the tissue to high shear or dithiothreitol (DTT) resulted in almost
complete solubilization.

It was noted however, that extraction of the tissue with deoxycholate left a very small residue although only some 20% of the mucin
was extracted. Thus, mucosal scrapings were suspended in 0.10M Na-
deoxycholate/Na-barbitone, pH 8.5 and left stirring for 5 h at 4°C.
After high-speed centrifugation the supernatant was discarded and the
gel-like pellet re-extracted with deoxycholate over-hight. The mucin
complex was again recovered by centrifugation, washed with barbitone
buffer followed by water. The material thus obtained (RI-DOC) was
free from nucleic acids and lipids. The yield was approximately 11mg
of lyophilized RI-DOC/rat, comprising about 80% of the total.

RI-DOC was solubilized with DTT in 6M GuHCl and the fragments

155

were subjected to isopycnic density-gradient centrifugation (4M
GuHCl/CsCl, initial density 1.39 g/ml). The carbohydrate-rich fragments were recovered at a density of 1.44 g/ml (Fig. 1a) and accounted
for more than 80% by weight of RI-DOC. Most of these glycopeptides
were excluded from Sepharose 2B (Fig. 1b). In contrast, solubilization by DTT without GuHCl produced much smaller fragments (Fig. 1c).
If RI-DOC that had been pretreated with diisopropyl fluorophosphate
(DFP) was exposed to DTT alone only partial solubilization was

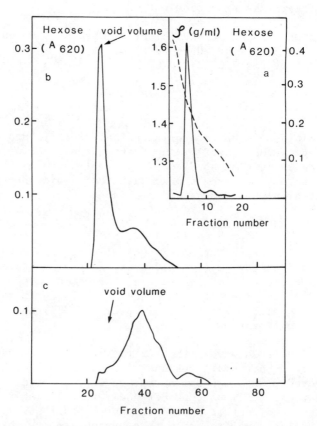

Fig. 1.   RI-DOC was solubilized by DTT in 6M GuHCl and the fragments
          were subjected to isopycnic density-gradient centrifugation
          in 4M GuHCl/CsCl, initial density 1.39 g/ml (a). The carbo-
          hydrate-rich fragments from the density gradients were
          chromatographed on Sepharose 2B after solubilization with
          DTT/6M GuHCl (b) or DTT alone (c).

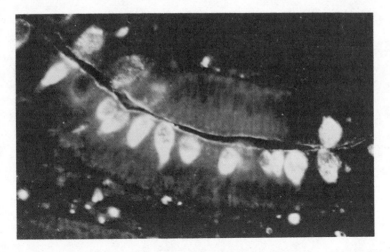

Fig. 2.   Goblet cells visualized by antibodies against RI-DOC using
          indirect immunofluorescence.

achieved, suggesting that an intrinsic proteolytic activity was working in concert with DTT. This suspicion was verified by the discovery
that RI-DOC catalysed the hydrolysis of N-Benzoyl-L-arginine ethyl
ester (BAEE) and that this activity was inhibited by boiling of
RI-DOC, by DFP or by GuHCl.

     Antibodies raised in rabbits against RI-DOC were localized to
the goblet cells in the rat small intestine by indirect immunofluorescence (Fig. 2).

CONCLUSIONS

     The major part of the goblet cell mucus in rat small intestine
constitutes an insoluble glycoprotein complex. The insoluble complex
can only be prepared if shear is avoided. The integrity of the gel
is dependent on disulphide bridges.

     A proteolytic activity is co-purified with RI-DOC. Control of
this activity is essential in order to obtain reproducible fragmen-
tation patterns in structural studies on these macromolecules. The
biological significance of this proteolytic activity is not apparent.
It might participate in the turn-over of the mucus gel.

COMPOSITION AND STRUCTURE OF RAT BRUNNER'S
GLANDS AND GASTRIC MUCOUS GLYCOPROTEIN

Henk L. Smits, P.J.M. van Kerkhof and M.F. Kramer

Department of Histology and Cell Biology
School of Medicine, University of Utrecht
3511 HG Utrecht, The Netherlands

INTRODUCTION

Previously, the isolation and partial characterization of rat gastric mucous glycoprotein (MG) has been described (Spee-Brand et al., 1980). We now have isolated a MG from Brunner's glands of the rat duodenum. The composition and size of the O-glycosidically linked oligosaccharide chains of both gastric and Brunner's glands MG are described.

RESULTS AND DISCUSSION

Homogenates of submucosal strips containing Brunner's glands were extracted twice with 0.5 M 2-mercaptoethanol and centrifuged at 1700 x g. The MG were purified by centrifugation at 150.000 x g for 60 hr on CsCl (initial density: 1.42 g/ml). A prominent peak of MG is found at 1.39 g/ml density, much lower than the 1.49 g/ml density of gastric MG. A pure MG was obtained after a second run either on CsCl or on CsBr. On gel electrophoresis it showed one band on top of a 2% (w/v) polyacrylamide -0.5% (w/v) agarose gel. No other (glyco-) proteins could be detected. The sugar composition is shown in Table I. Only fucose, galactose, galactosamine and glucosamine are present. Neuraminic acid is absent. Sulphate content is 2% by weight. Blood-group activity A is expressed by the purified glycoprotein but no B or H activity.

The MG of stomach (Spee-Brand et al., 1980) and of Brunner's glands were subjected to sodium borohydride degradation. MG (1 mg/ml) were dissolved in 0.05 N NaOH and 1 M NaBH$_4$ and incubated for 16 hr at 50°C. After neutralization and removal of the borate, the non-degradated residu was precipitated with alcohol. By this method 95% of the carbohydrate of Brunner's glands MG and 60% of the carbo-

TABLE I Sugar composition, as determined by gaschromatography

Stomach

	Glycoprotein (fractions 4-9 of 2nd CsCl gradient centrifugation	Oligosaccharides (after alkaline borohydride degradation of MG)	Oligosaccharide fractions		
			P_6	P_2	
			26–43	43–45	50–52
			G_v	G_8	G_7
gal	1.00	1.00	1.00	1.00	1.00
fuc	0.60	0.70	0.55	0.70	0.70
glcNAc	1.20	1.20	1.40	0.83	0.65
galNAc	0.60	0.25	0.45	0.35	–

Brunner's glands

	Glycoprotein (fractions 9-11 of 2nd CsCl gradient centrifugation)	Oligosaccharides (after alkaline borohydride degradation of MG	Oligosaccharide fractions			
			P_2			
			36–40	44–49	50–55	56–49
			B_v	B_8	B_7	B_6
gal	1.00	1.00	1.00	1.00	1.00	1.00
fuc	0.40	0.40	0.40	0.40	0.20	0.20
glcNAc	1.80	2.10	1.20	1.90	2.00	1.80
galNAc	1.80	1.00	0.80	0.70	0.40	0.60

hydrate of gastric MG were obtained. In the latter case the residu
was subjected to a second degradation and the recovered oligosaccharides were pooled. After desalting and removal of amino acids
by repeated ion exchange chromatography the oligosaccharides were
fractionated by chromatography on calibrated P_2 and P_6 columns.
Brunner's glands oligosaccharides eluted from the P_2 column in four
peaks, demonstrated by their hexose content, as measured by the
orcinol method. Two major peaks eluted with relative molecular
weights of an octamer (B_8) and a heptamer (B_7). A minor peak was
found in the hexamer fraction (B_6) and another minor peak eluted in
the void volume of the column (B_v). Gastric oligosaccharides eluted
in three peaks. Two of them eluted with relative molecular weights
of an octomer (G_8) and a heptamer (G_7). The major component, eluting
in the void volume of the P_2 column (G_v), was retarded on a P_6 column
and presumably contains oligosaccharides with 25 to 30 sugar residues. The sugar composition of the fractions, as analyzed by gaschromatography (Kamerling et al., 1975) is presented in Table I.
The structure of the carbohydrate chains is under study by NMR-
spectroscopy.

CONCLUSION

    Gastric MG of the rat has mainly long carbohydrate chains, while
Brunner's glands MG has shorter chains, mainly octamers and hepta-
mers. This difference confirms the autoradiographic data presented
earlier (Smits and Kramer, 1981).

REFERENCES

Smits, H.L., and Kramer, M.F., 1981, Glycoprotein synthesis in
        the mucous cells of the vascular perfused rat stomach.
        III. Mucous cells of the antrum and Brunner's glands.
        Am. J. Anat. (in press).
Spee-Brand, R., Strous, G.J.A.M., and Kramer, M.F., 1980,
        Isolation and partial characterization of rat gastric
        mucous glycoprotein. Biochim. Biophys. Acta, 621:104.
Kamerling, J.P., Gerwig, G.J., Vliegenthart, J.F.G., and Clamp,
        J.R., 1975, Characterization by gas-liquid chromatog-
        raphy-mass spectrometry and proton-magnetic-resonance
        spectroscopy of pertrimethylsilyl methyl glycosides ob-
        tained in the methanolysis of glycoproteins and glyco-
        peptides. Biochem. J., 151:491.

EFFECT OF LYSOLECITHIN ON THE CONSTITUENTS OF GASTRIC MUCUS

Bronislaw L. Slomiany,   George B. Jerzy Glass, Koichi
Kojima, Zofia Banas-Gruszka, Amilia Slomiany

Gastroenterology Research Laboratory, New York Medical
College, Research Center, Metropolitan Hospital
1900 Second Avenue, New York, N. Y. 10029

INTRODUCTION

The surface epithelium of gastric mucosa is covered by an extracellular and renewable layer of a viscous mucus gel. This, together with an intracellular apical layer of mucus has been referred
to by Hollander (1954) as "two layer mucous barrier". Yet its
ability to protect the underlying mucosa from the physical and enzymatic injury has been questioned by Heatley (1959), and later protective function has been rather assigned to the so-called "mucosal
barrier". This is supposed to consist of the plasma membranes and
the intracellular and basal membranes, as well as junctions of the
epithelium (Davenport, 1970). Only more recently, when the retardation of acid and pepsin diffusion by mucus gel has been demonstrated by newer technics (Allen, 1976; Pearson et al, 1980; Williams
and Turnberg, 1980) the potential role of the mucus in mucosal protection became plausible again. It has been also realized that the
solubilization, depolymerization and depletion of the mucus may be
contributory to the breaking of the "mucosal barrier" by anti-inflammatory agents, such as aspirin, indomethacin, phenylbutazone,
steroids, etc. or other agents, such as stress, bile acids or ethanol (Stremple et al, 1973; Glass and Slomiany, 1977; Glass et al,
1979).

Also lysolecithin, a lytic phospholipid produced by the action
of phospholipase A in pancreatic juice on the lecithin in the bile,
has been suspected to contribute to the damage of the gastric mucosa.
The reflux of bile into the stomach from the duodenum has been
incriminated as one of the pathogenetic factors in development of

163

the gastric erosions and ulcers in these patients (Davenport, 1970;
Rhodes, 1972; Orchard et al, 1977; Johnson and McDermott, 1974).
A consistent duodenal reflux of bile in these patients has been
also associated with an increased concentration of lysolecithin in
their stomachs, usually not seen in normal individuals (Slomiany
and Slomiany, 1979). The drop in the potential difference in these
patients with marked bile reflux into the stomach was suggestive
of the penetration of the refluxed lysolecithin through the viscous
mucus gel and injury to gastric mucosa. Although the lysolecithin
has been shown to modify the rheological properties of mucus in
vitro by reducing its viscosity and decreasing its elasticity (Martin
et al, 1978), little information exists on the mechanism of the
weakening of the mucous layer by lysolecithin and its adverse effect
on the constituents of the gastric mucous barrier.

     In this work we have attempted to clarify some of these effects
in vivo, by intragastrically instilling lysolecithin in the increasing concentrations into Ghosh-Lai preparations of 80 rats. The
effects of these instillations on the main components of the mucus
barrier were then examined, including the gastric glycoproteins,
protein-bound carbohydrates, and glycolipid constituents of the
gastric mucus, as well as their composition and contents. We have
also determined the effects of increasing concentration of lysolecithin on the solubilization of the main constituents of the gastric
mucus, which could be of important significance in the pathophysiology
of lysolecithin effects on gastric mucous barrier.

MATERIALS AND METHODS

     Male Sprague-Dawley rats of similar age and mean weight of 400g
were used for Ghosh-Lai stomach preparations (Balanzo and Glass,
1975). These were prepared under intraperitoneal anesthesia with
urethane (1.5ml/400g rat) as described previously. All animals were
fasted for 10 hrs. before surgery. The upper ligature was placed
over a polyester tubing introduced into the stomach through the
esophagus and ligated at the pyloric ring. This was used for filling the stomach. The lower catheter was used for draining. After
removal of the gastric preoperative content by washing with saline,
rats were placed in pairs, for the whole duration of the experiment
in a thermostatically controlled chamber at 37°C. The stomachs were
filled to capacity (4.5-6.2ml) with saline instillate which was then
used as control, each of the stomachs was washed twice with saline,
drained and filled to capacity by instillation with the solution
of lysolecithin in a concentration range from 0.02 up to 2.0 mg lysolecithin per ml saline (from Applied Science Lab State College, Pa).
The instillate was left in the Ghosh-Lai stomach for 1 Hour, and the
contents were completely recovered. At the conclusion of each experiment, rats were killed by an overdose of urethane, their stomachs
were removed, opened along the lesser curvature, washed gently with

saline, inspected for the presence of erosions and hemorrhagic
foci, and examined histologically.

In 36 pairs of Ghosh-Lai stomach preparations, lysolecithin
instillates and 2M NaCl instillates obtained from paired rats were
made into separate pools. Four pairs were used for each of the 8
lysolecithin concentrations and 4 pairs for 2M NaCl. The samples
were then dialyzed and lyophilized and assayed for protein and hemoglobin. Protein was determined by the procedure of Lowry et al,
(1951) using bovine serum albumin as standard, and hemoglobin was
measured by the method of Hanhs et al (1960). Stomachs were examined histologically with differential stains: Hematoxylin eosin,
periodic acid-Schiff, and Alcian blue, according to the standard
techniques (Slomiany et al, 1978).

Analytical methods

The dry samples were extracted with chloroform/methanol (2:1,
v/v), at room temperature for 24 hours, using 10 ml of solvent per
mg of lyophilizate. The extracts were percolated through silicic
acid columns (0.2 x 5cm) to retain the insoluble protein residue.
Each column was washed in sequence, with 15 ml aliquots of chloro-
form/methanol (2:1, v/v), chloroform/methanol (1:2, v/v), and methanol. The eluates were combined, concentrated to dryness, dissolved
in a small volume of chloroform and chromatographed individually on
silicic acid columns (0.2 x 15cm) into three major lipid fractions:
neutral lipids, eluted with chloroform; glycolipids, eluted with
acetone/methanol (9:2, v/v); and phospholipids,eluted with methanol
(Slomiany et al, 1977). The glycolipids present in the acetone/
methanol elutes were examined by thin-layer chromatography on silica
gel HR plates, from Analtech (Newark, Del.), and quantified by gas-
liquid chromatography. The content and composition of protein bound
carbohydrates was determined on the insoluble residues retained on
the top of the sialic acid columns (Slomiany et al, 1979).

Gas-liquid chromatography analyses of carbohydrates were performed on a Beckman GC-65 instrument equipped with glass columns
(180 x 0.2cm) packed with 3% SE-30 on Gas-Chrom Q (Slomiany and
Slomiany, 1978). Trimethylsilyl derivatives of methyl glycosides were
prepared according to Hughes and Clamp (1972). Glycolipids were
chromatographed on the thin-layer plates developed in chloroform/
methanol/water (65:25:4, v/v/v) and visualized with orcinol reagent
(Slomiany et al, 1978).

RESULTS

Exposure of the Ghosh-Lai rat gastric mucosa to lysolecithin
at concentrations of 0.08 mg/ml or more led to the removal of mucus
from gastric mucosa,as judged by the increased content of protein,
protein bound carbohydrate and glycolipids in the post lysolecithin

instillates. The extent of solubilization of the investigated mucus constituents was clearly dependent upon the lysolecithin concentration, and at high lysolecithin concentrations of 0.8-2.0 mg/ml it was comparable to that produced by 2M NaCl (Table 1).

Histological studies of the mucosa from stomachs instilled with the lysolecithin at concentrations 0.02-0.04 mg/ml showed that surface epithelial mucous lining was not affected at these low concentrations. However, 1 hour exposure to lysolecithin at the concentration range of 0.08-0.40 mg/ml led to the progressive removal of the visible mucus layer. At the very high concentrations of lysolecithin (1.6-2.0 mg/ml) in the instillates, most of the surface epithelial mucous lining and most of the preformed intracellular mucus were removed from the epithelial mucous cells. Yet, no disruption of the mucosal membranes was observed. This finding was further substantiated by absence of significant amounts of hemoglobin in the instillates, which confirmed relative integrity of vascular membranes.

Lysolecithin at concentrations of 0.02-0.04 mg/ml in the instillate had no effect on the mucus constituents: The contents of protein, protein bound glycoprotein and glycolipid in the instillates did not differ from that in saline controls. The higher lysolecithin concentrations (see Table 1) however, produced gradual enrichment of the instillates in the constituents of the gastric mucus layer. The enrichment of the instillate in protein was initially rather low (2-2.5 times) at 20 fold increases of lysolecithin concentrations in the instillate (from 0.04 to 0.8 mg/ml). At the further 2-2.5 increase in lysolecithin concentration (from 0.8 to 1.6-2 mg/ml) the increase of protein concentration was five fold, which was associated with a three fold increase in glycolipid and two fold increase in protein bound carbohydrate in the instillate.

Gas-liquid chromatography of the glycolipid fractions isolated from all the instillates revealed only the presence of glucose, glycerol ethers and fatty acids in all three types of samples. Neither fraction contained sphingosine or sugars other than glucose. Thus, it was concluded that glycolipids fraction of the gastric mucus represent not sphingolipids, but glyceroglucolipids, (Glass and Slomiany, 1977; Slomiany et al, 1980). The glyceroglucolipids contained in each type of instillate gave two major and two minor glycolipid bands on thin-layer chromatography. One of the major bands showed positive staining reaction with rhodizonate and was identical in its thin-layer mobility with the sulfated tetraglycosyl monoalkylmonoacylglycerol of the rat gastric surface epithelial mucous layer. The thin-layer chromatographic patterns of the remaining glyceroglucolipids of the mucous barrier of rat stomach (Slomiany et al, 1978) were similar to that of the neutral glyceroglucolipids.

The composition and content of protein bound carbohydrates found

Table 1. Protein, protein bound carbohydrate and glyceroglucolipid content in the instillates from Ghosh-Lai rat stomach

Lysolecithin (mg/ml instilled)	Protein		Protein bound carbohydrate		Glyceroglucolipid (as glucose)	
	Experiment	Control	Experiment	Control	Experiment	Control
			μg/ml of instillate			
0.02	71.1±13.2	65.7±12.8	3.89±0.45	3.87±0.41	0.60±0.10	0.59±0.11
0.04	69.6±14.1	68.1±11.4	3.93±0.42	4.00±0.38	0.63±0.12	0.60±0.10
0.08	118.9±19.8	68.5±13.1	4.48±0.40	3.93±0.42	0.81±0.13	0.62±0.12
0.16	120.2±15.3	67.7±14.2	4.55±0.45	3.79±0.47	0.79±0.10	0.58±0.11
0.40	168.0±14.6	68.1±12.0	5.56±0.83	3.85±0.36	1.02±0.21	0.57±0.13
0.80	185.2±15.7	65.3±11.4	6.59±0.91	4.01±0.49	1.37±0.25	0.61±0.12
1.60	328.5±28.6	67.1±10.8	8.07±1.03	3.90±0.40	1.82±0.23	0.59±0.14
2.00	330.7±32.8	71.3±12.4	8.43±1.02	3.88±0.50	1.81±0.26	0.65±0.10
2M NaCl	350.3±42.1	64.2±15.1	8.46±1.10	4.02±0.48	1.85±0.28	0.60±0.15

Values represent means ± SD from 4 pairs of rats for each lysolecithin concentration, saline controls, as well as for the 2M NaCl instillations.

in the lysolecithin, 2M NaCl and saline instillates is listed in
Figure 1. All three types of samples contained fucose, galactose,
N-acetylgalactosamine, N-acetylglucosamine, and sialic acid, the
carbohydrates native to gastric mucins, as well as mannose, which
is a characteristic component of serum glycoproteins (Glass and
Slomiany, 1977). At low lysolecithin concentrations (0.02-0.04 mg/ml
in instillate) the content of individual carbohydrates did not differ
from that of saline control. Increase in the concentration of
instilled lysolecithin led to the enrichment of the instillates in
carbohydrates typical to both the mucus and serum glycoproteins.
However, the rate of increase of the carbohydrates typical to mucus
glycoproteins was considerably greater than that of mannose. The
weight ratio of N-acetylgalactosamine to mannose was 1.67:1.0 in
saline controls, 2.25:1.0 in the instillates containing 0.4 mg/ml of
lysolecithin, and 2.63:1.0 in the instillates containing 2 mg/ml of
lysolecithin or 2M NaCl, therefore it became enriched in acetyl-
galactosamine by 50% as compared to mannose.

        The extent of enrichment of the instillates in protein bound
carbohydrates at the concentration of lysolecithin 2.0 mg/ml was comparable to that produced by 2M NaCl (117.0 mg NaCl/ml). Despite this
high concentration of lysolecithin the gastric mucosal or intracellular plasma membranes and the membranous constituents of the cells
were damaged by this exposure.

        The changes in some of the individual protein-bound sugars in
response to increase of the lysolecithin concentrations in the instillates varied significantly. The least changes were demonstrated
by protein bound mannose and fucose which moved almost in parallel
changes. On the other hand, protein bound galactose, N-acetylgalactosamine and N-acetylglucosamine demonstrated more that 2 to 3 fold
increase along the tracing with the 20-30 fold increase in the lysolecithin concentration. The levels of the protein bound sialic acid
content increased about twofold with the 20-30 fold increase in the
concentration of lysolecithin (Fig. 1).

DISCUSSION

        The data obtained indicate that solubilization of the mucous
layer constituents begins at concentration of lysolecithin above
0.04 mg/ml and proceeds in two phases (Fig. 2). One, at which lysolecithin (up to 0.16 mg/ml of instillate) does not break the gastric
mucous barrier, and the other (above 0.4 mg/ml) at which lysolecithin
breaks the gastric mucous layer. This interpretation is in agreement
with the results of histological examinations of the mucosa from
stomachs instilled with lysolecithin, which showed that lysolecithin
at concentrations above 0.4 mg/ml caused removal of the visible
surface epithelial mucus layer, and at concentrations above 0.8 mg/ml
removed also most of the preformed intracellular mucus. The extent
of gastric mucous barrier depletion by the high concentrations of

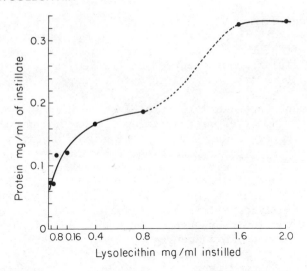

Fig. 1.   The effect of lysolecithin on solubilization of the protein-
          bound carbohydrates of rat gastric mucus.

          Values represent means from 4 pairs of rats for each lyso-
          lecithin concentration and saline controls.

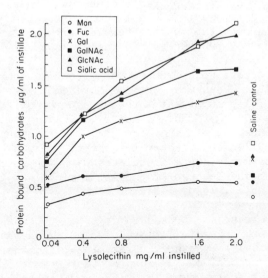

Fig. 2.   The effect of lysolecithin on solubilisation of the protein
          of rat gastric mucus.

          Values represent means from 4 pairs of rats for each lyso-
          lecithin concentration and saline controls.

lysolecithin was comparable to that produced by 2M NaCl, which is known to remove both the visible and preformed intracellular mucus constituents of the gastric mucous barrier. The removal of the pre-formed intracellular mucus occurred at lysolecithin concentrations that are known to break the gastric mucosal barrier.

Since at high concentrations of lysolecithin (1.6-2.0 mg/ml) no damage to even most exposed apical portions of the surface epithelial cells occurs and the vascular system of stomachs remained intact, one may infer that lysolecithin alone is not the direct single cause of the gastric erosions. However, this assessment does not take into account factors which might modify toxic effect of lysolecithin. One such factor is lowering of the pH. Reduction in mucus produced by lysolecithin would expose the gastric mucosa to the acidity of the lumen and peptic digestion and produce gastritis, and gastric erosions. This sequence of events leading to the gastri-tis and gastric ulcers is in agreement with the results of Orchard et al (1977), who showed that instillation of lysolecithin into the stomach causes macroscopic erosions and mucosal damage when proceded and followed by HCl exposure at pH 3.0.

The solubilization and partial removal of gastric mucus occurs at lysolecithin concentrations in the stomach which are equal or below those known to break the mucosal barrier. This suggests that the breaking of the mucus layers may play an important role in destruction of the integrity of gastric mucosa. Mucus removal by lysolecithin apparently facilitates the penetration of hydrochloric acid and pepsin through the mucosal barrier with all the consequences which its breaking may produce in the gastric mucosa.

This and previous work from our and other laboratories demon-strated solubilization and more or less advanced depletion of gastric epithelial mucus following treatment with hyperosmotic ethanol (Patkowska et al, 1978), 2M NaCl (Webster, 1967), various detergents and stressful agents, (Glass and Stremple, 1973; Rhodes, 1972; Stremple et al,1973). This injury to gastric "mucous barrier" was usually even at an early stage associated with the shedding of the surface epithelium and leakage of the serum albumin into the gastric lumen (Patkowska et al, 1978, 1981).

The mucus release as studied in dogs may procede by exocytosis, apical expulsion or cell exfoliation. Depending on the age and location of mucus cells, the species examined, and the integrity of or damage to gastric mucosa, one or more of these three mechanisms may enter into action. The exocytosis of the mucus usually occurs in younger and normal cells, which depending on species and cell loca-tion may release non-sulfated mucus. It is usually a slow and a continuous process which may involve only a part of an undamaged cell.

The most important mechanism of mucus discharge from both normal and damaged epithelial mucous cell is "apical expulsion". It occurs in various species including humans mostly in goblet cells and also in the human stomach, where it may be limited to individual cells or their small groups located often at crests of the pits. Here the intracellular mucus following a break in the plasma membrane is ejected as a suddened release of the whole apical mucus package. This occurs in the older cells in the area and leaves degenerated sites of the cells after the mucus loss. The apical expulsion of the mucusis accelerated following damage to the cells and results in formation of a honeycomb-like appearance of the cells. This has been described after gastric stress bleeding (Lucas et al, 1971; Baskin et al,1976). We have seen it after removal of mucus by hyperosmotic solutions of 2M NaCl (Stremple et al, 1973) or 20-30% ethanol (Patkowska et al, 1978, 1981) and (with Konturek et al, 1981) after injury by aspirin at low pH. We, therefore, share the opinion of Pfeiffer (1970) and Zalewsky and Moody (1979) that apical expulsion is a frequent mode of releasing mucus seen most often in response to injury to mucous barrier, or aging of the cell.

The third mechanism of mucus release is the "cell extrusion" which includes the loss of the entire mucus surface cell. This mechanism called also "cell exfoliation" is rare in dogs but seen in fasted rats; it usually involves the most aged mucus cells but is also seen in various animal stomachs as results of injury, including stress (Croft et al, 1963).

Each of the mechanisms of release of mucus following injury to mucus membrane may be called "breaking of the mucus barrier" which has to be sharply differentiated from the process of "breaking of mucosal membrane". This developes as a result of the more extensive and prolonged injury to gastric mucosa resulting in the back diffusion of H^+ ions through the unstirred layer of water within mucus and its structural interstices, and perhaps also reduced secretion of the bicarbonates by gastric epithelium into mucus which wakens the so called mucus-bicarbonate barrier (Williams and Turnberg, 1980). One may look at the breaking of the mucus barrier as the initial stage of a more advanced injury to the gastric mucosa which results in the breaking of the mucosal barrier (Davenport, 1970). The latter comprises the basal and plasma membranes as well as the intracellular membranes and junction of the gastric mucosa. At this stage the marker of the mucosal damage caused by the injurious agent is the drop in the mucosal potential difference and the breaking of the mucosal and submucosal vasculature. This in turn will be followed by isolated and then more confluent hemorrhages in the lamina propria, leading to gastric erosions and subsequently ulcerations.

Thus, as seen, here the solubilization of the surface mucus layers by detergents and reducing agents which break the disulfide bridges between the naked protein segments of the human protein core

(Schrager, 1969) are responsible for the weakening the polymer structure of native mucus. The solubilization that occurs more frequently
is due to proteases digesting some large chunks of the glycoprotein
and also breaking the peptide links of the partly digested porteins
(Allen, 1978; Pearson et al, 1980). This is the important mechanism
of the breaking of the mucous barrier with all its consequences, which
include breaking of the mucosal membranes and vasculature and damage
and injury to gastric mucosa.

ACKNOWLEDGEMENT

This work was supported by United States Public Health Service Grants
AM#21684-03 and AM#25372-02 from National Institute of Arthritis,
Metabolism and Digestive Diseases, National Institutes of Health.

SUMMARY

     The effect of lysolecithin on the protein, glycoprotein and
glycolipid constituents of the rat gastric mucous barrier was investigated. Eighty Ghosh-Lai rat stomach preparations were instilled
in vivo with 0.9% NaCl (control) for 1 hour followed by various
concentrations of lysolecithin or 2M NaCl, and the contents of protein, protein bound carbohydrate and glyceroglucolipids in the
recovered instillates were determined. The low lysolecithin concentrations (0.02-0.04 mg/ml of instillate) had no effect on the solubilization of the investigated components of the gastric mucous
barrier. The higher lysolecithin concentrations, however, produced
gradual enrichment of the instillates in protein, protein bound
carbohydrate and glyceroglucolipid. This enrichment occurred at two
rates. One, at low lysolecithin concentrations of 0.08-0.40 mg/ml,
at which protein increased about twice over the control level. The
other, at high lysolecithin concentrations above 0.8 mg/ml, at which
fivefold increase of protein, threefold increase of glyceroglucolipid
and twofold increase of protein bound carbohydrate developed. The
extent of gastric mucous barrier depletion by the high concentrations
of lysolecithin was comparable to that produced by 2M NaCl which is
known to remove both the visible and preformed intracellular mucus
from gastric mucosa.

     The changes in some of the individual protein-bound sugars in
response to increase of the lysolecithin concentrations in the instillates varied significantly. The least changes were demonstrated
by protein bound mannose and fucose which moved almost in parallel.
On the other hand, protein bound galactose, N-acetygalactosamine and
N-acetylglucosamine demonstrated more than 2 to 3 fold increase along
the tracing with the 20-30 fold increase in the lysolecithin concentration. The levels of the protein-bound sialic acid content increased about twofold with the 20-30 fold increase in the concentration of lysolecithin.

All our data indicate that lysolecithin at low concentrations exerts topical solubilization action on the surface mucus layer and at high concentrations causes the depletion of the preformed intracellular mucus from mucus cells of gastric epithelium, as well. The subsequent injury to mucosal membranes and vasculature by lysolecithin with injury to gastric mucosa, from all what we know about the protective role of the mucus, is most probably facilitated by damage to the mucous layers, resulting in their partial solubilization and removal.

REFERENCES

Allen, A. (1977) in, Mucus in Health and Disease, (Ed.) Elstein, M. and Parke, D.V., pp. 283-299, Plenum Press, New York and London.
Allen, A. (1978). Br. Med. Bull., 34, 28.
Balanzo, J.T. and Glass, G.B.J. (1975). Digestion, 13, 290.
Baskin, W.N., Ivey, K.J., Krause, W.J. (1976). Ann. Intern. Med., 95, 2990.
Clamp, J.R., Allen, A., Gibbons, R.A. and Roberts, G.P. (1978). Br. Med. Bull., 34, 28.
Croft, N.D. (1963). Brit. Med. J., 2, 897.
Davenport, H.W. (1970). Proc. Soc. Exp. Biol. Med., 126, 657.
Glass, G.B.J. (1976) in Mucus and Gastric Injury, North Amer. Symp. on Carbenoxolone, (Ed.) Beck, J.T., pp. 29-43, Excerpta Medica, Amsterdam.
Glass, G.B.J. and Stremple, J.F. (1973). Lancet, 1, 1506.
Glass, G.B.J., Slomiany, B.L. (1977) in Mucus in Health and Disease, (Ed.) Elstein, M. and Parke, D.V., pp. 311-347, Plenum Press, New York and London.
Glass, G.B.J., Slomiany, B.L. and Slomiany, A. (1979) in, Biochemistry and Pharmacology of Ethanol, (Ed.) Majchrowicz, E. and Noble,E.P., Vol. 1, pp. 551-586, Plenum Press, New York and London.
Hanks, C.E., Cassell, M., Ray, R.N. and Chaplin, H. (1960). J. Lab Clin. Med., 56, 486.
Heatley, N.G. (1959). Gastroenterol., 37, 313.
Hollander, F. (1954). Arch. Intern. Med., 93, 107.
Hughes, K.W. and Clamp, J.R. (1972). Biochim. Biophys. Acta. 264, 418.
Johnson, A.G. and McDermott, S.J. (1974). Gut, 15, 710.
Konturek, S., Swierczyk, P., Glass, G.B.J. (1981) in press.
Lowry, O.H., Rosebrough, N.J., Farr, A.L. and Randall, R.J. (1951). J. Biol. Chem. 193, 265.
Lucas, C.E., Sugawa, C., Riddle, J., (1971). Arch. Surg., 102, 266.
Martin, G.P., Marriot, C. and Kellaway, I.W. (1978). Gut, 19, 103.
Orchard, R., Reynolds, K., Fox, B., Andrews, R., Parkins, R.A. and Johnson, A.G. (1977). Gut, 18, 457.
Patkowska, M., Yano, S. and Glass, G.B.J. (1978). Gastroenterol., 74, 1075.
Patkowska, M., Yano, S. and Glass, G.B.J. (1981). Gastroenterol., in press.

Pearson, J., Allen, A. and Venables, C. (1980). Gastroenterol., 78, 709.

Pfeiffer, C.J. (1970). Exp. Mol. Pathol., 13, 319.

Rhodes, J. (1972). Gastroenterol., 63, 171.

Schrager, J. (1969). Digestion, 2, 73.

Slomiany, B.L., Slomiany, A. and Glass, G.B.J. (1977). Eur. J. Biochem. 78, 33.

Slomiany, A., Yano, S., Slomiany, B.L. and Glass, G.B.J. (1978). J. Biol. Chem. 253, 3785.

Slomiany, B.L., and Slomiany, A. (1978). Eur. J. Biochem. 90, 39.

Slomiany, B.L., Galicki, N.I., Kojima, K. and Slomiany, A (1980). Eur. J. Biochem., 111, 259.

Slomiany, B.L. and Slomiany, A. (1979). IRCS Med. Sci. 7, 373.

Slomiany, A., Patkowska, M.J., Slomiany, B.L. and Glass, G.B.J. (1979). Int. J. Biolog. Macromol., 1, 165.

Slomiany, A., Slomiany, B.L. and Glass, G.B.J. (1980). J. Applied Biochem., 2, 366.

Stremple, J.F., Mori, H., Lev, R. and Glass, G.B.J. (1973) in, Current Problems in Surgery, (Ed.), Ravitch, M., pp. 1-64, Chicago.

Webster, D.R. (1967) in, Gastric Secretion: Mechanisms and Control, (Ed.), Shnitka, T.K., Gilvert, J.A. and Harrison, R.C., 215-224, Pergamon Press Ltd., Oxford.

Williams, S. E., Turnberg, L.A. (1980). Gastroenterol., 79, 299.

Zalewsky, C.A., and Moody, F.G. (1979). Gastroenterol., 719.

SYNTHESIS AND SECRETION OF GLYCOPROTEINS BY ISOLATED

RAT SMALL INTESTINAL EPITHELIAL CELLS

Tony Howe and Peter J. Winterburn

Department of Biochemistry
University College
Cathays Park, Cardiff CF1 1XL, U.K.

INTRODUCTION

Mucins are synthesised by the goblet cells of the small intestinal mucosa and secreted into the lumen.   Subcellular fractions
from rat small intestinal mucosa incorporate glucosamine into brush
border glycoproteins and mucins, which can also be labelled in vivo
(Forstner, 1970).   Human small intestinal mucosa biopsy samples
have been cultured and shown to produce mucins (Browning and Trier,
1969).

Several groups of workers have used isolated small intestinal
epithelial cells for various metabolic studies (Dawson and Bridges,
1979; Porteous, 1979; Shirkey et al., 1979; Watford et al., 1979).
Here we are reporting the use of such isolated cells for studying
the synthesis and secretion of glycoproteins with a view to the
development of a model system with which to investigate the effects
of compounds that modify such synthesis and secretion.

EXPERIMENTAL

Small intestinal cells were prepared by a modification of the
method of Watford et al. (1979).   Rats were killed by cervical
dislocation.   The small intestine was removed from 5cm below the
pylorus to 10cm before the junction with the colon and flushed
with 40ml ice-cold calcium-free Hanks' medium (buffer A).   One
end was ligated and the lumen was partially filled with calcium-
free Hanks' medium containing 5mM-EDTA and 0.25% (w/v) dialysed
bovine serum albumin.   The other end was ligated and the intestine

incubated in 100ml buffer A at 37°C for 15 min in a shaking water
bath.   The lumen was emptied and flushed with 20ml ice-cold Hanks'
medium containing 2.5% (w/v) dialysed bovine serum albumin. After
refilling with the same medium, the intestine was placed on an
ice-cold glass plate and massaged for 1 min.   The contents, containing isolated cells, clumps of cells and mucus, were removed
and dispersed by two passages through a 15cm length of polyethylene
cannula tubing (Portex pp240).   The suspension was centrifuged
at 500g for 3 min, the pellet washed three times with 40ml Williams
medium E containing 2.5% (w/v) dialysed bovine serum albumin and
0.05mg/ml gentamicin at 37°C, and finally resuspended in the
desired volume of the same buffer.   Up to four intestines were
processed concurrently.   All solutions were gassed with 95%
O_2/5% CO_2 and all glassware which came into contact with the cells
was siliconised.

     Portions (12ml) of the cell suspension were incubated in 50ml
siliconised conical flasks on a shaking water bath at 37°C, and
gassed continuously with 95% O_2/5% CO_2.   After 30 min, either
L-[3-3H]serine (14 Ci/mmol) or D-[6-3H]glucosamine hydrochloride
(34Ci/mmol) was added to give final concentrations of 2.5 Ci/ml
or 2.0 Ci/ml respectively.   Portions (2ml) were removed at
intervals, centrifuged to separate the cells from the medium, and
the cell pellets washed with 2ml medium.   The washed pellet was
homogenised in 1ml ice-cold 10% trichloroacetic acid/1% phospho-
tungstic acid containing either 0.1M-glucosamine or 0.1M-serine.
The precipitate was washed five times with 5ml of the same solution,
twice with 5ml chloroform/methanol (1:1, v/v) air dried and
dissolved in 2ml NCS (Amersham-Searle) containing 10% (v/v) water.
Radioactivity was determined by liquid scintillation counting.
The media were dialysed three times against 0.15M-NaCl, the first
change containing unlabelled glucosamine or serine, twice against
6M-urea, and twice against 0.15M-NaCl.   As noted by Yeomans and
Millar (1980), non-specific binding of label to protein occurred
from either glucosamine or serine.   The dialysis against urea
reduced this non-specific binding.   Aliquots of the dialysed
media were determined for radioactivity by liquid scintillation
counting or used for characterisation of the product.

RESULTS

     The cell suspension consisted largely of small rafts of cells
which were able to exclude Trypan Blue.   Approximately 10% of the
cells in the rafts stained with Alcian Blue, indicating the
presence of goblet cells; the remainder appeared to be brush
border cells.   Dispersion methods involving the use of proteolytic enzymes or more vigorous mechanical methods gave higher
yields of isolated cells, but these were less viable than cells
in rafts, as also reported by Porteous (1979).   Most of the
protozoa were removed during the washing procedure.

A number of different media and colloid osmotic supports were tested for their ability to maintain cell viability. Williams medium E supplemented with 2.5% (w/v) dialysed bovine serum albumin was able to preserve viability for at least a 6h period and this combination also gave the highest incorporation rates of glucosamine and serine into secreted material. After 6h, 82% of the incorporated serine was present in the medium and 18% in the cells. The corresponding figures for glucosamine incorporation were 78% and 22% respectively. The glucosamine-labelled material in the cell pellet was presumably mainly glycoproteins of the cell brush border, glycocalyx and mucin awaiting secretion.

Glucosamine-labelled material from the medium was subjected to gel chromatography on a column (45cm x 3cm) of Sepharose 4B and eluted with 0.15M-NaCl. Two labelled components accounted for 90% of the eluted material. The larger component had a K_{av} value of 0.43 corresponding to an approximate molecular weight for a globular protein of 7×10^5 whereas the smaller component (K_{av} 0.66) had an apparent molecular weight of 10^5 and may be related to Component II described by Forstner (1970).

Incubation of the secreted glucosamine-labelled material with 1M-NaBH$_4$ in 0.5M-NaOH for 15h at 45°C followed by chromatography on Sephadex G-25 revealed that 74% of the radioactivity eluted from the column was at the included volume. This suggested that most of the carbohydrate chains were O-linked to the protein core. A sample of the secreted glucosamine-labelled material was hydrolysed with formic acid (pH2, 70°C for 60 min, according to Schauer (1978). 22% of the radioactivity was liberated suggesting that the glucosamine had been incorporated into the sialic acids that occur in this mucin.

ACKNOWLEDGEMENT

We thank Boehringer Ingelheim Ltd. for financial support.

REFERENCES

Browning, T.H. and Trier, J.S., 1969, J. Clin. Invest., 48, 1423-1432.
Dawson, J.R. and Bridges, J.W., 1969, Biochem. Pharmacol., 28, 3299-3305.
Forstner, G.G., 1970, J. Biol. Chem., 245, 3584-3592.
Porteous, J.W., 1979, Environmental Health Perspectives, 33, 25-35.
Schauer, R., 1978, Methods in Enzymology, 50, 64-89.
Shirkey, R.J., Kao, J., Fry, J.R. and Bridges, J.W., 1979, Biochem. Pharmacol., 28, 1461-1466
Watford, M., Lund, P. and Krebs, H.A., 1979, Biochem. J., 178, 589-596.
Yeomans, N.D. and Millar, S.J., 1980, Digestive Diseases and Sciences, 25, 295-301.

ULTRACENTRIFUGATION OF SALIVARY MUCINS

P.A.Roukema, C.H.Oderkerk, A.V.Nieuw Amerongen

Dept.of Oral Biochemistry, Faculty of Dentistry, Vrije
Universiteit, P.O.Box 7161, Amsterdam

INTRODUCTION

The salivary mucins contribute to the protection of enamel
against the attack by bacteria and their metabolic products among
others by the formation of a protein pellicle on its surface.
To study the interaction of salivary mucins with enamel and bacteria, pure and well characterized mucins should be available.
A widely used method for the isolation of salivary mucins is based
on clot formation with a quaternary ammonium compound, followed by
fractional precipitation of the mucin from the resolved clot by
ethanol (1). This method is only convenient if the glycoproteins
are rich in sialic acid or sulphate. Another effective method used
heating at $100^{\circ}C$ to remove contaminating proteins, followed by
further purification on a Biogel P-300 column (2,3). However this
method could not be applied for the preparation of a mucin from
the human sublingual glands. We therefore replaced the Biogel step
by ultracentrifugation and purified the human sublingual mucin
(HSL), the human submandibular mucin (HSM), the mouse sublingual
and submandibular mucins (MSL and MSM) in this way.
HSM was studied in more detail, because it has been described (3)
that the incorporation of proteolytic digestion and/or -S-S-bond
reduction in the purification procedure results in preparations
with very low protein values.

EXPERIMENTAL

Glands were homogenized in 5-10 parts distilled water (w/v)
and centrifuged at 25 000 g for 20 min. The supernatant, adjusted
to pH=6.0, was heated at $100^{\circ}C$ for 10-15 min.
The soluble fraction, obtained after centrifugation at 100 000 g
for 60 min was dialyzed, freeze-dried and dissolved in the follo-

wing buffer: 0.1052 M NaCl + 0.0144 M Na_2HPO_4 + 0.0563 M NaH_2PO_4,
pH=7.5.
Centrifugation at 100 000 g (HSM, HSL, MSL) or 170 000 g (MSM) for
24 h resulted in a sediment, which contained most of the mucin
(3x10 ml swing-out rotor, ± 3 mg material/ml). Ultracentrifugation
of the bottom fractions was repeated, till no further purification
was obtained.
To detect the effect of -S-S-bond reduction, the purified materials
were preincubated with the centrifugation buffer with (a) 1% SDS
(w/v) overnight at 20°C and (b) in addition with 0.05% N-acetyl-
cystein (NAC), pH 8.0 at 37°C for 1 h. Centrifugation on a analytical
or preparative scale (only HSM) was then performed at 40 000 rev./
min.

RESULTS AND DISCUSSION

As judged from the amino acid composition, polyacrylamide gel
electrophoresis (PAGE) and $S_{20,w}$ values, HSM, MSL and MSM yielded
comparable products by preparative ultracentrifugation as were
formely obtained by chromatography on Biogel P-300 (3,4). HSL was
comparable to HSM in its $S_{20,w}$ value (resp. 13.7 and 13.6 at 3 mg/
ml), but had a much lower protein content (resp. 16.4 and 28.1%).
Some amino acid values are given in table I.

Table I. Amino Acid Composition of Salivary Mucins.

Mol %	MSL ref.7	MSL centr.	HSL centr.	HSM ref.5	HSM centr.	HSM comp.I	comp.II
Thr	19.3	21.3	15.7	14.2	13.5	24.7	5.8
Ser	20.3	21.9	7.8	13.2	12.5	13.1	7.9
Glx	5.4	4.4	8.7	9.8	9.3	6.1	16.6
Asx	3.8	2.6	5.0	7.3	5.9	3.7	8.8
Protein g/100 g	34.1	27.5	16.4	28.1	29.2	16.0	90.5

Analytical ultracentrifugation in SDS + NAC indicated that HSM
($S_{20,w}$ 12.1 and 2.3) and HSL ($S_{20,w}$ 12.5 and 2.7) consisted of two
components. The $S_{20,w}$ values of MSM (5.5) and MSL (10.9) did not
change.
Preparative ultracentrifugation gave two components, one sedimenting to the bottom (I), the other remaining in solution (II).
Component I (table I) was characterized by PAGE (migration rate 0.3
strong PAS reaction) and blood group activity as the mucin component.
Compared to the original HSM the protein content was much lower
(28.1 → 16.1%), Thr higher (14.2 → 24.7%). The amino acid composition was comparable to a mucin obtained by Schrager and Oates (5)
after proteolytic digestion and NAC treatment.
Component II consisted of 90.5% protein. Glx and Asx were high
compared to component I, Thr and Ser were low; 6.2% phosphate was

present. The migration rate in PAGE was 0.6.

The results indicate, that ultracentrifugation as described is a fast and convenient method for the preparation of mucins in a high yield. In that way the time-consuming chromatographic procedure on Biogel P-300 can be circumvented. That is especially of advantage with low amounts or high viscosity (e.g. MSL) of the material.

HSM and HSL yielded two components in the analytical ultracentrifuge with SDS + NAC (compare 6 and 7 for other mucins).

Dunstone and Morgan (6) used proteolytic treatment or S-S-bond reduction. We used solely that latter method. In the present concept a mucin may consist of two covalently bound protein cores (8), one of which (the naked peptide) can be released by trypsin or broken down by pepsin. Component II, released by NAC/SDS, must be bound in a different way. If component I contains a naked peptide has still to be investigated. The present results show that both component I and II are firmly bound to each other, resisting purification procedures. In our opinion, such a situation holds also in the oral cavity, where different proteins (mucins, phosphoproteins) may adsorb together, contributing to the protection of the teeth.

REFERENCES

1. Pigman, W. and Tettamanti, G., Arch.Biochem.Biophys. 124: 41-50 (1968).
2. Rölla, G. and Jonsen, J., Caries Res. 2: 306-316 (1968).
3. Oemrawsingh, I. and Roukema, P.A., Archs.Oral Biol. 19: 753-759 (1974).
4. Roukema, P.A., Oderkerk, C.H. en Salkinoja-Salonen, M.S., Biochim.Biophys.Acta 428: 432-440 (1976).
5. Schrager, J. and Oates, M.D.G., Archs Oral Biol. 16: 287-303 (1971).
6. Dunstone, J.R. and Morgan, W.T.J., Biochim.Biophys.Acta 101: 300-314 (1975).
7. Snary, D. and Allen, A., Biochem. J. 123: 845-853 (1971).
8. Bhushana Rao, K.S.P. and Mason, P.L., Adv. in Exp.Med. and Biol. 89: 275-282 (1976).

THE GASTRIC "MUCUS-BICARBONATE" BARRIER: EFFECT OF LUMINAL ACID ON

HCO$_3^-$ TRANSPORT BY AMPHIBIAN FUNDIC MUCOSA IN VITRO

Jon R. Heylings, Andrew Garner and Gunnar Flemström

I.C.I. Research Laboratories
Alderley Park, Macclesfield
Cheshire SK10 4TG, U.K.

INTRODUCTION

Transport of HCO$_3^-$ by surface epithelial cells has been demonstrated in amphibian isolated gastric mucosa (Flemström, 1977) and in mammalian stomach in vivo (Garner and Flemström, 1978). These studies together with work on surface mucus gel have led to the proposal that the two secretions provide a first line of defense against intraluminal H$^+$ ions (Allen and Garner, 1980). Although the stimulatory pathways controlling acid secretion and HCO$_3^-$ transport are different, titratable alkalinity of fluid perfusing the canine Heidenhain pouch increased following instillation of HCl into the gastric remnant (Garner and Hurst, 1980). Thus regulation of gastric HCO$_3^-$ output by intraluminal acid may involve a humoral mechanism. The present study was designed to investigate the effect of luminal acidification on alkaline secretion by the isolated mucosa.

METHODS

The technique for measuring alkaline secretion by amphibian fundic mucosa was as described in detail previously (Flemström, 1977) with the exception that the in vitro chamber was modified to incorporate two separate mucosae. Pairs of frogs (Rana temporaria) were killed, the stomachs removed and fundic mucosae separated from underlying muscle layers by blunt dissection. The two mucosae were mounted as membranes (surface area 1.8 cm^2) in a chamber with their serosal surfaces bathed by the same frog Ringer solution buffered at pH 7.20 and gassed with 95% O$_2$ and 5% CO$_2$. Luminal surfaces were bathed by separate unbuffered solutions and gassed with 100% O$_2$. Spontaneous acid secretion was inhibited and a titratable alkalinity induced in both tissues by addition of the histamine H$_2$ receptor

183

antagonist ICI 125211 (10^{-4} M) to the common serosal side solution.
Luminal alkalinization was determined by pH-stat titration (end
point 7.40) with solutions containing 5 mM HCl. Transmucosal PD
across each tissue was measured via matched calomel electrodes.
When stable rates of alkaline secretion and PD had been recorded for
a period of 60 min, the luminal side of one mucosa was acidified by
addition of small volumes of concentrated HCl. Consecutive periods
of acidification (final concentrations 5 and 50 mM HCl) were each of
60 min duration. Experiments were performed at $20 \pm 0.1^{\circ}$C.

RESULTS AND DISCUSSION

In 12 control mucosae there was a slight fall in the rate of
alkaline secretion (0.36 ± 0.05 to 0.28 ± 0.04 µeq cm^{-2} h^{-1}, mean
$\pm$ SE) over 3 h whereas PD remained stable (42 ± 1.7 to 43 ± 1.6 mV)
during this period. Exposure of one mucosa to increasing concentrations of acid (5 and 50 mM) was associated with a dose-related
increase in the rate of alkaline secretion by the other tissue in
6 out of 10 preparations. The remainder showed a maximal response
at 5 mM and no further increase in secretion was observed from the
non-acidified mucosa at 50 mM. As shown in Fig. 1, the maximum
increase in secretory rate was $27.4 \pm 6.1\%$ at a mean pH of 1.85
(calculated from molarity) and occurred 20 min after exposure to
acid. The PD of these tissues remained constant throughout.

Since the tissues are totally separate in the dual mucosal
preparation described here, it seems likely that gastric HCO$_3^-$ trans-

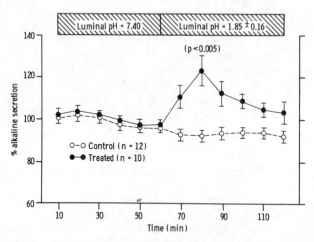

Fig. 1. Effect of luminal pH on alkaline secretion by amphibian
         fundic mucosa in vitro. The percent change in secretory
         rate (mean $\pm$ SE) in control and acidification experiments
         is shown. In ten fundic mucosal pairs, reduction of luminal
         pH from 7.40 to 1.85 caused a significant increase in the
         rate of alkaline secretion by the non-acidified tissue.

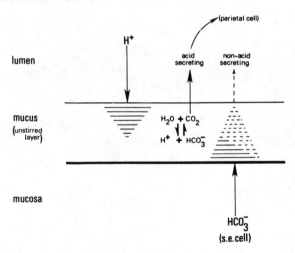

Fig. 2. Model for neutralization of H^+ by HCO_3^- in a mucus gel layer
         lining the surface of gastric epithelium. In acid secreting
         stomachs, HCO_3^- appears in the luminal solution as CO_2 (solid
         lines); some CO_2 may be utilized by parietal cells. Free
         HCO_3^- only appears in the lumen in acid-inhibited stomachs.

port is under humoral control from a transmitter released on exposure
of the luminal membrane to acid. Although acid per se produces an
increase in alkaline secretion, HCO_3^- output is nevertheless small
($< 10\%$) compared with maximum H^+ output. However, its effectiveness
in protecting the mucosal surface would be markedly enhanced if
neutralization was confined to a boundary zone of low turbulence
(Fig. 2). The recent demonstration of a pH gradient from 2.31 to
7.26 across the mucus layer of rabbit fundic mucosa in vitro
(Williams and Turnberg, 1981) supports this concept. A defect in
the output of HCO_3^- and/or mucus would reduce the effectiveness of
this barrier and may be one component in the pathogenesis of gastric
mucosal damage.

REFERENCES

Allen, A., and Garner, A., 1980, Mucus and bicarbonate secretion in
      the stomach and their possible role in mucosal protection, Gut,
      21:249.
Flemström, G., 1977, Active alkalinization by amphibian gastric
      fundic mucosa in vitro, Am. J. Physiol., 233:E1.
Garner, A., and Flemström, G., 1978, Gastric HCO_3^- secretion in the
      guinea pig, Am. J. Physiol., 234:E535.
Garner, A., and Hurst, B. C., 1980, Stimulation of bicarbonate
      secretion in the Heidenhain pouch dog by exogenous acid, Proc.
      Int. Union Physiol. Sci., 14:427.
Williams, S. E., and Turnberg, L. A., 1981, The demonstration of a
      pH gradient across mucus adherent to rabbit gastric mucosa:
      evidence for a "mucus-bicarbonate" barrier, Gut, 22: in press.

STUDIES OF THE PROTECTIVE PROPERTIES OF GASTRIC MUCUS

S.E. Williams, L.A. Turnberg

Department of Medicine, University of Manchester

Hope Hospital, Eccles Old Road, Salford M6 8HD. U.K.

It is often assumed that the mucus layer protects gastric mucosa from damage by intraluminal acid and pepsin. However the mechanism by which it might achieve this is uncertain. We examined therefore the ability of gastric mucus to resist the diffusion across it of hydrogen and other small ions. Pig gastric mucus was gently scraped off freshly slaughtered pigs' stomachs. The mucus was gently washed with isotonic saline to remove adherent food particles and small samples were layered between gauze supports in a perspex flux chamber. The layer of mucus, approximately 2.5 mm thick, separated equimolar solutions of HCl, on one side, and NaCl on the other (154 mM). The solutions were bubbled with 100% oxygen to ensure adequate mixing and were maintained at 20°C. The pH on the saline side was monitored with a pH probe placed within the solution and samples were taken simultaneously from both sides for estimation of sodium concentrations.

Compared with the gauze supports alone, mucus markedly delayed the rate at which the saline solution became acidified. Thus, the hydrogen ion concentration rose to 15 mEq/l after 120 minutes whereas with the mesh alone this rise occurred within 3 minutes. Even after 180 minutes hydrogen ion concentration only reached 36% of equilibration values. Similar restrictions to sodium movement occurred. Using tritiated water and isotopic sodium, it was confirmed that mucus considerably delayed equilibration across it.

These studies confirm that mucus effectively retards the passage of hydrogen and other small ions across it. To determine whether this was simply the result of the maintenance of an unstirred layer of water or the result of another property of mucus

which delayed diffusion, we measured diffusion coefficients for
hydrogen ions in mucus and in a similar thickness of isotonic
unstirred saline. This technique involved the use of 1 mm and 2
mm thick perspex discs perforated by varying sized holes into which
mucus or saline was inserted. The rate at which hydrogen ion
concentration changed in saline on one side of this disc when HCl
was placed on the other in the chamber was measured accurately
over a short time period. Using this technique the diffusion
coefficient for hydrogen through mucus was calculated to be 1.75
$\pm$ 0.23 x 10^{-5} cm^2/sec (n = 27) compared with 6.65 $\pm$ 0.33 x 10^{-5}
cm^2/sec (n = 20) in saline. Thus, diffusion was retarded by
mucus, probably by the ability of this material to immobilise
some water in its interstices making it unavailable for diffusion,
a characteristic which is probably dependent on its viscosity.

These investigations provide some indication of a mechanism
by which mucus may protect the gastric mucosa. Delay of diffusion
of hydrogen ions from lumen towards the epithelium must contribute
to its protective properties. It is likely to provide a medium in
which bicarbonate secreted by surface mucosal cells can neutralise
hydrogen entering from the lumen and thus provide one element of a
'mucus-bicarbonate barrier'.

STUDIES OF THE pH GRADIENT ACROSS THE MUCUS ON RAT GASTRIC

MUCOSA IN VIVO AND ACROSS MUCUS ON HUMAN GASTRIC MUCOSA IN VITRO

I.N. Ross, H.M.M. Bahari, L.A. Turnberg

University Department of Medicine

Hope Hospital, Eccles Old Road, Salford. M6 8HD, U.K.

The gastric mucosal barrier to hydrogen ion back diffusion is thought to consist of the apical membrane of the surface epithelial cells and the tight junctions of adjacent cells. An alternative hypothesis is that gastric mucus acts as an unstirred layer allowing hydrogen ions diffusing from the lumen, to be neutralised by bicarbonate secreted from the surface epithelial cells (1). If this hypothesis is correct, it should be possible to demonstrate a pH gradient across the mucus layer. In this study the presence of such a mucus pH gradient was looked for in rat fundic mucosa, in vivo and the effects on this pH gradient of potential damaging agents such as aspirin and N-acetyl-1-cysteine (NAC) examined.

A pedicle of fundic stomach was exteriorised from the anaesthetised rat and placed in a chamber such that the luminal surface was bathed with pH 2 HCl. With the antimony microelectrode in this luminal fluid the mean ($\pm$ ISD) callibrated pH was 1.87 $\pm$ 0.40 (n = 30). Just prior to entry into the mucus layer a slight rise of 0.1 to 0.3 pH units could be obtained; this rise was abolished by gentle turbulence in the luminal fluid. The electrode was winched through the mucus layer in 42.3 um steps and the pH recorded at each step. The pH gradually rose to a maximum point after which there was a sudden fall in pH as the electrode probably entered tissue. The mean maximum pH in 30 rats was 6.68, with a range varying from 5.10 to 7.95. The mean distance travelled by the electrode to reach this maximum point was 695 $\pm$ 202 um.

A constant pH reading could be maintained at any step along the pH gradient, for at least 120 mins, with a luminal pH of 2.0. However substituting the luminal solution with one at a pH of 1.4 or less resulted in a gradual fall in the pH gradient to the luminal pH value.

Instillation of the mucolytic agent NAC as a 5% solution in pH 2.0 HCl onto the luminal side caused a significant fall in a constant pH value in 12 of 18 rats. After 30 minutes exposure the mean pH of 5.75 had fallen to 4.44 ($p < 0.001$), while the maximum pH obtainable in those rats who had shown a fall in constant pH was reduced to 4.18 ± 0.92 ($p < 0.001$).

Instillation of 10 mM aspirin in pH 2.0 HCl onto the luminal side caused a significant fall in a constant pH value in 12 of 24 rats. After 30 minutes exposure the mean pH of 5.67 had fallen to 4.84 ($p < 0.001$). While the maximum pH obtainable in those rats who had shown a fall in constant pH was reduced to 4.18 ± 0.92 ($p < 0.001$).

The effect of low pH, NAC and aspirin appeared to be only local as in those rats who showed a reduction in maximum pH values after treatment, other untreated areas of the same stomachs showed normal maximum pH values. Why only some of the rats showed a fall in constant pH could not always be explained. However in certain animals thick alkaline mucus was produced due to irritation by the cuff forming the luminal chamber. In others the same cuff caused bleeding, while haemorrhage beneath the electrode could actually cause a rise in pH.

This study confirms the existence of a pH gradient within the mucus layer in vivo. Presumably under physiological conditions this 'mucus-bicarbonate' barrier acts to neutralise any back diffusing hydrogen ions. The ability of the mucus layer to maintain this pH gradient was overcome by a luminal pH of 1.4 or less, equivalent to a hydrogen ion concentration of approximately 40 mM. The rapid fall in pH after application of NAC was probably due to a reduction in the mucus layer thickness which would then allow more acid to diffuse through. The precise cause for the fall in pH after application of aspirin is unknown, but is probably due to an effect on epithelial cell bicarbonate secretion and/or shedding of adherent mucin. The results support a role for the 'mucus-bicarbonate' barrier in gastric mucosal protection in the rat.

Similar studies were performed on macroscopically normal human stomachs which were obtained from gastric resections performed for carcinoma or peptic ulceration. The specimens were transported in oxygenated bicarbonate free Ringer's lactate solution and used within 1 hour of removal. The tissue was

mounted in a specially prepared chamber so that the serosal side
was bathed with bicarbonate free glucose Ringer's solution and
the luminal side with pH 2.0 HCl.

With an antimony microelectrode in the luminal solution the
callibrated pH was 2.25 ± 0.29 (n = 10). As the electrode was
winched into the mucus layer in 42.3 um steps, a gradual rise in
pH was recorded. The maximum pH value found in 10 stomachs was
6.96 ± 0.88. A stable, constant pH value could be obtained at
any point along this pH gradient.

Substituting the luminal fluid with pH 1.4 HCl resulted in
a fall in a constant pH value in 6 stomachs tested, so that by 35
min the mean constant pH value of the group had fallen from 6.19
to 3.69 ($p < 0.001$). The maximum pH obtainable at that time was
reduced to 4.27 ($p < 0.001$).

Application of 5% NAC in pH 2.0 HCl to the luminal side
resulted in a fall in a constant pH value in 6 of 7 stomachs. By
35 min the mean constant pH value of the 7 stomachs was reduced
from 6.34 to 5.05 ($p < 0.01$). The maximum pH obtainable 35 min
after application of NAC was reduced to 4.69 ± 1.08 ($p < 0.001$).

Application of 10 mM aspirin in pH 2.0 HCl to the luminal
side, again resulted in a fall in constant pH values in 6 of 7
stomachs. The mean constant pH value of the 7 stomachs was
reduced from 6.53 to 5.38 by 35 minutes ($p < 0.01$). After 35 min
exposure to aspirin the mean maximum pH was reduced to 4.90
± 1.01 ($p < 0.001$).

These results confirm the presence of a pH gradient within
the mucus layer adherent to human stomach, in vitro. The gradient
is compromised by acid of pH 1.4 and by NAC and aspirin. This
'mucus-bicarbonate' barrier may play an important physiological
role in protecting gastric tissue from intraluminal acid.

REFERENCES

1.   Heatley NG (1959) Mucosubstance as a barrier to diffusion.
        Gastroenterology, 37, 313-317.

RESTRICTION OF HYDROGEN AND SODIUM ION DIFFUSION IN PORCINE

GASTRIC MUCIN: A CONCENTRATION DEPENDENT PHENOMENON

Michael Lucas

Institute of Physiology
University of Glasgow
Glasgow G12 8QQ
Scotland UK

Gastric mucin has long been thought of as having a protective role
against the high levels of acidity that the mucosa are often exposed
to. With the advent of the concept of an 'unstirred layer' next
to all membranes, mucin has been envisaged as providing a structural
framework for maintaining and extending an external unstirred layer.
Such an unstirred layer within the mucus is thought to reduce access
to the mucosa by increasing the mean diffusion path length (1) and
the possible ability of mucus to retard hydrogen ion access by re-
stricting the free diffusion is often discounted (2) Since mucus
is an exceptionally viscous substance, a restriction of the diffusion process might be expected, as predicted by the Stokes-Einstein
equation (3).

Consequently, studies were undertaken to investigate the effect
of various porcine mucus concentrations on the diffusion of hydrogen and sodium ion in a three compartment diffusion cell. Two experimental variants of this method were used, steady state fluxes
of hydrogen ion were measured to assess the effect on the hydrogen
ion diffusion coefficient, and in contrast, an approach to equilibrium
method was used for the determination of the sodium ion diffusion
coefficient. In any diffusion cell experiments which utilise filters to keep mucus in place, the overall diffusion coefficient will
be the sum of three components i.e. the mucus, filter and unstirred
layer components. Additionally, where well stirred end compartments
are used, the effect of mucus may be to impose an additional zone
of restricted turbulence as well as any effects on diffusion itself.
The following experiments describe the effects of gastric mucus on
the movement of sodium and hydrogen ion after taking these factors
into account.

METHODS:Prior to studies with mucus, the effect of various numbers
of millipore filters on the rate of transfer of cation was tested
in a two compartment chamber. For hydrogen ion,the time taken for
the pH to change in phthalate buffer from pH 9 to 10 was measured
in one compartment by pH-electrode,with the other compartment filled
with 1 N HCl. For sodium ion, the transfer rate constant was measured
for the movement of ion from a solution of 1000 mM to one of 15 mM,
using a sodium-ion sensitive electrode.Extrapolation of either the
diffusion time or the rate constant to 'zero'millipore filters then
allowed calculation of the unstirred layer component and the effect
of two millipore support filters on the measured diffusion coefficient.

In subsequent three compartment studies,mucus was mounted in
a central well of 0.5 cm path length and the experiments were carried out as described.For hydrogen ion,the time taken to effect the
standard pH-change in a well stirred end compartment was noted: for
sodium ion a transfer rate constant from the sigmoid appearance of
sodium ion was measured.This was calculated by assuming that the
resultant sigmoid curve could be described by a three compartment
model with rate constants of equal value at the two membranes (4)
and deriving the best estimate by BMDP. The apparent diffusion coefficient was calculated from these experimentally determined parameters and from this the specific component due to the inclusion
of gastric mucus was calculated (5) since:-

$$\left(\frac{D}{\delta}\right)_{obs} = \left\{\frac{\delta_1}{D_1} + \frac{\delta_2}{D_2} + \frac{\delta_3}{D_3}\right\}^{-1}$$

1 = unstirred layer
2 = millipore filter
3 = mucus layer

RESULTS: When mucus was included in the central well of the diffusion
chamber,the time taken for hydrogen ion to permeate the composite
membrane increased considerably. This effect was concentration dependent and statistically very highly significant. The diffusion
coefficient was calculated from this time using Fick's first Law.
Similarly,(Table) the transfer rate for sodium ion permeation was
greatly reduced in the presence of mucus,was concentration dependent
also and highly statistically significant.

Table:Effect of varying concentrations of porcine mucin on the
apparent diffusion coefficient of sodium and hydrogen ion.

Mucus conc. (%g/ml)	Hydrogen ion Time taken (minutes)	$D(10^5)$ (cm^{2}sec^{-1})	Sodium ion Rate constant (minutes (10^3)	$D(10^5)$ (cm^{2}sec $^{-1}$)
0	27.9 + 0.7	10.50	9.8 ± 0.10	1.85
10	49.8 + 2.2	4.49	4.7 ± 0.1	0.88
20	81.2 + 3.9	2.49	2.7 +. 0.05	0.53
30	128.0 + 9.8	1.47	1.1 ± 0.1	0.21

Results given as mean + standard error of mean for n = 5

DISCUSSION:When considering the effects of mucus on hydrogen and sodium ion movement,the possible contribution of buffering and ion binding to the measurements must be assessed.In the steady state however,exit from the centre well should equal entry from the compartment of higher concentration and any buffering should cause some deviation from linearity.Since this was not the case,although there was a delay in establishing the steady state,it is unlikely that buffering within the central compartment had much effect after the steady state is achieved.With sodium ion,the total amount of ion that gastric mucus can bind was far exceeded by the total flow of ion in the diffusion chamber.

An additional problem is the extent to which well-stirred end compartments induce a degree of turbulence in the centre well,which might be only partially unstirred and which might be made less turbulent by the addition of viscous mucus.There is undoubtedly a degree of turbulence in the centre well since the calculated diffusion coefficients are slightly higher than the free-solution values (6) when no mucus is present.However,addition of mucus caused the values to fall below their free solution value showing clearly that the effect of mucus is not only to reduce turbulence but to additionally restrict the process of free diffusion.

The present experimental data demonstrating a reduction in the observed diffusion coefficient through mucus,after taking into account unstirred layer and filter effects,agree with other published data (7). These observations add experimental support to the hypothesis that mucus assists in cytoprotection by slowing down the rapid rate of appearance of hydrogen ion at the surface cells such that neutralisation can be achieved (8). A second area of interest may be in intestinal absorption studies based on unstirred layer modelling (9).If the present findings can be generalised,it may be that intestinal mucus also restricts diffusion and that the unstirred layer estimates and derived parameters using the value for diffusion coefficients in free solution may be substantially in error.

REFERENCES

1) Allen,A.(1978) Br med Bull 34:28-33
2) Carter,D.C.(1980) Scientific Foundation of Gastroenterology, Eds.Sircus,W. & Smith,A.N.,William Heinemann,London. p344.
3) Einstein,A.(1905) Ann der Physik 17: 549-560
4) Brown,B.M.(1961) The Mathematical Theory of Linear Systems John Wiley,New York. p7
5) Jacobs,M.H. (1935) Ergb Biol 12: 70-71
6) Robinson,R.A. & Stokes,R.H.(1965) Electrolyte Solutions. Butterworth,London,UK.
7) Williams,S.E.& Turnberg,L.A. (1980) Gastroenterol 79: 299-304
8) Allen,A. & Garner,A. (1980) Gut 21: 249-262
9) Dietschy,J.M. et al.,(1971) Gastroenterol 61: 932-934

MUCUS AND GASTRIC ACID-BICARBONATE INTERACTION

J.B. Elder and A.R. Hearn

University Department of Surgery

Manchester Royal Infirmary, Oxford Road, Manchester U.K

Normally the mucosa of the stomach has a unique ability to resist digestion by the acidic proteolytic gastric secretion within its lumen. This property is impaired in peptic ulceration whether it occurs in the stomach or in the duodenum. The role of the layer of mucus on the surface of gastric and duodenal epithelial cells is poorly understood but it is believed that there is a degree of cyto-protection in the form of a barrier (1). Recently others have shown that substances with anti-ulcer properties stimulate gastric bicarbonate secretion (2) which is probably dissolved in mucus. There thus exists the possibility that interaction takes place between bicarbonate ions and hydrogen ions within the layer of mucus itself at the surface of the epithelial cells. Any such interaction would result in the formation of CO_2 which would be likely to diffuse rapidly through the mucus layer into the gastric lumen. Using a new flexible silicon-coated teflon catheter (BOC) with a semi-permeable membrane 1 cm from its tip measurements of the intragastric pCO_2 by a mass spectrometer were made in a total of 17 human subjects. Six normal subjects in the fasting state and 11 patients with either gastric, duodenal ulcer or combined duodenal and gastric ulcers and one patient with gastric carcinoma were studied. pCO_2 values in the resting stomach ranged from 9 to 30 mmHg in normal subjects and from 23 to 80 mmHg in the patients. The pattern was not a plateau in that some patients had sharp brief peaks of up to 170 mmHg pCO_2 lasting for a few seconds at a time. The pH of the resting gastric juice was measured by a glass electrode and the intragastric bicarbonate concentration calculated using the Henderson Hasselbach equation. Intragastric bicarbonate concentrations (calculated) ranged from 5_1 to 136 nmol l^{-1} in normal subjects and 12 nmol - 960 mmol l^{-1} in those subjects with

197

diseased stomachs.

In five patients with dyspepsia and in one with cancer of the stomach, Carbenoxolone sodium 100 mgs 3 times a day were given for 72 hours with informed consent and measurements of intra-gastric pCO_2 repeated. Carbenoxolone sodium is effective clinically in gastric and duodenal ulceration and is believed to exert its effect by stimulation of gastric mucus secretion (3). The fasting pCO_2 in the treated patients decreased from a mean of 34.6 ± 4.89 sem. to 26.6 ± 2.4 sem. mmHg ($p < 0.05$) with corresponding changes in the calculated bicarbonate concentration.

These results suggest firstly that interaction between bicarbonate and hydrogen ion has taken place in the mucus layer covering the epithelial cells with the resulting evolution of CO_2 and in some circumstances very high intragastric pCO_2 values were found. The gastric mucosa itself is relatively impermeable to the CO_2 and this is likely therefore to be a local phenomenon. Secondly it appears that carbenoxolone may have increased the thickness of the unstirred layer of mucus in the treated patients with the reduction of hydrogen ion penetration into the mucus layer and consequent reduction of interaction between hydrogen and bicarbonate resulting in lowering of the pCO_2 in the lumen of the stomach.

REFERENCES

1.   Davenport, H.W. (1972), Sci. Ameri. 226, 86-93

2.   Garner, A, and Heyling , J.R. (1979), Gastroenterology, 76, 497-503

3.   Parke, D.V. (1978), Medical Bull. 34, 89-94

CLINICAL ASPECTS OF GASTROINTESTINAL MUCUS

G. Forstner, A. Wesley, J. Forstner

Kinsmen Cystic Fibrosis Research Centre, Dept. of
Paediatrics and Biochemistry, The Hospital for Sick
Children and the University of Toronto, Toronto, Canada

In his Croonian Lecture of 1964,[1] Sir Howard Florey attributed
three significant functions to gastrointestinal mucus: protection
of the underlying mucosa from chemical and physical injury, lubrication of the mucosal surface to facilitate passage of luminal
ingredients and the removal of parasites by binding and entrapment.
In re-reading Florey's seminal contribution to this field, one is
struck by the little our understanding of the fundamental functions
of mucus has expanded in the intervening years. To a great extent
even the three "Floreyian" roles for this "slimey secretion" remain
articles of faith, buttressed at best, by indirect and circumstantial
pieces of evidence which lack convincing bite. If I can use a phrase
turned by a Torontonian philosopher, Marshall McLuhan, it may be as
true of mucus as of modern communication that "the medium is the
message". If so, we will probably never understand function fully
until the structural and organizational complexities of mucus are
unravelled.

In 1954 the major cause for ignorance of structure was the
"difficulty of preparing in a pure form the substances which make
up the ill-defined class of mucins".[1] This problem has been largely
overcome, with the result that numerous investigators have characterized highly pure mucin preparations from the stomach,[2,3] intestine,
[4,5] and colon,[6,7] in several species, and a considerable amount of
of data exists with regard to molecular weight and other physical
characteristics as well as structure. These features are reviewed
in detail by others in this symposium and need not be repeated here.
For the purpose of our discussion, however, one can generalize by
noting that all non-proteolytically derived mucins from these areas
behave as extremely large molecules with estimated molecular weights
ranging upward from 1.5 to 15.0 millions.[6] At least 80% of the dry

weight is carbohydrate, and whenever oligosaccharides have been examined following their release by alkaline borohydride reduction, a complex mixture of short and long chains with branching has been found.[8,9,10,11,12] Although at first glance the complexity of reported structures is bewildering, it is not endless. In fact, similar sequences tend to appear in most analyses, and on present evidence it seems most unlikely that regional mucins will differ by virtue of unique oligosaccharide chains.

Sialic acid is of particular interest because of reported variations in disease. In both salivary and colonic mucins it is linked α 2-6 to the core Gal NAC.[8,12] Slomiany et al also found it linked α 2-6 to the penultimate and ultimate galactose residues in elongated chains in rat colonic mucin.[8] The sialyl transferase which carries out this function has not been described but the linkage has been reported in other mucins.[14] Depending upon the presence of the two sialyl transferases and other competing transferases,[14] one might expect to find that sialation of individual mucins would vary with the degree of glycosylation. We approached this question by subfractionating a pure human intestinal mucin preparation by passage through a DEAE Sephacel column eluted with buffers of increasing ionic strength and alkalinity. As one can see in Table 1, several fractions with increasing proportions of sialic acid were obtained. The two sialic acid containing fractions II and III were less well glycosylated and contained shorter chains than fraction I. These results seem to indicate that poorly glycosylated, smaller mucins with short oligosaccharide chains, contain the sialic acid. In contrast to sialic acid, fucose was preferentially increased in the most highly glycosylated fraction of the human mucin (fraction I). Although the fucosyl α 1-2 gal transferase probably transfers fucose to galactose when adjacent to the core GalNAC quite well,[14] it would appear, in human intestine at least, that the majority of the fucose is added to somewhat longer chains. Our impression

Table 1. Composition of Three Fractions of Human
          Intestinal (non secretor) Mucin Removed
          From DEAE Sephacel

	I	II	III
		nmol/µg protein	
Total carbohydrate	43.3	19.2	19.6
GalNAC	7.1	4.9	4.7
GlcNAC	14.9	4.3	4.0
Gal	13.1	3.8	4.4
Sialic acid	1.3	5.5	5.3
Fucose	6.9	0.7	1.2
Total CHO/GalNAC	6.1	3.9	4.2

from this and other studies (see Cystic Fibrosis, infra) is that sialated mucin species have on average more short chains than non-sialated species, while highly fucosylated species are more highly glycosylated with longer chains.

The changing proportions of mucin species with different glyco-sylation and charge might be expected to have a marked influence on the ultimate physical properties of the secreted product. The entire mucin from the human small intestine is markedly polydisperse when examined by techniques such as band ultracentrifugation, which limits self interaction by reducing concentration to a minimum, or analytical Cs Cl density gradient ultracentrifugation, which reduces ionic bonding to a minimum,[15] (Fig. 1). In contrast, a more homogeneous species such as fraction I behaves in both systems in a more uni-modal manner. As shown in Fig. 1 for band ultracentrifugation, reduction and alkylation had a major effect on the entire mucin, greatly increasing the amount of small 14 S material, but no effect on the large 33-37 material in fraction I. It seems therefore, that disulphide bonds play a role in holding large and small species together, although partial disentangling is evident even before reduction.

Self interaction or aggregation of mucin species in vivo is probably due to a mixture of covalent and non-covalent forces which may be extremely important in determining the characteristics of the final mucus product. In addition to mucin species, of course, these interactions may also include other proteins, nucleic acids and smaller molecules, such as drugs. In a study using proteolytically digested (commercial) hog gastric mucin we showed that viscosity was greatly enhanced by the addition of albumin at low shear (< 45 S^{-1}).[16] Electrostatic interaction seemed relatively unimportant but both urea and heat enhanced viscosity, suggesting that anything which reduced hydrogen bonding and permitted linear entangling of the molecules would enhance their interaction. Maximum viscosity occurred with an albumin-mucin ratio of 2 (W/W). Normal mucus secretions contain little albumin but this is not true of in-flammatory states where the concentration of albumin rises toward that of serum, or of meconium in Cystic Fibrosis where very high albumin-mucin ratios are found.[17] Under these circumstances the interaction between the two molecules may create very high viscosity.

Recently Allen and colleagues have described specific proteins of 70,000 daltons in porcine gastric, and 90,000 daltons in porcine small intestinal mucin, which are released by reduction.[18] The intestinal mucins we have worked with are comparatively resistant to reduction but we have also been able to show recently, by reduction under more drastic conditions in the presence of SDS, that there is a release of a protein species with a molecular weight of approximately 100,000 daltons in rat intestinal mucin. The protein cannot be labelled by the galactose oxidase method and

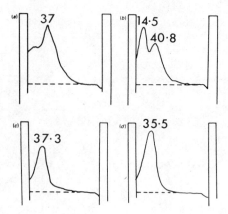

Fig. 1.   Band ultracentrifugation of entire mucin, (a) before and
          (b) after reduction and alkylation, compared with fraction
          I mucin, (c) before and (d) after reduction and alkylation.
          Numbers equal S^o values.   (For details see ref. 15).

therefore probably does not contain carbohydrate. A covalently
linked naked peptide chain of 70-100,000 daltons may be an important
and universal cross-linking agent in many intestinal mucins. Naked
peptide segments are much more likely to be cleaved by proteases
than their carbohydrate-decorated counterparts, and are rapidly
destroyed by pronase in porcine gastric mucin. Thus intestinal
mucins, which are often thought to be quite impervious to luminal
proteases, may have a very susceptible chink in their armour.

    In addition to structural similarities, intestinal mucins also
share considerable antigenic cross-reactivity. Immunofluorescent
microscopic techniques show quite clearly that antibody raised
against rat small intestinal mucin binds to mucins of the rat sublingual gland, stomach and colon.[4] A similar broad cross-reactivity
has been shown for antibody raised against other intestinal mucin
antigens.[19,20]

    Figure 2 illustrates a radio-immunoassay for human intestinal
mucin, which we developed principally as a means of identifying and
measuring mucins in mixed solutions.[21] Mucins present in post
mortem surgical tissues from many cases of the intestinal tract
were compared with respect to their ability to compete with a radio-
actively labelled small intestinal mucin antigen. For identical

antigens the slope of the displacement curve should always be the
same, whereas totally unlike antigens will not displace the standard
at all. Using these criteria it is evident that the colon and
intestine contain identical mucin antigens. Stomach and esophageal
mucins have shallower displacement slopes which indicate that they
are not identical, although sharing some antigenic determinants.
When mucins from the same region, but from several individuals,
bearing variable ABH and Lewis specificity, were compared the
slopes were always parallel. Mucins from particular areas of the
intestinal tract appear, therefore, to possess common antigenic
determinants which override individual specificities. Interestingly,
species cross-reactivity was limited. The human antibody reacted
with dog, monkey, and rabbit mucins but not with rat, porcine, toad
and oyster mucins.

    Immunology is a potentially powerful tool for detecting minor
differences between mucins. Thus far, its application to diseases
of the intestinal tract has been limited almost entirely to tumour
antigens with indifferent results. There is an urgent need to
develop monoclonal antibodies for narrow antigenic specificities,

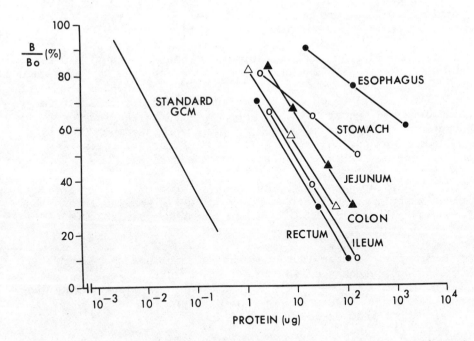

Fig. 2.   Radio-immunoassay for human intestinal goblet cell mucin
           (GCM) applied to mucins from various segments of the
           human intestinal tract.

so that mucins can hopefully be characterized and distinguished in
peptic ulcer disease, inflammatory bowel disease and neoplasia,
where there is considerable evidence that relatively subtle changes
in mucin structure occur.

PERMEABILITY

One accepts that mucus must limit the permeability of noxious
substances such as hydrochloric acid and bile salts if it is to
function as a protective skin, but it has not been easy to prove if
or how this is done. Heatley[22] could not show that crude gastric
mucins limited H^+ ion diffusion. In fish, snails and various
vertebrates, removal of external mucus leads to tissue oedema,[23,24]
as if mucus formed a barrier to the free flow of water. Lukie[25]
has shown that rat intestinal mucus greatly retards water permeability in vitro, in direct proportion to its thickness. A recent
paper by Nimmerfall and Rosenthaler[26] suggests that surface mucus
secretion constitutes a diffusion barrier which is identical in
properties and thickness to the unstirred water layer. They showed
that the rate of diffusion of 7 small molecular weight compounds
through, and retention by, expressed rat intestinal mucus correlated
very well with their rate of absorption in vivo. Molecular weight
and electrostatic interactions were the most important factors in
determining these responses. Nakamura et al[27] attempted to
correlate binding to intestinal mucus in vitro with absorption of
water soluble and lipophilic dyes through intact mucosa in vivo,
at first denuded of mucus by washing and then replete with mucus
after a recovery period. The presence of mucus resulted in reduced
absorption of the lipophilic drug, quinine, but enhanced the absorption of four water soluble cationic and anionic dyes. All five
compounds bound mucus, but this was unrelated to the rate of
absorption. Other factors, such as the changing composition of the
mucus components during resting and hypersecretory states, were
possibly more important in determining the outcome.

The antibiotic tetracycline binds to proteolytically treated
porcine gastric mucin, causing it to cross-link and undergo a
viscous transformation.[28,29] This in return appears to limit bioavailability of the drug.

Absorption of phenylbutazone, an anti-inflammatory drug, has
been shown to be enhanced in rat intestine after interaction with
model mucins.[30] The mechanism of the enhancement is unknown. Low
concentrations of bile salts increased absorption further, possibly
by improving the water-solubility of the drug.

Rat intestinal mucin bound divalent cations and solubility
decreased as cation concentration increased.[31,32] Coleman et al[33]
have shown that Ca^{2+} becomes distributed in a patchy way through

intestinal surface mucus, and have suggested that there are
specialized anionic sites to which Ca^{2+} and Mg^{2+} ions are normally
attracted. Possibly the binding enhances passive ion diffusion
through the epithelium. On the other hand, toxic heavy metals may
be trapped and held tightly by mucin for later excretion in the
faeces. Cations may also favour secondary binding reactions to
mucin, as in the case of tetracycline which forms a drug-Ca-mucin
complex.[34]

These studies suggest that mucus is an important barrier to
water, electrolytes and small molecules, which varies in its
properties with thickness, secretory state, ionic environment and
the presence of cross-linking or viscotropic agents. In almost all
cases however, the evidence is inferential. Model mucins, or mucus
of varying purity and physical states, have been used and there is
little certainty that physiology rather than artifact has been
described. The answer may lie with model in vivo systems, such as
that described by Nakamura,[27] which at least permit one to measure
function in the presence or absence of a naturally produced surface
coat. Standardization is an obvious problem which will have to be
solved before progress can be made.

PEPTIC ULCER DISEASE

Agents which inhibit glycoprotein synthesis such as aspirin,[35,36] phenylbutazone,[37] and steroids,[38] appear to be ulcerogenic
while agents which appear to stimulate mucus synthesis and secretion
such as carbonoxolone,[39,40] and prostaglandins[41] are cytoprotective.[42,43] Although these facts satisfactorily implicate mucus as a
mucosal protective agent, some caution needs to be exercised since
the inhibitory experiments have not dealt specifically with mucins
and it is doubtful that the secretion experiments do so either.
Dye binding techniques[41] do not distinguish mucins from acid mucopolysaccharides or other glycoproteins. The results shown in
Table 2, taken from Nozamis and Robert,[44] are illustrative of the
problem created by the restricted measurement of either sialic acid
or hexosamine. The NANA:hexosamine molar ratio for the gastric
contents of control rats was 4.4 and 4.1 in rats stimulated with
16.16 dimethyl PGE_2 for 30 minutes. NANA accounted for approximately
30% of the dry weight of the mucosa in the latter case. When one
considers the actual structure of mucins which have been isolated,
and the specificities of the transferases which produce them, it is
difficult to conceive of a structure with a NANA:hexosamine ratio
greater than 2. Even then, the only structure which might qualify
(shown in Table 2) has not been described in mucins and must be
extremely rare. Data of this type suggests that saccharides derived
from other sources are an important, perhaps even dominant part of
the secretory response to prostaglandins.

Table 2.   Gastric Secretion After Prostaglandin, Modified from
            Ref. 44.

	Controls	Oral Prostaglandin	Proposed Structure NANA/Hexosamine=2.0
	mg		
Dry wt	8.5	27.0	Gal — GalNAC — {ser / thr
NANA	1.84	9.45	
Hexosamine	0.3	1.65	
NANA/Hexosamine (molar)	4.4	4.1	NANA NANA

   Guslandi et al[45] suggested that cimetidine reduces the amount
of neutral dissolved glycoprotein in basal gastric contents while
increasing acid glycoprotein. Cathcart et al[46] found that mucus
in the surface epithelium and sulphated mucus in the chief cells
was decreased in stress ulcer and hemorrhagic gastritis. These
observations seem to suggest that qualitative changes may occur in
secreted mucins during the inflammatory stage or as a consequence
of drugs used in treatment.

   Sulphated mucins, especially those of the gastric epithelium,
have gained a reputation for being antiulcerogenic, through their
inhibitory action on pepsin digestion of protein substrates.[47]
Mikuni-Takagaki and Hotta[48] have shown that sulphated gastric mucins
form an insoluble ionic complex with albumin at low pH, thereby
rendering albumin resistant to pepsin attack. The authors have
suggested that the physiologic role of sulphated gastric glyco--
proteins is to coat and protect the gastric and duodenal mucosal
surface constituents from peptic auto-digestion. In humans, unlike
rodents, small intestinal mucins are poorly sulphated,[49] which may
account for the vulnerability of the human duodenum to peptic ulcer
formation in the presence of gastric pepsin and acid. It may also
account for the therapeutic usefulness in peptic ulcer disease of
sulphate-containing polymers such as amylopectin sulphate.[50]

MUCUS AND BACTERIA

   Bacteria grow best where conditions are static. In the upper
intestine these conditions can only be achieved by adherence to,
and colonization of, the luminal cell surface. Mucus may retard
this process physically by forming a viscous barrier to organisms,
which entraps them and bundles them down the tract, chemically by
providing false binding sites which compete with those of the
cellular surface, and organizationally by providing a semi-stable
matrix in which immunoglobulins and intestinal enzymes are embedded
so as to confront invading organisms. The mechanisms which control
mucus penetration are probably complex. Virulent cholera strains

are motile and penetrate to the base of the crypts.[51] E. coli K88
strains are not very motile, but bind microvilli tightly. They end
up plastered to the sides of the villus but are apparently unable
to penetrate more deeply.[52] Bacterial glycosidases or proteases may
aid penetration but these have not been studied systematically.
Williams and Gibbons[53] showed that salivary mucin competes with
buccal epithelium for receptors on certain pathogenic streptococci.
Strombeck and Harrold described similar competition between gastric
mucin and intestinal epithelium for cholera toxin.[54] Mucins also
bind viruses more or less specifically. Sialic acid is a well known
ligand for influenza virus, but sulphate groups appear to be critical
for some human enteroviruses.[55]

     In the colon, mucins interact with some anaerobic bacteria by
serving as a major nutrient.[56] Hoskins and Boulding showed that the
glycosidic spectrum of colonic anaerobes is controlled genetically
by the blood group specificity of the host mucin.[57,58]

     Other roles for mucins are possible through physical inter-
action with specific substrates. Rat caecal mucins, for example,
inhibited the normal bacteriocidal activity of serum for E. coli.[59]
Both carbohydrate and protein moities seemed to be necessary for
inhibition, but it is not clear whether trapped small molecular
weight inhibitors may have been present.

PARASITIC INFECTIONS

     Mucus appears to play a central role in the elimination of at
least one intestinal parasite, Nippostrongylus brasiliensis, a
helminth which invades the small intestine of the rat following the
subcutaneous injection of larvae. The worm is commonly expelled in
a sudden immune dependent manner 10-20 days later. Shortly before
expulsion the number of villus goblet cells increases dramatically.[60]
The timing of both the goblet cell hyperplasia and expulsion are
accelerated by the adoptive transfer of immune thoracic duct lymphocytes of immune serum.[61] These events can be prevented by steroids
or reserpine,[60] both of which interfere with mast cell function.
In experiments performed in our laboratory in collaberation with
Dr. Dean Befus of McMaster University, we found that intestinal
slices from rats secreted high concentrations of both immunoreactive
mucin and DNA at this period, and in much more elegant experiments,
Miller et al have also shown that increased mucus secretions are
associated with increased goblet cells (H.R.P. Miller, pers. commun.)
Thus, antibody appears to act indirectly, perhaps via a mast cell
dependent mechanism such as histamine release, to stimulate both
goblet cell differentiation and increased mucin secretion. It is as
yet too early to determine whether differentiation and secretory
stimulation are coupled or independent events, or whether the
secretory stimuli involves antibody-antigen interaction at the cell

surface as described by Walker and Wu.[62] The intermediate role of
mast cells is also in doubt since in some rat species mast cell
hyperplasia and the appearance of increased mucosal histamine follow,
rather than coincide with, expulsion.[63] In normal rats, chronic
administration of reserpine stimulates glycoprotein synthesis and
leads to increased accumulation of mucosal mucin.[64] It seems likely
therefore that the prevention, by reserpine, of goblet cell differentiation in the parasite infestation model is mediated indirectly
via the mast cell.

The interaction of a non-specific mucosal mechanism of defence
such as mucus secretion also seems necessary to explain the intestinal
response to Giardia Muris in mice. Underdown et al[65] have shown
that C3H/He mice, which develop a chronic infestation of Giardia
Muris after ingestion of cysts, do so in spite of the fact that
they are quite similar to resistant strains, in elaborating
specific IgA antibodies to the parasite, developing resistance to
reinfection and having the capacity to transmit resistance to off-
spring. Their capacity to elaborate and secrete mucin has as yet
not been examined, but the mouse Giardia model is attractive for
studying the possible protective role of mucus, since unlike the
Nippostrongylus model, there is very little inflammatory response
associated with the eventual rejection of the parasite.

MUCINS, LECTINS AND ANTIGENS

The stimuli which control mucus secretion are not well understood. In the bronchus,[66] cat trachea,[67] and rat small intestine,[68]
cholinergic agents stimulate deeper mucus glands, but superficial
cells seem to respond primarily to local stimuli. Cell surface
receptors, possibly glycoproteins, may play a significant role.
Lectins stimulate blast transformation in lymphocytes by binding
cell surface carbohydrates and are potentially useful probes for
uncovering surface mediated events. Freed and Buckley[69] fed
concanavalin A by gastric tube to rats and observed depleted goblet
cells in rat jejunum four days later, in association with considerable gas and diarrhoea. Whether goblet cell depletion was a primary
lectin mediated event or secondary to fluid secretion is not clear,
but this observation deserves further investigation in more easily
controlled systems. Many common foods contain lectins which bind
components of saliva,[77] and probably other intestinal glycoproteins.
Mucins could protect the underlying mucosa by intersecting these
agents before they reach the plasma membrane.

Walker and Wu[62] showed that immune complexes of bovine serum
albumin and rat antibodies to bovine serum albumin caused goblet
cell emptying in rats when injected into segments of intestine.
Increased ^{35}S-labelled glycoprotein secretion was also seen. In a
later paper from the same laboratory, Lake et al[70] showed that the

[35]S label was present in a high molecular weight glycoprotein excluded by Sepharose 4B. Antigen was able to stimulate increased secretion in immunized animals but only when they had been immunized by the oral route. The immune complexes could interact with surface receptors, or more classical intermediates such as activated components of the complement system could mediate mucus release through local receptors, but the exact mechanism is presently unknown. The observation is extremely important, however, because it provides a mechanism through which an immune reaction at the cell surface might stimulate mucus secretion in order to entrap and wash away the offending antigen. If these surface reactions are mediated through carbohydrate recognition one suspects that they may also be triggered by bacterial lipopolysaccharide antigens such as those which activate the indirect complement pathway.

CHOLERA TOXIN

A number of bacterial toxins cause severe diarrhoea through a mechanism which involves increased secretion of water and electrolytes mediated by cyclic AMP. The "rice water" stool of cholera is a mixture of fluid and mucus. The issue is whether mucus and electrolyte secretions are co-stimulated or whether the mucus is simply a consequence of the diarrhoeal state. Cyclic AMP stimulates glycoprotein synthesis in the intestine,[71] but it is not clear whether mucus secretion is affected. Several investigators have found morphological evidence of goblet cell emptying following exposure to crude cholera filtrate.[72,73,74,75] Sherr at al[76] reported that protein bound hexosamine secretion increased in rabbit ileal loops when exposed to either crude filtrate or pure cholera enterotoxin. We have recently studied the effect of crude and pure cholera toxin on mucin secretion in rat intestinal slices using precursor labelling and mucin immunoassay techniques.[78] The toxin stimulated both synthesis and a two-fold increase in secretion of labelled glycoprotein. The increase in the secretion of mucin was much more dramatic however. As shown in Fig. 3, a dose dependent increase in secretion was evident at 30 minutes, rising to a four to five-fold increase over control slices at 90 minutes. Interestingly, although both stimulate mucin secretion, the crude filtrate was more effective as a secretagogue than the pure enterotoxin. One therefore wonders whether the enhancement of mucus and electrolyte secretion is triggered by exactly the same stimulus. When tissue glycoprotein was prelabelled with ^{14}C glucosamine by injection 3 hours before sacrifice, and labelled again during the incubation in vitro with the ^{3}H analogue, there was a consistent fall in the ^{14}C/^{3}H label of material secreted in response to cholera toxin as if the toxin stimulated the preferential secretion of the most recently labelled material. This material may not be mucin since we measured label in total glycoprotein, but if it is, one might usefully ask whether there is a differential release of newly

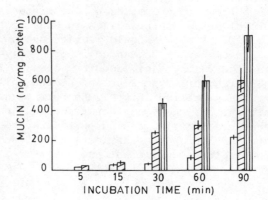

Fig. 3.   Secretion of mucin by rat intestinal slice, unstimulated
           (clear bars) or stimulated by 12.5 mg (hatched bars) or
           25 mg cholera toxin (crude filtrate), (vertical stripe
           bars). Mucin was analyzed by RIA.

synthesized and stored mucins. The recent work of Specian and
Neutra[79] adds special point to the speculation. They have shown
that microtubules or microfilaments control the outward movement of
newly labelled mucus along the lateral margins of the granule mass
within the goblet cell. Acetylcholine initiates the release of
central, less well labelled, mucus by rapid exocytosis at the center
of the cell apex, and does so independently of the microtubule-
microfilament pathway. Stored and continuously secreted mucus seem
therefore to be handled quite differently within the goblet cell.
Cholera toxin might be an example of an agent which preferentially
stimulates the microtubule-microfilament pathway.

CYSTIC FIBROSIS

    Thick, viscid mucus is the hallmark of this inherited disease
which is characterized by blockage of epithelium-lined ducts throughout the body. The intestine is involved through the obstruction of

intralobular ducts in the pancreas, which eventually leads to exocrine gland atrophy and the obstruction of the intestinal lumen by
inspissated meconium. Altered pupillary responses and salivary
hypersecretion suggest that the basis for the pathology of the
disease may lie in secretory aberrations produced by some form of
altered autonomic responsiveness. Rats chronically stimulated by
repeated injection of reserpine may constitute a suitable model for
the disease since PAS positive material accumulates in the cells
and ducts of the salivary,[80] tracheobronchial,[81] and pancreatic
glands.[82] Glycoprotein secretion from the trachea shows increased
sulphation in both the animal model[83] and CF patients.[84] Experiments conducted in our laboratory indicate that chronic reserpine
treatment produces increased concentrations of immunospecific goblet
cell mucin in intestinal tissue and significantly greater mucin
secretion by tissue slices over a 90 minute period (Fig. 4). These
changes were associated with enhanced incorporation of radioactive
glucosamine by tissue glycoproteins, both in vivo, following intraperitoneal injection and in vitro, during incubation of slices, but
no evidence of increased secretion of labelled glycoprotein. We
have tentatively concluded that, unlike cholera, the majority of
the secreted mucin was derived from a stored pool which was not well
labelled. In this case the characteristics of the secretion are
those which might be expected from triggering of the acetylcholine
pathway.[79] It was also noted that a great deal of the increased
isotope incorporation occurred in small molecular weight, non-mucin
fractions, suggesting that reserpine stimulates the synthesis of
glycoproteins in general rather than mucin in particular. Net
synthesis and secretion of mucus glycoprotein is nevertheless
enhanced and this probably accounts for the histological changes.

In man, autoradiographic study of rectal biopsies, pulse-
labelled with ^{3}H glucosamine and maintained in organ culture for
periods up to 24 hours, failed to reveal any abnormality in the
intracellular transport of label in cystic fibrosis patients.[85]
As well, the pattern of incorporation of ^{3}H fucose, ^{3}H-N-acetyl-
mannosamine and ^{35}S-sulphate was not different from that of biopsies
from controls. These experiments do not exclude a model in which
net synthesis is increased, as with reserpine, since one would not
necessarily expect to find an increase in transport rate. It might
be objected that greater net synthesis should result in fatter, or
more numerous, goblet cells, but a small fractional increase might
be difficult to identify by histology.

Morrisey and Tymvios[86] studied goblet cells in the duodenum,
jejunum and ileum in infants with and without CF by quantitative
microdensitometry. Sulphomucins were increased in CF patients,
particularly in the ileum, whereas total acid mucins were not
different from controls. Thus there seems to be reasonably good
evidence that CF intestine, like the lung, contains increased
amounts of sulphated, as compared with other acid mucins.

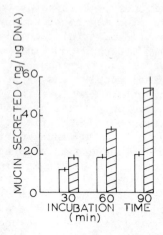

Fig. 4.   Secretion of mucin by rat intestinal slices from control
           rats (clear bars) and chronically reserpinized rats
           (hatched bars). Mucin determined by RIA.

Dische first reported that mucoproteins from CF duodenal
secretions had high fucose/sialic acid ratios in CF in 1959,[87]
and later stressed the reciprocal nature of these two sugars in a
variety of mammalian glycoproteins.[88] Clamp and Gough[17] recently
invigorated this theme by demonstrating that mucus glycopeptides
from a pool of three blood group A patients with cystic fibrosis had
greater amounts of fucose and lesser amounts of N-acetylneuraminic
acid than normals. A number of CF oligosaccharides released by
reductive cleavage with alkaline borohydride had a rather large
amount of fucose, which suggests that there could be an underlying
transferase abnormality. We have recently completed a comparison
of 10 purified mucins isolated from post mortem human small intestine

six from non CF and four from CF subjects. The outstanding result
from this study, in our minds, is the fact that CF mucin was more
highly glycosylated than the mucins from controls (Table 3). Glycosylation correlated nicely with increased amounts of gal, fucose
and GlcNAC, suggesting chain elongation, while neither GalNAC nor
sialic acid increased. CF mucins were also denser. Thus we would
conclude that CF is associated with the accumulation of very large,
highly glycosylated mucins, which are poorly sialated simply because
sialic acid is not added to elongating chains. We also see no evidence of enhanced fucose content beyond that which is to be expected
for the degree of glycosylation. At the moment one can only
speculate as to whether these changes result from altered control
of chain elongation, which appears to be a transferase function in
mucin, or from altered rates of synthesis and storage within the
goblet cell.

ADENOCARCINOMA

     A great deal of evidence, particularly in the colon, suggests
that mucins produced by adenocarcinomas, and even transitional
mucosa in the neighbourhood of the tumour, are abnormal. By histochemical staining sialomucins are increased at the expense of sulphomucins.[89,90] Similar changes have been found in carcinoma induced
by dimethylhydrazine in rats.[91] It is not always easy to demonstrate the exact counterpart for histochemical observations in
extracted mucins but Gold and Miller have shown that sialic acid is
increased in reasonably pure mucins isolated from tumours.[92] These
observations have been pushed somewhat further by Filipe's group
and Culling and Reid and associates who have shown that the neoplastic tissue produces relatively less O-acetyl substituted sialic
acid.[93,94,95] The histochemical methodology involves a rather
clever combination of oxidation, reduction and saponification which
is summarized in Fig. 5. The explanation for these changes is not
yet resolved. Neoplastic tissue could code for a different mucin
altogether. Some support for this proposition comes from the work
of Gold and Miller who extracted mucin from 12 tumours and found

Table 3.   Total Carbohydrate Sialic Acid in Human
            Intestinal Goblet Cell Mucin

	CF (4)	N (6)	P
	nmols per µg protein $\pm$ SD		
Total CHO	22.87 ± 3.41	14.36 ± 4.20	< 0.01
Sialic acid	1.68 ± 0.36	2.16 ± 0.99	N.S.

that they had a lower threonine and higher aspartic acid content than normal mucins, as well as greatly decreased carbohydrate.[92] Alternatively, transferase activity could be abnormal. Kim and Isaacs[96] indicated that the activity of the core GalNAC transferase was reduced, for example, as one might expect with poor glycosylation, but they also found decreased sialyl transferase activity which does not fit with increased staining for sialomucins. One should not ignore the possibility that many of the current observations may not be related to mucins at all, since neither the histological nor bio-chemical techniques have been that selective or specific. Cancers of the colon often produce much less mucin than normal tissue,[95] while other tumours have been shown to contain fewer blood group antigens.[97] Mucin depletion may account for the fact that decreased incorporation of carbohydrate precursors by tumour tissue was found by two groups of investigators.[98,99] However, O'Gorman and LaMont found that a marked decrease in particulate glycoprotein synthesis accounted for the difference rather than the incorporation into the more soluble mucins.[98]

Hakkinen[100] has suggested that sulphated glycoproteins are significantly increased in gastrointestinal, particularly gastric, tumours. Glycoproteins which seem to be mucins have been isolated

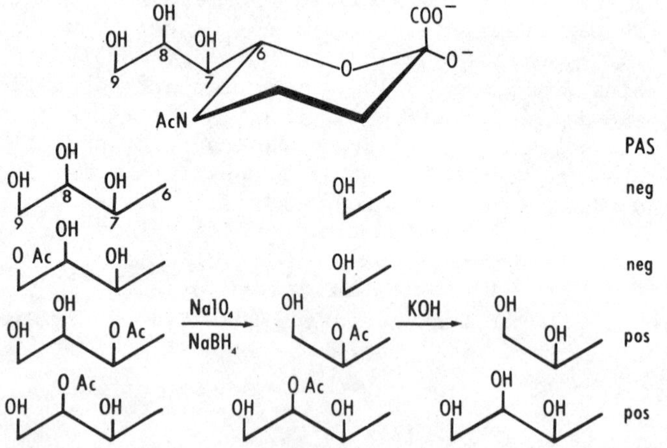

Fig. 5.   Reaction sequence of periodate oxidation, followed by reduction and saponification in the determination of free and substituted hydroxyls in sialic acid with the PAS reagent.

from gastric tumours in man, and experimental tumours in rat.[101]
Bara et al[19] found 7.5% sulphate and a sialic acid:fucose molar
ratio of 5 in the gastric material which also cross-reacted by antibody immunofluorescence with goblet cell contents in intestine and
colon. This is an extraordinarily high sialic acid:fucose ratio.
Could the glycoprotein have been simultaneously rich in sialic acid
and sulphate?

Specific Mucoprotein Tumour Markers

Mucins produced by tumour cells share antigenic determinants
with mucus glycoproteins from normal tissue.[20,102,103,104] If
neoplastic mucins are different, however, even in degree, one might
expect to find antigenic determinants shared poorly or not at all
with normal mucins. Gold and Miller[103] extracted human mucoproteins
from three adenocarcinomas by identical techniques used for their
colonic mucoprotein, which appears to be a well characterized mucin.
Antisera to the colonic mucoprotein precipitated antigens from all
three tumour extracts, while antisera raised against the tumour
extracts precipitated only tumour antigens and failed to react with
the normal mucoprotein. None of the antigens reacted with goat anti-
CEA antisera. Although the human extracts had carbohydrate and
amino acid characteristics of mucin glycoproteins, no direct evidence
was given that the tumour marker was necessarily a mucus glycoprotein.
This subject deserves more attention than it is getting. It is
somewhat disappointing that the initial report has been neither
confirmed nor rejected. Bara and Burtin[105] have recently shown
that a mucin antigen derived from ovarian cyst material can be
localized within goblet cells of transitional mucosa adjacent to
colonic adenocarcinomas by immunofluorescence. This antigen is
present in the fetal and newborn colon but not in adult colon, thus
suggesting a reverse into a more immature type of colonic goblet
cell with tumour growth.

INFLAMMATORY BOWEL DISEASE

Mucosal mucin is markedly decreased in ulcerative colitis.
Indeed, the absence of goblet cells forms an important clue to the
diagnosis. Sialomucins are less affected than sulphomucins.[106]
Culling, Reid and Dunn have recently applied their saponification
and have found that substitution is decreased with the severity of
the lesion in both Crohn's disease and ulcerative colitis.[107]
These changes are very similar to those seen in adenocarcinoma;
dedifferentiation of rapidly proliferating mucosal epithelium may
be the common denominator. In contrast to neoplastic tissue, glycoprotein synthesis appears to be enhanced in ulcerative colitis.[108]
An analysis of carbohydrate in a Sepharose 2B excluded fraction
from extracts of surgically removed colonic material taken from

patients with ulcerative colitis was compared with that of similar material derived from post mortem bowel of control patients by Fraser and Clamp.[109] The preparations contain less than .15 mols of mannose per mol of galactose, and were therefore relatively free of contamination by non mucin glycoproteins. Unlike the histochemical findings there was no evidence of increased sialic acid.

GALLSTONES

The healthy gallbladder secretes relatively little mucin but as gallstones form in experimental animals there is a pronounced hyperplasia of mucus-secreting cells with greatly increased secretion. Lee, LaMont and Carey[110] have investigated this phenomenon in the prairie dog which is particularly susceptible to cholesterol gallstone formation and have concluded that the mucoprotein may act as a nucleating agent for stone formation. They were able to demonstrate that aspirin, in sufficient dose to inhibit glycoprotein synthesis and reduce mucus secretion, was able to prevent the formation of stones or crystals, in spite of bile highly saturated with cholesterol.

SUMMARY

In such a bird's eye view of a very complicated and complex literature it is inevitable that significant contributions, particularly from earlier investigators, will have been overlooked. We have tried, however, to provide a reasonable framework for the many presentations and discussions which will take place at this conference. As in the past, it is evident that much needs to be done to reconcile the excellent histochemical studies of gastrointestinal mucus in many diseases with the increasing knowledge of mucin structure and composition. This will only be achieved by extraction of highly pure mucin from diseased bowel. In this regard, post mortem material provides an opportunity for mucin researchers which is not available to workers in other fields because of relative freedom from decomposition. The wedding of mucin technology with immunology is also a major priority. Immunoassay techniques provide the answer to quick and accurate product identification in secretion studies. Specific monoclonal antibodies will provide the route to structural differentiation of antigens in complex mixtures. It is also evident that we must seek to develop a variety of controllable models for the study of functional parameters of mucus in physiological conditions, parasite rejection, neoplasia and inflammatory states. Promising initiatives have been touched upon in this review, but these must only be the beginning. One must conclude, however, by recognizing that advances in knowledge have been truly remarkable since the last conference in 1976. One senses that some of the keys which will one day

unlock the gates of this "ill-defined" kingdom are already in hand,
while the remainder are at least within reach.

REFERENCES

1. H. Florey, Mucin and the protection of the body, Proc. Roy.
    Soc. London, Series B, 143:147 (1955).
2. J. Schrager, and M. D. G. Oates, The isolation and composition
    of the major glycoprotein from human gastric aspirates,
    Gut 12:559 (1971).
3. D. Snary, and A. Allen, Studies on gastric mucoproteins. The
    isolation and characterization of the mucoprotein of the
    water-soluble mucus from pig cardiac gastric mucosa,
    Biochem. J. 123:845 (1971).
4. J. Forstner, N. Taichman, V. Kalnins, and G. Forstner, Intestinal
    goblet cell mucus: isolation and identification by immuno-
    fluorescence of a goblet cell glycoprotein, J. Cell. Sci.
    12:585 (1973).
5. I. Jabbal, D. I. C. Kells, G. Forstner, and J. Forstner. Human
    intestinal goblet cell mucin, Can. J. Biochem. 54:707 (1976).
6. T. Marshall, and A. Allen, The isolation and characterization
    of the high-molecular-weight glycoprotein from pig colonic
    mucus, Biochem. J. 173:569 (1978).
7. V. L. N. Murty, F. Downs, and W. Pigman, Rat colonic, mucus
    glycoprotein, Carb. Res. 6:139 (1978).
8. B. L. Slomiany, V. Murty, and A. Slomiany, Isolation and
    characterization of oligosaccharides from rat colonic mucus
    glycoprotein, J. Biol. Chem. 255:9719 (1980).
9. B. L. Slomiany, and K. Meyer, Oligosaccharides produced by
    acetolysis of blood group A (A+H) sulfated glycoproteins
    from hog gastric mucin, J. Biol. Chem. 248:2290 (1973).
10. W. Newman, and E. A. Kabat, Immunological studies on blood
    groups. Structures and immunochemical properties of nine
    oligosaccharides for β-active blood group substances of
    horse gastric mucosae, Arch. Biochem. Biophys. 172:535
    (1976).
11. V. Derevitskaya, N. Arbatsky, and N. Kochetkov, The structure
    of carbohydrate chains of blood group substance, Eur. J.
    Biochem. 86:423 (1978).
12. M. D. G. Oates, A. Rosbottom, and J. Schrager, Further inves-
    tigations into the structure of human gastric mucin. Carb.
    Res. 34:115 (1974).
13. D. Carlson, Structures and immunochemical properties of oligo-
    saccharides isolated from pig submaxillary mucins, J. Biol.
    Chem. 243:616 (1968).
14. T. Beyer, J. Rearick, J. Paulson, J. P. Prieels, J. E. Sadler,
    and R. L. Hill, Biosynthesis of mammalian glycoproteins.
    Glycosylation pathways in the synthesis of the nonreducing
    terminal sequence, J. Biol. Chem. 254:12531 (1979).

15. J. F. Forstner, I. Jabbal, R. Qureshi, D. Kells, and G. G. Forstner, The role of disulphide bonds in human intestinal mucin, Biochem. J. 181:725 (1979).

16. S. J. List, B. P. Findlay, G. G. Forstner, and J. F. Forstner, Enhancement of the viscosity of mucin by serum albumin, Biochem. J. 175:565 (1978).

17. J. R. Clamp, and M. Gough, Study of the oligosaccharide units from mucus glycoproteins of meconium from normal infants and from cases of cystic fibrosis with meconium ileus, J. Path. 126:197 (1978).

18. A. Allen, M. Mantle, and J. Pearson, The structure and properties of mucus glycoproteins, in: "Proceedings of the 8th International Congress on Cystic Fibrosis," J. M. Sturgess, ed., Imperial Press Ltd, Toronto (1980).

19. J. Bara, A. Paul-Gardias, F. Loisillier, and P. Burtin, Isolation of a sulfated glycopeptide antigen from human gastric tumors, Int. J. Cancer 21:133 (1978).

20. D. V. Gold, and F. Miller, A Mucoprotein with colon-specific determinants, Tissue Antigens 11:362 (1978).

21. R. Qureshi, G. G. Forstner, and J. F. Forstner, Radioimmunoassay of human intestinal goblet cell mucin, J. Clin. Invest. 64: 1149 (1979).

22. N. G. Heatley, Mucosubstance as a barrier to diffusion, Gastroent. 37:313 (1959).

23. J. Machin, Osmotic gradients across snail epidermis. Evidence for a water barrier, Science 183:759 (1974).

24. V. Negus, Mucus, Proc. Roy. Soc. Med. 60:75 (1967).

25. B. E. Lukie, Studies of mucus permeability, Mod. Prob. Paed. 19:46 (1977).

26. F. Nimmerfall, and J. Rosenthaler, Significance of the goblet-cell mucin layer, the outermost luminal barrier to passage through the gut wall, Biochem. Biophys. Res. Comm. 94:960 (1980).

27. J. Nakamura, K. Shima, T. Kimura, S. Muranishi, and H. Sezaki, Intestinal mucus in the absorption of quinine and water-soluble dyes from the rat small intestine, Chem. Pharm. Bull. 26:857 (1978).

28. M. P. Braybrooks, B. W. Barry, and E. T. Abbs, The effect of mucin on the bioavailability of tetracycline from the gastro-intestinal tract, in vivo, in vitro correlations, J. Pharm. Pharmacol. Commun. 27:508 (1975).

29. I. W. Kellaway, and C. Marriot, The influence of mucin on the bioavailability of tetracycline, J. Pharm. Pharmacol. Commun. 27:281 (1975).

30. E. G. Lovering, and D. B. Black, Drug permeation through membranes III: Effect of pH and various substances on permeation of phenylbutazone through everted rat intestine and poly-dimethyl-siloxane, J. Pharmac. Sci. 63:671 (1974).

31. J. F. Forstner, and G. G. Forstner, Calcium binding to intestinal goblet cell mucin, Biochim. Biophys. Acta 386:283 (1975).

32. J. F. Forstner, and G. G. Forstner, Effects of calcium on intestinal mucin. Implications for cystic fibrosis, Pediat. Res. 10:609 (1976).

33. J. R. Coleman, and L. B. Young, Metal binding by intestinal mucus, in: "Scanning Electron Microscopy II", A. M. F. O'Hare, ed., S. E. M. Inc., Ill. (1979).

34. A. Albert, and C. W. Rees, Avidity of the tetracyclines for the cations of metals, Nature 177:433 (1956).

35. P. W. Kent, and A. Allen, The biosynthesis of intestinal mucins, Biochem. J. 106:645 (1968).

36. B. E. Lukie, and G. G. Forstner, Synthesis of intestinal glyco- proteins. Inhibition of (I-^{14}C) glucosamine incorporation by sodium salicylate in vitro, Biochim. Biophys. Acta 283: 380 (1972).

37. B. E. Lukie, and G. G. Forstner, Synthesis of intestinal glyco- proteins. Inhibition of (I-^{14}C) glucosamine incorporation by phenylbutazone in vitro, Biochim. Biophys. Acta 338:345 (1974).

38. R. Menguy, and A. E. Thompson, Regulation of secretion of mucus from the gastric antrum, Ann. N.Y. Acad. Sci. 140:797 (1967).

39. W. Domschke, S. Domschke, M. Classen, and L. Demling, Some properties of mucus in patients with gastric ulcer. Effects of treatment with carbonoxolone. Scand. J. Gastroent. 7:647 (1972).

40. J. S. Shillingford, W. E. Lindup, and D. V. Parke, The effects of carbonoxolone on the biosynthesis of gastric glycoproteins in the rat and ferret, Biochem. Soc. Trans. 2:1104 (1974).

41. J. P. Bolton, D. Palmer, and M. Cohen, Stimulation of mucus and non parietal cell secretion by E2 prostaglandins, Am. J. Dig. Dis. 23:359 (1978).

42. A. Robert, Antisecretory antiulcer cytoprotective and diarrheo- genic properties of the prostaglandins, Adv. Prostag. Thrombox. Res. 2:507 (1976).

43. W. Domschke, S. Domschke, J. Hagel, L. Demling, and D. N. Croft, Gastric epithelial cell turnover, mucus production and healing of gastric ulcers with carbonoxolone, Gut 18:817 (1977).

44. J. E. Nozamis, and A. Robert, Gastric mucus may mediate the cytoprotective effect of prostaglandins, Gastroent. 78:1228 (1980).

45. M. Guslandi, M. Cambielli, L. Bierti, and A. Tittobello, Relapse of duodenal ulceration after cimetidine treatment, Brit. Med. J. 1:718 (1978). (Letter).

46. R. S. Cathcart, C. T. Fitts, J. McAlhany, and S. S. Spicer, Histochemical changes in gastric mucosubstances with acute and chronic ulcer disease, Surgery 180:1 (1974).

47. N. B. Roberts, and W. H. Taylor, The inactivation by carbon- oxylone of individual human pepsinogens and pepsins, Clin. Sci. Mol. Med. 45:213 (1973).

48. Y. Mikuni-Takagaki, and K. Hotta, Characterization of peptic inhibitory activity associated with sulphated glycoproteins

isolated from gastric mucosa, Biochim. Biophys. Acta 584:288, (1979).

49. M. I. Filipe, and C. Fenger, Histochemical characteristics of mucins in the small intestine. A comparative study of normal mucosa, benign epithelial tumours and carcinoma, Histochem. J. 11:277 (1979).

50. Y. S. Kim, A. Bella, Jr., J. S. Whitehead, R. Isaacs, and L. Remer, Studies on the binding of amylopectin sulfate with gastric mucin, Gastroent. 69:138 (1975).

51. M. N. Guentzel, L. H. Field, E. R. Eubanks, and L. J. Barry, Use of fluorescent antibody in studies of immunity to cholera, Inf. Immun. 15:539 (1977).

52. H. U. Bertschinger, J. W. Moon, and S. C. Whipp, Association of E. coli with intestinal epithelium, Inf. Immun. 5:595 (1972).

53. R. C. Williams, and R. J. Gibbons, Inhibition of streptococcal attachment to receptors on human buccal epithelial cells by antigenically similar salivary glycoproteins, Inf. Immun. 11:711 (1975).

54. D. R. Strombeck, and D. Harrold, Binding of cholera toxin to mucins and inhibition by gastric mucin. Inf. Immun. 10:1266 (1974).

55. R. di Girolamo, J. Liston, and J. Matches, Ionic binding: the mechanism of viral uptake by shellfish mucus, Appl. Environ. Microbiol. 33:19 (1977).

56. D. C. Savage, Factors involved in colonization of the gut epithelial surface, Am. J. Clin. Nutr. 31:131 (1978).

57. L. E. Hoskins, and E. T. Boulding, Degradation of blood group antigens in human colon ecosystems, J. Clin. Invest. 57:63 (1976).

58. L. E. Hoskins, and E. T. Boulding, Degradation of blood group antigens in human colon ecosystems. J. Clin. Invest. 57:74 (1976).

59. R. Winsnes, T. Midtveldt, and A. Trippestad, Rat intestinal glycoprotein lowering bactericidal activity of serum on ^{32}P-labelled E. Coli, Acta Path. Microbiol. Scand. Sect. C, 84: 77 (1976).

60. H. R. P. Miller, and Y. Nawa, Immune regulation of intestinal goblet cell differentiation, Nouv. Rev. Fr. Hematol. 21:31 (1979).

61. H. R. P. Miller, and Y. Nawa, Nippostrongylus brasiliensis: intestinal goblet cell response in adoptively immunized rats, Exper. Parasit. 47:81 (1979).

62. W. A. Walker, and M. Wu, Stimulation by immune complexes of mucus release from goblet cell of the rat small intestine, Science 97:370 (1977).

63. A. D. Befus, N. Johnston, and J. Bienemstock, Nippostrongylus brasielensis: mast cells and histamine levels in tissues of infected and normal rats. Exper. Parasit. 48:1 (1979).

64. J. Forstner, B. Maxwell, and N. Roomi, Intestinal secretions of mucin in chronically reserpinized rats. Am. J. Physiol.

(In press).

65. B. J. Underdown, I. C. Roberts-Thomson, R. F. Anders, and G. F. Mitchell, Giardiasis in mice: studies on the characteristics of chronic infection in C^3H/He mice, J. Immunol. 126:669 (1981).

66. S. Coles, Regulation of the secretory cycles of mucous and serous cells in the human bronchial gland, in: "Mucus in Health and Disease," M. Elstein and D.V. Parke, eds., Plenum Press, London (1977).

67. J. T. Gallagher, P. W. Kent, R. Phipps, and P. Richardson, Influence of pilocarpine and ammonia vapour on the secretion and structure of cat tracheal mucins, in: "Mucus in Health and Disease," M. Elstein and D.V. Parke, eds., Plenum Press, London (1977).

68. R. D. Specian, and M. R. Neutra, (personal communication).

69. D. L. Freed, and C. H. Buckley, Mucotractive effect of lectin, Lancet 1:585 (1978).

70. A. M. Lake, K. J. Block, M. R. Neutra, and W. A. Walker, Intestinal goblet cell mucus release, J. Immunol. 112:834 (1979).

71. G. Forstner, M. Shih, and B. Lukie, Cyclic AMP and intestinal glycoprotein synthesis: the effect of β-adrenergic agents, theophylline, and dibutyryl cyclic AMP, Can. J. Physiol. Pharmacol. 51:122 (1973).

72. H. L. Elliot, C. C. J. Carpenter, R. B. Sack, and J. H. Yardley, Small bowel morphology in experimental canine cholera. A light and electron microscopic study, Lab. Invest. 22:112 (1970).

73. H. W. Moon, S. C. Whipp, and A. L. Baetz, Comparative effects of enterotoxin from Escherichia coli and Vibrio cholerae on rabbit and swine small intestine, Lab. Invest. 25:133 (1971).

74. J. H. Yardley, T. M. Bayless, E. H. Leubbers, C. H. Holland, and T. R. Hendrix, Goblet cell mucus in the small intestine. Findings after net fluid production due to cholera toxin and hypertonic solutions, Johns Hopkins Med. J. 130-131:1 (1972).

75. S. E. Steinberg, J. G. Banwell, J. H. Yardley, G. T. Keusch, and T. R. Hendrix, Comparison of secretory and histological effects of Shigella and cholera enterotoxins in rabbit jejunum, Gastroent. 68:309 (1975).

76. H. P. Sherr, B. R. Mertens, and R. Broock, Cholera toxin-induced glycoprotein secretion in rabbit small intestine, Gastroent. 77:18 (1979).

77. R. J. Gibbons, and I. Dankers, Lectin-like constituents of foods which react with components of serum, saliva and streptococcus mutans, Appl. Environ. Bacteriol. (In press).

78. J. F. Forstner, N. W. Roomi, R. E. F. Fahim, and G. G. Forstner, Cholera toxin stimulates secretion of immunoreactive mucin, Am. J. Physiol. 3:G10 (1981).

79. R. D. Specian, and M. R. Neutra, Mucous granule transport and secretion: effects of colchicine and cytochalasin B, J. Cell

        Biol. 87:278 (1980).
80. J. R. Martinez, E. Adelstein, D. Quissell, and G. Barbero, The
        chronically reserpinized rat as a possible model for cystic
        fibrosis. 1. Submaxillary gland morphology and ultrastructure,
        Pediat. Res. 9:463 (1975).
81. T. P. Mawhinney, M. S. Feather, I. Q. Martinez, and G. J.
        Barbero, The chronically reserpinized rat as an animal model
        for cystic fibrosis: acute effect of isoproterenol and pilo-
        carpine upon pulmonary lavage fluid, Pediat. Res. 13:760
        (1979).
82. M. E. Setser, S. S. Spicer, J. A. V. Simpson, M. Adamson, and
        J. R. Martinez, The effects of reserpine on the ultra-
        structure and secondary response of rat exocrine pancreas,
        Exper. Mol. Pathol. 31:413 (1979).
83. T. P. Mawhinney, J. R. Martinez, M. S. Feather and G. J.
        Barbero, Effects of kinin, peptides and prostaglandins on
        glycoprotein release by the perfused trachea of control and
        reserpine treated rats, presented at the 21st annual meeting
        of the Cystic Fibrosis Club, San Antonio, Texas, April 29,
        1980 (abstract).
84. R. Frates, J. Last and T. Kaizu, Mucus glycoproteins secreted
        by cultured airway tissue from human subjects with and with-
        out cystic fibrosis, presented at the 20th annual meeting of
        the Cystic Fibrosis Club, Atlanta, Georgia, May 1, 1979
        (abstract).
85. M. R. Neutra, R. J. Grand, and J. S. Trier, Glycoprotein syn-
        thesis, transport and secretion by epithelial cells of human
        rectal mucosa, Lab. Invest. 36:535 (1977).
86. S. M. Morrissey, and M. C. Tymvios, Acid mucins in human intes-
        tinal goblet cells, J. Path. 126:197 (1978).
87. Z. Dische, P. Di Sant'Agnese, C. Pallavicini, and J. Youlos,
        Composition of mucoprotein fractions from duodenal fluid of
        patients with cystic fibrosis of the pancreas and from
        controls, Paed. 24:74 (1959).
88. Z. Dische, Reciprocal relation between fucose and sialic acid
        in mammalian glycoproteins, Ann. N.Y. Acad. Sci. 106:259
        (1963).
89. M. I. Filipe, Value of histochemical reactions from muco-
        substances in the diagnosis of certain pathological conditions
        of the colon and rectum, Gut 10:577 (1969).
90. M. I. Filipe, and A. C. Branfoot, Abnormal patterns of mucus
        secretion in apparently normal mucosa of large intestine
        with carcinoma, Cancer 34:282 (1974).
91. M. I. Filipe, Mucous secretion in rat colonic mucosa during
        carcinogenesis induced by dimethylhydrazine, Br. J. Cancer
        32:60 (1975).
92. D. V. Gold, and F. Miller, Comparison of human colonic muco-
        protein antigen from normal and neoplastic mucosa, Cancer
        Res. 38:3204 (1978).
93. C. M. Rogers, K. B. Cooke, and M. I. Filipe, Sialic acids of

human large bowel mucosa: O-acylated variants in normal and malignant states, Gut 19:587 (1978).

94. C. F. Culling, P. E. Reid, J. Worth, and W. L. Dunn, A new histochemical technique of use in the interpretation and diagnosis of adenocarcinoma and villous lesions in the large intestine, J. Clin. Path. 30:1056 (1977).

95. P. A. Dawson, J. Patel, and M. I. Filipe, Variations in sialo-mucins in the mucosa of the large intestine in malignancy: a quantimet and statistical analysis, Histochem. J. 10:559 (1978).

96. Y. S. Kim, and R. Isaacs, Glycoprotein metabolism in inflammatory and neoplastic diseases of the human colon, Cancer. Res. 35: 2092 (1975).

97. I. Davidson, S. Kovarik, and L. Y. Ni, Isoantigens A, B and H in benign and malignant lesions of the cervix, Arch. Pathol. 87:306 (1969).

98. T. A. O'Gorman, and J. T. LaMont, Glycoprotein synthesis and secretion in human colon cancers and normal colonic mucosa Cancer Res. 38:2784 (1978).

99. P. A. Dawson, and M. I. Filipe, A comparison of (^{3}H) galactose and (^{3}H) fucose uptake with the morphological and histo-chemical changes observed in mucous secretion in chemically induced rat colonic carcinoma, Histochem. J. 12:23 (1980).

100. I. P. Hakkinen, FSA-foetal sulphoglycoprotein antigens associated with gastric cancer, Transplant. Rev. 20:61 (1974).

101. H. Munakata, and Z. Yosizawa, Isolation and characterization of a sulfated glycoprotein from a transplantable colorectal adenocarcinoma of rat, Biochim. Biophys. Acta 623:412 (1980).

102. D. M. Goldenberg, C. A. Pegram, and J. I. Vazquez, Identification of a colon specific antigen in normal and neoplastic tumor, J. Immunol. 114:1008 (1975).

103. D. V. Gold, and F. Miller, Chemical and immunological differ-ences between normal and tumoral colonic mucoproteins, Nature 255:85 (1975).

104. W. Rapp, M. Windisch, P. Peschke, and W. Wurster, Purification of a human intestinal goblet cell antigen (GoA). Its immuno-logical demonstration in the intestine and mucus producing gastrointestinal adenocarcinomas, Virchows Arch. (Pathol. Anat.) 382:167 (1979).

105. J. Bara, and P. Burtin, Mucus-associated gastrointestinal anti-gens in transitional mucosa adjacent to human colonic adeno-carcinomas: their "fetal type" association, Eur. J. Cancer 16:1303 (1980).

106. M. I. Filipe, and I. M. P. Dawson, The diagnostic value of muco-substances in rectal biopsies from patients with ulcerative colitis and Crohn's disease, Gut 11:229 (1970).

107. C. F. Culling, P. A. Reid, and W. L. Dunn, A histological com-parison of O-acylated sialic acids of the epithelial mucins in ulcerative colitis; Crohn's disease and normal controls, J. Clin. Path. 32:1272 (1979).

108. R. P. MacDermott, R. M. Donaldson, and J. S. Trier, Glycoprotein
     synthesis and secretion by mucosal biopsies of rabbit colon
     and human rectum, J. Clin. Invest. 54:545 (1974).
109. G. Fraser, and J. R. Clamp, Changes in human colonic mucus in
     ulcerative colitis, Gut 16:832 (1975).
110. S. P. Lee, J. T. LaMont, and M. Carey, The prevention of
     cholesterol gallstones with aspirin, Gastroent. 78:1205
     (1980).

ULTRASTRUCTURAL ALTERATIONS IN THE COLONIC MUCUS LAYER DURING

CARCINOGENESIS : A SCANNING ELECTRON MICROSCOPY STUDY

O. J. Traynor, C. B. Wood and N. Costa

Department of Surgery, Royal Postgraduate Medical
School, London, and Life Sciences E M Unit, Department
of Botany, Imperial College, London

The colonic mucus layer is considered to have several functions, including lubrication, waterproofing, and protection of the mucosal epithelium (Forstner, 1978). The precise extent of its protective role is not certain but it seems likely that mucus affords some protection against noxious agents in the bowel lumen, possibly including carcinogens. The functions of mucus are intimately related to its structure (Allen, 1977) and minor alterations in its chemical composition may impair its protective functions. Histochemical and biochemical alterations in colonic mucus in malignancy have previously been reported (Filipe, 1969; Culling et al, 1977; Filipe and Cooke, 1974) but the effect of these alterations on the structure and functions of mucus have not been described. This study was designed to examine changes in the physical appearance of the colonic mucus layer during carcinogenesis, using the high magnifications which can be attained by scanning electron microscopy.

MATERIALS AND METHODS

The appearances of the mucus layer were studied in experimental animals during chemical carcinogenesis and also in humans with colorectal cancer and colonic polyps.

A: Animal studies:

Seventy five Sprague-Dawley rats were given weekly subcutaneous injections (20 mg/kg body weight) of 1,2 dimethylhydrazine (DMH) for 16 weeks. Groups of 6 treated animals plus 2 controls were sacrificed at 2 week intervals until the 16th week, and then at 4 week intervals until frank tumours appeared (28-36 weeks). The mucus layer was then examined by SEM.

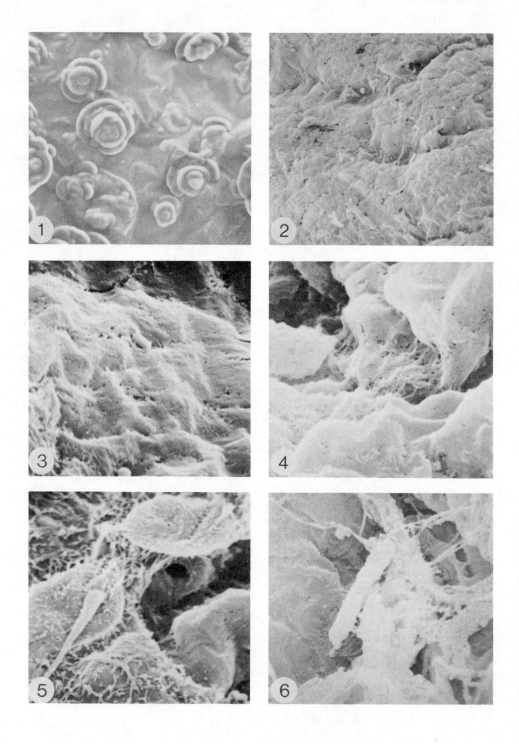

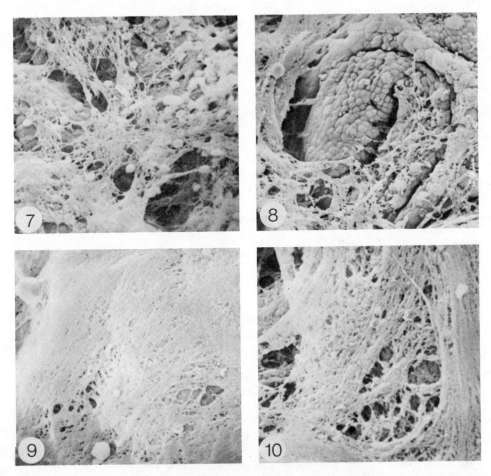

FIGS 1, 2, 7, 9 and 10 are reproduced with kind permission of the
Editor, British Journal of Surgery
FIG 1. (x320) Normal colonic mucus layer, (human). Strands of
mucus streaming from underlying crypts. FIG 2. (x 2500) Normal
mucus (human). Tiny fenestrations seen. FIG 3. (x 2500) DMH
treated rat (6 weeks). Increase in size and number of fenestrations. FIG 4. (x 2500) DMH rat (10 weeks). Mucus layer breaking
into strands. FIG 5. (x 2500) DMH rat (14 weeks). Advanced disintegration of mucus layer. FIG 6. (x 2500) DMH rat (18 weeks).
Mucus forms clumps on mucosa. FIGS 7 and 8. (x 640) Advanced
disintegration of mucus layer on surface of human carcinoma,
exposing underlying epithelial cells. FIG 9. (x 640) Large
defects in mucus layer adjacent to human polyp. FIG 10. (x 640)
Mucus layer adjacent to human carcinoma. Large defects in its
continuity.

B: Human studies:

     Fifteen patient specimens were also examined by SEM (5 patients
had colorectal carcinoma, 5 patients had colonic polyps and 5
patients had normal colons with no evidence of any colonic disease).
The specimens were obtained by surgical resection, except in the
case of normal patients, when sigmoidoscopic biopsies were used.

C: Specimen preparation:

     Preparation for SEM was similar for both human and animal
specimens. Fixation in 10% formol-saline for a minimum of 48 hours
was followed by dehydration through a series of graded acetone
immersions. The specimens were then critical point dried, using
liquid carbon dioxide, and thinly coated with gold prior to examination in the scanning electron microscope (Philips PSEM 500).

RESULTS

     The normal mucus layer had a similar appearance in control
animals and normal patients. This was a dense homogeneous layer
(Fig 1) which completely covered and prevented visualisation of
the underlying epithelium. In places, strands of mucus were seen
streaming from the underlying crypt orifices. At high magnification tiny fenestrations were seen in the mucus layer (Fig 2).

     During carcinogenesis with DMH changes were detected as early
as 4-6 weeks after the first injection. The fenestrations increased
in number and size (Fig 3) and later coalesced to cause focal
defects in the integrity of the mucus layer (Fig 4). By 10-14
weeks the mucus layer began to disintegrate into strands of mucus
(Fig 5) and eventually formed thick clumps rather than a continuous
layer (Fig 6) thus exposing large areas of the mucosal epithelium.

     Patients with colorectal carcinoma had advanced changes in the
appearance of the mucus layer, which were similar to those seen
in rats with fully developed tumours. Breakup of the mucus layer
was seen (Figs 7 and 8) not only on the surface of tumours but also
over the apparently normal mucosa adjacent to the tumours (Fig 10).
In patients with colonic polyps changes were seen which resembled
the preneoplastic appearances of the experimental animal. The
mucus was abnormal in these patients, not only on the surface of
polyps, but also in the immediately adjacent area where it showed
a tendency to break into strands (Fig 9) causing defects in its
continuity.

DISCUSSION

     This study has shown that characteristic ultrastructural
changes occur in the colonic mucus layer in malignancy which

result in failure to provide a continuous covering for the colonic
epithelium. In the experimental animal these changes begin at an
early stage of carcinogenesis and culminate in a complete breakdown
in the structural integrity of the mucus layer. Similar mucus
layer disintegration is seen in humans with colorectal cancer and
patients with colonic polyps display changes which resemble the
intermediate stages of carcinogenesis seen in the rat.

The explanation for the observed changes in the mucus layer is
uncertain. However, sialic acid is considered to play an important
role in determining the viscoelastic properties of mucus (Litt and
Khan, 1976; Gottschalk, 1960) and it is possible that the alterations in sialic acid content of mucus which occur in malignancy
(Filipe and Cooke, 1974) may be responsible for the inability of
mucus to remain as a cohesive layer.

In the rat, changes were seen throughout the entire length of
the colon during carcinogenesis. The changes were not so widespread
in man, but, nevertheless, marked abnormalities were seen in several
apparently normal areas of mucosa remote from tumours and polyps,
suggesting that the colon undergoes a field-change in malignancy.
These findings are in keeping with histochemical studies on "normal"
mucosa adjacent to and distant from colonic tumours (Filipe and
Branfoot, 1974) which have shown an increase in sialomucins with a
decrease or absence of sulfomucins in these areas.

It is not certain whether the changes we have observed in the
mucus layer are a primary phenomenon, which results in exposure of
the mucosal epithelium to the action of carcinogens, or a secondary
phenomenon resulting from abnormal glycoprotein synthesis in epithelial cells which are already neoplastic. However, the changes
occur at an early stage of neoplastic transformation in the colon
and are not confined to sites of frank tumour development. These
changes may have useful applications in earlier detection of malignancy in the colon. Further studies on the origins and extent of
mucus abnormalities in colonic malignancy may assist in the detection of carcinoma and may give valuable clues to the actual
mechanism of carcinogenesis in the colon.

REFERENCES

Allen, A., 1977, Adv. Exp. Med. Biol. 89:283
Culling, C. F. A., Reid, P. E., Worth, A. J., and Dunn, W. L.,
      1977, J. Clin. Path. 30:1056
Filipe, M. I., 1969, Gut 10:577
Filipe, M. I., and Cooke, K. B., 1974, J. Clin. Path. 27:315
Filipe, M. I., and Branfoot, A. C., 1974, Cancer 34:282
Forstner, J. F., 1978, Digestion 17:234
Gottschalk, A., 1960, Nature 186:949
Litt, M., and Khan, M. A., 1976, Biorheology 13:27

GALLBLADDER MUCIN GLYCOPROTEIN HYPERSECRETION IN EXPERIMENTAL CHOLELITHIASIS: ROLE OF MUCIN GEL IN NUCLEATION OF CHOLESTEROL GALLSTONES

J. Thomas LaMont

Chief, Section of Gastroenterology, Evans Memorial
Department of Medicine, University Hospital, Boston
University Medical School, Boston, MA 02118 USA

The emphasis of gallstone research in the past decade has focused on the role of cholesterol supersaturation in bile and reduction in bile salt pool size in the pathophysiology of cholelithasis. However it is now recognized that cholesterol supersaturation is a frequent finding in normal individuals (1), and that a nucleation factor (2) is present in lithogenic gallbladder bile. Womack (3) first suggested that gallbladder mucus glycoproteins could serve as a nucleating agent for cholesterol stones, and noted that nucleation of cholesterol crystals occured in mucus aggregates in the gallbladder of hamsters fed a lithogenic diet. Increased concentrations of gallbladder mucus glycoproteins has been described in experimental cholelithiases in rabbits (4) and dogs (5) and in gallbladder bile from patients with cholesterol stones. (6) Analysis of human gallstones reveals the presence of mucus glycoproteins which may form a matrix for the stones. (7)

In order to investigate further the possible relationship between mucus secretion and cholesterol gallstone formation, we chose the cholesterol-fed prairie dog model. These animals rapidly and reproducibly form cholesterol gallstones when fed a diet containing 1.2% cholesterol. We correlated the appearance of cholesterol crystals with mucus synthesis and secretion in gallbladder organ culture using (3H)-glucosamine as a precursor.

Cholesterol fed prairie dogs had visible accumulation of mucus gel in the gallbladder starting on the fifth day of the lithogenic diet. As shown in Table I, cholesterol feeding was associated with a marked stimulation of mucus synthesis in the biopsies, and release of labelled glycoproteins into the organ culture medium.

The peak of secretion was reached on the fifth day, and regressed somewhat thereafter. Histologic examination of gallbladders

did not reveal infiltration with inflam atory cells or evidence
of bacterial infection.

Table 1.   Incorporation of (3H)-Glucosamine
Into Gallbladder Glycoproteins

Days on Diet	Tissue Glycoproteins	Secreted Glycoproteins
	(dpm/mg protein x 10^{-4})	
0	38.1*	11.8*
3	47.5**	18.1**
5	82.5**	70.9**
8	57.5**	37.0**
14	56.0	28.9

(*p < 0.05; **p < 0.01)

     Gel filtration of the secreted glycoprotein on Sepharose 4B in-
dicated that the majority of radioactivity was present in a glycoprotein of greater than 1 million molecular weight. The increased
secretion of gallbaldder mucin was organ specific, in that incorporation of (3H)-mannose into gallbladder membrane glycoproteins
was not altered by cholesterol feeding. The rate of glycoprotein
synthesis and secretion returned to normal upon withdrawal of the
cholesterol diet, and ligation of the cystic duct prior to cholesterol feeding prevented gallbladder mucin hypersecretion. Both results indicate that the stimulus to mucin secretion was a constiuent
of bile.
     In order to further study the relationship of gallbladder mucus
and crystal formation, we monitored the effect of mucus gel on cholesterol crystal formation in saturated bile in vitro. Mucus glycoproteins were purified by gel filtration from human gallbladder epithelium obtained at elective cholecystectomy for gallstone disease.
Gallbladder mucin was concentrated to produce a viscous gel and then
mixed with samples of hepatic bile from cholesterol-fed prairie dogs
which was saturated with cholesterol but did not contain crystals.
Purified human gallbladder mucin gel was shown to induce nucleation
of lecithin-cholesterol liquid crystals from supersaturated hepatic
bile within 2 to 4 hours. These grew in number and gave rise to
cholesterol monohydrate crystals after 18 hours of incubation. Control supersaturated hepatic bile could not be nucleated by the addition of other proteins (albumin, pig gastric mucus), and was stable
for days upon standing. These results suggested that the increase
in cholesterol content of bile in cholesterol-fed prairie dogs stimulated gallbladder mucus hypersecretion, and that gallbladder mucus gel served as a nucleating agent for cholesterol microprecipita-
tion from bile.

Because non-steroidal anti-inflammatory agents are known to inhibit mucus secretion in other epithelial organs, we tested the effects of aspirin and indocin on gallbladder mucus synthesis. Both agents strongly inhibited gallbladder mucus secretion in vitro with MIC_{50} of 5×10^{-3} for aspirin and 10^{-4} for indocin. Indocin caused a similar dose dependent inhibition of release of 6-keto-PGF_1, a breakdown product of prostacylin, into the organ culture medium. Aspirin at a daily oral dosage of 100 mg/kg/day was administered to cholesterol-fed prairie dogs in order to determine its effect on cholesterol crystal formation.

Aspirin caused a marked inhibition of gallbladder mucus synthesis and mucus gel accumulation, and completely prevented gallstone crystal and stone formation at 14 days of cholesterol feeding. (Table 2) In contrast, 100% of control animals pair-fed the same amount of cholesterol without aspirin had thick mucus gel and numerous stones and crystals in the gallbladder. Aspirin had no effect on biliary lipid secretion, as evidenced by a cholesterol saturation index of 123% in the aspirin plus cholesterol animals (N=15) versus 125% in the group fed cholesterol alone (N=15).

Table 2.   Effect of Aspirin on Experimental
Cholelithasis

Diet	Crystals and stones	Mucus Gel	% Cholesterol Saturation
Control (N=20)	0	0	48*
1.2% Cholesterol (15)	15	15	125*
1.2% Cholesterol + Aspirin (15)	0	0	123*

(*$p < 0.001$)

Conclusions: 1) Lithogenic hepatic bile causes hypersecretion of gallbladder mucus in cholesterol-fed prairie dogs. 2) Cholesterol crystals form in a mucus gel which accumulates in the gallbladder lumen beginning at 5 days of cholesterol feeding. 3) Gallbladder mucus gel can nucleate cholesterol crystals from saturated bile. 4) Aspirin and indocin inhibit gallbladder mucus secretion and aspirin prevents stone formation and mucus accumulation. These studies suggest that the formation of human gallstones might be prevented by drugs which inhibit gallbladder mucus secretion.

References:

1) Carey, M.C., and D.M. Small. 1978. The physical chemistry of cholesterol solubility in bile: relationship to gallstone formation and dissolution in man. J. Clin. Invest. 61: 998-1026.

2) Holan, K.R., R.T. Holzbach, R.E. Hermann, A.M. Cooperman,
    and W.J. Claffey. 1979. Nucleation time: a key factor
    in the pathogenesis of cholesterol gallstone disease.
    Gastroenterology 72: 611-617.

3) Womack, N.A. 1971. The development of gallstones. Surg.
    Gyn. Obstet. 133: 937-945.

4) Hofmann, A.F., E.M. Mosbach. 1964. Identification of al-
    lodeoxylcholic acid as the major component of gallstone in-
    duced in the rabbit by 5 α-cholestan-3β-ol. J. Biol. Chem.
    239: 2813-2821.

5) Englert, E., C.G. Harman, J.W. Freston, R.C. Straight, and
    E.E. Wales. 1977. Studies on the pathogenesis of diet-
    induced dog gallstones. Am. J. Dig. Dis. 22: 305-312.

6) Bouchier, I.A.D., S.R. Cooperband, and B.M. El Kodsi. 1966.
    Mucous substances and viscosity of normal and pathological
    human bile. Gastroenterology. 49: 343-353.

7) Lee, S.P., T.H. Lim, and A.J. Scott. 1979 Carbohydrate
    moieties of glycoproteins in human hepatic and gallbladder
    bile, gallbladder mucosa and gallstones. Clin. Sci. Mol.
    Med. 56: 533-538.

GASTRIC MUCUS DEGRADATION IN VIVO IN PEPTIC ULCER PATIENTS

AND THE EFFECTS OF VAGOTOMY

Fekry Younan[*], Jeffrey P. Pearson[*], Adrian Allen
& Christopher W. Venables.
Department of Physiology and Surgery[*]
The University
Newcastle upon Tyne NE1 7RU, U.K.

Pepsin and other proteolytic enzymes solubilise mucus gel. Studies in vitro on human and pig gastric mucus have shown that this can be explained in molecular terms by the production of degraded glycoprotein subunits (molecular weight 5×10^5) which no longer retain the gel-forming properties of the native glycoprotein (molecular weight 2×10^6) (Scawen & Allen, 1977, Pearson et al., 1980). These studies imply that any increase in mucus glycoprotein in gastric juice could (Allen & Garner, 1980) reflect an increased peptic erosion of the surface gel and not increased mucus secretion as is often stated. We have investigated this possibility by examining the mucus glycoprotein present in gastric washouts following insulin (I.V. 0.2 u/Kg) stimulation in duodenal ulcer patients.

The amounts of native and proteolytically degraded mucus glycoprotein in gastric juice were analysed by gel filtration on Sepharose 2B (Mantle & Allen, 1978). Pepsin and acid concentrations were determined in aliquots of the washout whilst the remainder was immediately neutralised to prevent breakdown of the glycoprotein by pepsin. Insulin stimulation produced a significant threefold rise in luminal glycoprotein content from an average of 26 mg/h. to 87 mg/h. On further analysis of this rise, the pepsin degraded component had risen from an average of 6 mg/h. to 68 mg/h. (78% of the total material) and this was associated with a significant rise in the average level of pepsin from 83 mg/h. to 273 mg/h. In contrast, vagotomised patients showed no rise in the total luminal glycoprotein or pepsin activity following insulin stimulation (Fig. 1). These results show that pepsin erosion of the mucus gel occurs in vivo to produce degraded, lower molecular weight, glycoprotein in the lumen and that the rise in total glycoprotein after insulin stimulation reflects such an erosion.

235

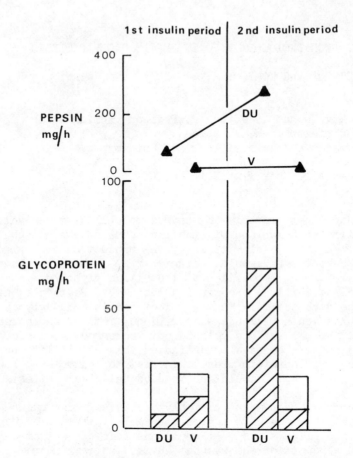

Fig. 1 Mucus glycoprotein and pepsin content of gastric washouts
from duodenal ulcer and vagotomised patients.   The first and second
insulin periods were for the successive 30 minutes and 45 minutes
respectively following insulin injection.
Lower molecular weight glycoprotein ▨; native glycoprotein ☐ ;
pepsin ▲ .

Analysis was also made of the surface mucus gel from mucosal scrapings of antral gastric mucosa resection specimens. All the samples analysed contained lower molecular weight material in addition to 'native' glycoprotein, the amount varying with the disease. Gastric mucus from normal mucosa obtained by resection of the antrum during removal of a pancreatic tumour, duodenal ulcer patients and gastric ulcer patients contained 33.4 ± 5.1, 50.2 ± 3.3 and $65.1 \pm 2.8\%$ respectively of the lower molecular weight material. These results show that in peptic ulcer patients the mucus contained a significantly greater amount of lower molecular weight glycoprotein which has reduced viscosity and gel-forming properties. Further taken in bulk the surface mucus gel of the gastric ulcer group (and to a lesser extent duodenal ulcer group) is potentially of a weaker structure, with presumably less resistance to erosion by shear and solubilisation, than mucus gel from a healthy antral mucosa.

Allen, A. & Garner, A., 1980, Gastric mucus and bicarbonate secretion and their possible role in mucosal protection. Gut 21: 249-262.

Mantle, M. & Allen, A., 1978, A colorimetric assay for glycoproteins based on the periodic acid/Schiff stain. Biochem. Soc. Trans., 6: 607-609.

Pearson, J.P., Allen, A. & Venables, C.W., 1980, Gastric mucus: isolation and polymeric structure of the undegraded glycoprotein: its breakdown by pepsin. Gastroenterology 78: 709-715.

Scawen, M. & Allen, A., 1977, The action of proteolytic enzymes on the glycoprotein from pig gastric mucus. Biochem. J. 163: 363-368.

HISTOCHEMICAL CHANGES IN MUCUS IN DUODENAL ULCERATION

Shelagh M. Morrissey, P.M. Ward, A.P. Jayaraj* and F.I.
Tovey*
Department of Physiology, Queen Elizabeth College
University of London, London, W8 7AH
* Department of Surgery, University College Hospital
   Medical School, London, WC1

INTRODUCTION

Mucosubstances in the human duodenum are normally secreted from
goblet cells in the epithelium of the villi as well as from Brunners
glands lying in the submucosa. Histochemical studies have been
carried out on epithelial mucins in the digestive tract of human
subjects (Lev, 1966), and consist of both neutral and acidic mucins.
Electron microscopical studies on mucins in duodenal ulceration have
been carried out by Moshel (unpublished data) but quantitative investigations in this field are not documented. This collaborative work
was carried out in order to get quantitative values for any observed
changes in the mucosubstances secreted by the normal duodenum both
before and after ulceration.

MATERIALS AND METHODS

Endoscopy biopsy tissue samples were obtained from Basingstoke
District Hospital. Duodenal control tissue (15 cases) are those
samples which after presentation for routine endoscopy were found not
to have any visible duodenal ulcer. These are described as Non-ulcer
group.

There were 13 cases with active ulceration and 9 with healed
ulcers. In those with ulceration the biopsy was taken from mucosa
as close as possible to the ulcer.

All specimens obtained were fixed in formalin, embedded in paraffin wax and sectioned at 4 nm. Sections were dewaxed and stained
with Alcian Blue (AB) (8GX) at pH 2.6 for acidic mucins and periodic

acid-Schiff (PAS) reagent for neutral mucins. Other sections were
stained with AB alone or PAS alone.

Quantitative estimation of mucins

A Vickers M85/0010 Scanning and Integrated Microdensitometer was
used on Alcian Blue sections and PAS stained sections. Goblet cell
mucins were estimated by a random count method independently of surface epithelium. Surface epithelial cells showing staining with PAS
were also scanned.

Goblet cell counts

35 mm slide photographs were taken from the stained sections at
a uniform magnification and projected at a distance of 5.5 metres. A
grid of 1600 square centimetres was placed over the projected picture
and the total number of cells falling within the area was counted.

Surface epithelial staining in relation to total length

A planimeter was used to find the total length of the surface
epithelium on a given slide. The length of PAS stained epithelium
was expressed as a ratio of the total length measured.

RESULTS

(1) Goblet Cell I.D. readings

	Mean PAS/550 nm	Mean AB nm/610 nm
Non Ulcer	45.0 ± 2.3 SEM	38.9 ± 2.5 SEM
Ulcer	44.2 ± 5.4 SEM	29.3 ± 2.4 SEM
Healed	45.0 ± 4.8 SEM	35.6 ± 2.5 SEM

(2) Goblet Cell Count

	Mean Goblet Cell Count PAS	Mean Goblet Cell Count AB
Non Ulcer	32.3 ± 3.6 SEM	27.3 ± 2.0 SEM
Ulcer	10.9 ± 2.5 SEM	14.0 ± 4.8 SEM
Healed	21.0 ± 4.7 SEM	17.1 ± 2.4 SEM

(3) Surface Epithelium I.D. Readings

	Mean PAS/550 nm	
Non Ulcer	50.1 ± 8.0 SEM	4 out of 10 cases (40%)
Ulcer	58.0 ± 6.6 SEM	7 out of 8 cases (87.5%)
Healed	63.6 ± 5.4 SEM	4 out of 8 cases (50%)

(4) Length of PAS stained surface epithelium related to total surface length

	Mean % of surface length	
Non Ulcer	0.058 ± 0.04 SEM	4 out of 10 cases (40%)
Ulcer	0.445 ± 0.11 SEM	7 out of 8 cases (87.5%)
Healed	0.325 ± 0.18 SEM	4 out of 8 cases (50%)

DISCUSSION

A Vickers Scanning and Integrating Microdensitometer was used to estimate quantitatively surface epithelial mucins in human duodenum and to study changes due to ulceration. The three groups investigated show that major changes occur in the quantity and distribution of neutral mucins during ulceration. The goblet cell I.D. readings for PAS show no significant difference between the three groups but PAS mucus appears in the surface epithelium with the progression of duodenal ulceration. In cases in which this PAS mucus appears goblet cells have almost disappeared and a greater part of the surface epithelium shows marked PAS staining. These changes are progressive during ulceration and return towards the normal in healing. During healing there is an increase in goblet cell counts coinciding with the disappearance of surface PAS staining. Although the AB goblet cell counts fluctuate there is no evidence of AB staining in the surface epithelium. The villi also become stunted in ulceration and return to normal length in healing. The histological appearance suggests that the metaplasia on the villi starts from the crypts and spreads towards the surface, so that in advanced ulceration PAS staining may be seen in almost the whole of the surface epithelium.

The three groups chosen show an overlap without any sharp boundaries, the changes being most marked in the ulcer group. The most significant change is in the increase in surface PAS staining mucus. The changes in the duodenal villi during ulceration suggest a metaplasia towards fundic mucosa which may have a protective function.

We are grateful to Charing Cross Hospital, Chelsea College and Professor P. Gahan at Queen Elizabeth College for the use of their Vickers M86 and Zeiss photomicroscopes.

P.W. was supported by a grant from Smith, Kline & French.

PROTECTION AGAINST NEMATODES BY INTESTINAL MUCUS

H.R.P. Miller and J.F. Huntley

Department of Pathology,
Animal Diseases Research Association, Moredun Institute
408 Gilmerton Road, Edinburgh EH17 7JH

INTRODUCTION

Several lines of evidence suggest that intestinal mucus is in-volved in protection against nematode infections (Miller and Nawa, 1979). For example, Trichinella spiralis larvae become entrapped in mucus both in vivo in immune rats and in vitro (Lee and Ogilvie, 1980). Similarly, the immune expulsion of Nippostrongylus brasiliensis from rat intestine is accompanied by goblet cell hyperplasia and increased turnover of mucin (Miller et al., 1981). The experiments described here provide further information on the role of mucus in protection against N. brasiliensis.

RESULTS

Wistar rats immunized by infection with 5,000 Nippostrongylus larvae 18 days previously, together with nonimmune controls were challenged intraduodenally with 250 immature adult worms and the distribution of the parasites in the small intestine was examined 2 and 4 hours later. More than 50% of the challenge infection in immune rats was in the distal half of the intestine at 2 hours and at 4 hours 92% of the parasites had been expelled (Table I). In controls, the worms established in the proximal half of the intestine and remained there throughout the course of the experiment (Table I).

Previous studies of this expulsion phenomenon had indicated that the parasites, unable to penetrate between the villi, became enveloped in mucus in the intestinal lumen, whereas in nonimmune rats they rapidly established between the villi (Miller et al., 1981). In order to quantitate these events, rats challenged intraduodenally with 450 worms were killed 1 and 2 hours later and

243

Table I.   Rapid Expulsion of N. Brasiliensis

Time after Challenge	Group	Worm Burden ± SE Proximal Intestine	Distal Intestine	% Worm Expulsion
2 h	Control	172 ± 13	1 ± 0.6	–
	Immune	60 ± 21	96 ± 10	10
4 h	Control	180 ± 10	1 ± 1	–
	Immune	2 ± 1	13 ± 5	92

the small intestines removed.   Each unopened intestine was flushed
with 20 mls of saline followed by 20 mls of air.   The intestinal
washings were tipped into gauze suspended in saline at 37°C.   Less
than 40% of the parasites in immune intestines remained adherent.
to the mucosa, 50-56% were free in the gut lumen and able to migrate
from the gauze, and 6-20%, trapped in viscous globules of mucus, were
retained in the gauze (Table II).   In controls, 78-79% of the
worms were attached to the mucosa and 21-22% were free in the lumen,
none was mucus-trapped (Table II).   These results indicate that,
in immune intestine, the worms are excluded from the mucosa and
some of them become trapped in mucus.

Since it seemed likely that the mucus layer was involved in the
exclusion of the parasites from the mucosa, the effect on worm expulsion of agents designed to breach the integrity of this layer,
was examined.   Immune rats were injected intraduodenally either
with 1 ml of mustard powder (20% w/v) in olive oil (Lee and Ogilvie,
1980) or with 2 mls of 1M cysteine containing 10 mgs papain (cyst/
papain - Table III) and were challenged intraduodenally 1.5 h later
with 350-400 parasites.   Immune rats given saline intraduodenally
1.5 h previously and nonimmune rats were also infected as controls
Both agents were effective in blocking worm expulsion (Table III),

Table II.   Immune exclusion and mucus-trapping of N. brasiliensis

Time after Challenge	Group	Worm Burden ± SE	% Attached To Mucosa ± SE	% Free In Lumen ± SE	% Mucus-Trapped ± SE
1 h	Control	356 ± 19	78 ± 3[a]	22 ± 3[a]	0 [a]
	Immune	355 ± 26	38 ± 5[c]	56 ± 4[c]	6 ± 2[b]
2 h	Control	361 ± 18	79 ± 3[a]	21 ± 3[a]	0 [a]
	Immune	294 ± 36	29 ± 5[c]	50 ± 4[c]	21 ± 4[c]

Student's t test   a vs b p < 0.05   a vs c p < 0.001

Table III.   Effect of Mucolytics on Parasite Expulsion

Group	Treatment	Worm Burden $\pm$ SE	% Expulsion	Mucosal Damage
Control	–	394 ± 30	–	–
Immune	Saline	31 ± 9^c	92	–
Immune	Mustard Oil	352 ± 28^a	8	–
Control	–	300 ± 9	–	–
Immune	Saline	47 ± 33^b	84	–
Immune	Cyst/papain	224 ± 11^a	25	+ +

Student's t test   a vs b p < 0.01   a vs c   p < 0.001

mustard oil did so without histological evidence of mucosal damage, whereas cysteine/papain caused epithelial shedding.

DISCUSSION

During the expulsion of N. brasiliensis from immune intestine the majority of the parasites were unable to attach to the mucosa and a significant proportion became trapped in mucus.   Pretreatment of immune intestine with mustard oil which strips off the mucus layer (Lee and Ogilvie, 1980), or with cysteine/papain which is mucolytic, permitted the establishment of the parasites.   These findings, together with evidence of increased turnover of intestinal mucin (Miller et al., 1981) and of immunologically-mediated intestinal goblet cell hyperplasia during primary N. brasiliensis infections (Miller and Nawa, 1979) lend further support to the concept that mucus serves an important function in protection against intestinal nematodes.

REFERENCES

Miller, H.R.P., and Nawa, Y., 1979, Immune regulation of goblet cell differentiation, specific induction of nonspecific protection against helminths?   Nouv. Rev. Fr. Hematol., 21:31
Lee, G.B., and Ogilvie, B.M., 1980, The mucus layer in intestinal nematode infections, in: 'Mucosal immune system in health and disease, report of 81st Conference on pediatric research' Ogra, P.L. and Bienenstock, J., Ed., (in press).
Miller, H.R.P., Huntley, J.F., and Dawson, A.McL., 1981, Mucus secretion in the gut, its relationship to the immune response in Nippostrongylus infected rats, in: 'Current Topic in Veterinary Medicine and Animal Science', Bourne, F.J., ed., Martinus and Nijhoff, London and The Hague (in press).

THE INTESTINAL MUCUS BARRIER TO PARASITES AND BACTERIA

Drs. G.B. Lee and B.M. Ogilvie

Depts. of Immunology and Gastroenterology, Clinical
Research Centre, Harrow, Middlesex and Dept. of
Parasitology, National Institute for Medical Research
Mill Hill, London NW7

Introduction

The intestinal mucus layer has been recognised as a layer in
which both antibodies are contained[1] and microorganisms are
trapped[2,3,4]. We have examined the interrelationship between
these three (i.e. intestinal mucus layer, antibodies and micro-
organisms) using nematode parasite infections of the rat and the
resident bacteria of the intestines of man.

If 3000 infective larvae of the nematode Trichinella spiralis
are administered orally or intra-duodenally to rats not previously
exposed to the parasite, at least 50% penetrate through the mucus
layer to embed in the intestinal epithelium. However using
techniques designed to remove the mucus layer, less than 25% of
administered larvae can penetrate through the layer in rats
previously exposed to T.spiralis. The majority of the remainder
get trapped in the mucus to be carried away when its superficial
layers (containing the larvae) get rolled up and propelled
distally by peristalsis.

This property of the mucus layer of immune animals can be
transferred to some extent to naive animals by administering
immune serum intraperitoneally concurrently with the larval
infection.

Furthermore an in vitro test using intestinal mucus from
rats, immune or control rat serum or bile (a rich source of IgA)
and infective T.spiralis larvae was developed. The results show
that in the presence of uninactivated control serum or immune
serum, the mucus can entrap about 40% of the infecive larvae, a

247

figure that falls below 5% when the control serum is inactivated
or when bile (with or without protein enrichment) is used. This
would suggest that in the T.spiralis model anyway, the complement-
conglutinin or non-IgA antibodies are involved in the mucus trap-
ping.

    In man it has previously been demonstrated that resident
bacteria cultured from homogenates of the small intestine are
contained in the mucus layer and are not directly in contact with
the epithelium[5]. These findings are reinforced by our findings
using frozen sections taken from both small and large intestine.
Immunofluorescence performed on such sections would suggest that
this containment of bacteria to the mucus layer is antibody dependent.

Conclusion

    This work, the studies referred to above and other studies
e.g. that of Miller[6], all suggest that the intestinal mucus layer
is a fundamental and integral part of the immune system of the gut.
Furthermore, we believe that in combination with antibodies, it is
capable of manifesting a highly specific form of immunity.

REFERENCES

1. W.A. Walker, K.J. Isselbacher and K.J. Bloch. Immunologic
    control of soluble protein absorption from the small intestine
    - a gut-surface phenomenon. Am. J. Clin. Nutr. 27, 1434 (1974)

2. H.W. Florey. Mucin and the protection of the body. Proc. Roy.
    Soc. London. 143 (B) 147 (1955)

3. D.C. Savage, R. Dubos and R. Schaedler. The gastrointestinal
    epithelium and its autochthonous bacterial flora. J. Exp.
    Med. 127, 67 (1968).

4. G.D. Schrank and W.F. Verwey. Distribution of cholera organisms
    in experimental Vibrio cholerae infections. Inf. & Imm.
    13, 195 (1976).

5. A.G. Plaut et al. Studied of intestinal microflora. III.
    The microbial flora of human small intestinal mucosa and
    fluids. Gastroenterology 53, 868 (1967).

6. H.R.P. Miller and Y. Nawa. Immune regulation of intestinal
    goblet cell differentiation. Nouv. Rev. Fr. Hematol, 21, 31
    (1979).

CERVICAL MUCUS

STRUCTURE AND FUNCTION OF CERVICAL MUCUS

Eric Chantler

Department of Obstetrics & Gynaecology
University of Manchester
Manchester M20 8LR

The cervical mucus is a complex secretion produced within the cervical canal and crypts and containing compounds from different areas of the female genital tract. The major site of production of the mucus is the secretory epithelium of the cervix which is rich in dense secretory bodies containing the characteristic viscoelastic mucus. However, the mucus obtained by aspiration of the cervix will also contain proteins and low molecular weight compounds from serum transudation, soluble proteins secreted by the uterine epithelium and probably peritoneal fluid from the pouch of Douglas. Additionally the possibility of contamination with seminal plasma must not be overlooked.

The physical properties of mucus show a cyclic variation during the menstrual cycle. Oestrogen dominance, at the time of ovulation, results in a copious secretion of mucus when 0.7 - 1.5 ml may be aspirated from the external os. This clear mucus has a low viscoelasticity and is readily penetrated by sperm. In contrast, only 200 µl or less can be obtained during the luteal phase when progesterone is dominant. This mucus is characteristically viscid and cellular, has a high viscoelasticity and resists penetration by sperm.

Sperm can only penetrate the cervical mucus to any extent during the periovulatory period when the effect of oestrogen is maximal. At this time sperm rapidly traverse the cervix showing different swimming characteristics compared to those shown in free solution (Katz, Mills and Pritchett, 1978). Under conditions of gestagen dominance sperm cannot penetrate the mucus. This regulatory role of the mucus on sperm colonisation of the upper genital tract is most significant in species in which vaginal insemination occurs; when a

low volume of ejaculate with a high sperm count is usual. Under
these circumstances the cervix and its mucus secretion exert a positive regulatory effect on sperm penetration. The hormonally dependent
changes in the rheological properties may be manipulated by the
administration of exogenous steroids. A rapid response is observed
in response to single doses of oestrogens or gestagens (Kesseru 1973)
showing the dynamic nature of mucus biosynthesis. Observations of
the rate of ^{3}H fucose fixation by the human cervix in tissue culture
has shown that glycosylation (and presumably synthesis of the protein
core) occurs very rapidly. Autoradiography using ^{3}H fucose shows the
appearance of label in the cellular secretory bodies after 90 minutes
and its appearance at the cell surface,incorporated into glycoprotein
after 2-3 hours (Chantler & Harris, unpublished observation).

PHYSICAL PROPERTIES

     Cervical mucus, in common with other mucus secretions, is a visco-
elastic gel and consequently exhibits non-Newtonian viscosity. It is
important to define carefully the parameters used in describing changes
in the physical properties of cervical mucus as these can be ascribed
to either of two variables, i.e. those properties relateable to the
recoverable elastic response of the material, the dynamic storage
modulus (G') or those related to the unrecoverable viscous loss, the
dynamic loss modulus (G'')

     The rheological properties of the mucus gel are solely a property
of the mucus glycoprotein (Gibbons 1969) and variation in the protein
concentration or low molecular weight components of the mucus probably
do not affect the viscoelastic properties in vivo.

     There are two models for the macromolecular arrangement of the
mucus glycoprotein. Odeblad (1973) suggests, on the basis of nuclear
magnetic resonance (NMR) studies, that the high viscosity component
of mucus exists as a miscelle with bound water in association. Within
the miscelle is a low viscosity phase containing freely diffusible
water. The high viscosity phase is cross-linked and three components
of this system are visualised as changing in response to the steroid
hormonal stimulation of the cervix e.g. under gestagen dominance.

   1) the degree of cross linking of the high viscosity phase
      increases.

   2) there is an exchange of water between the free and bound
      states resulting in an increase of about 33% in bound water.
      However, the depth of the bound water surrounding the
      glycoprotein strands decreases from 0.4 μm to 0.03 μm
      (Odeblad 1973).

   3) there is a shift from predominantly hydrogen bonds to "dipole"
      bonds.

Sperm migration occurs through the low viscosity phase of the miscelle during oestrogen dominance and the decrease in the volume of this domain when gestagens are active results in a physical barrier to sperm penetration.

In contrast to this model Blandau and Lee (1977) using light scattering measurements suggest that the high viscosity phase of cervical mucus is a random coiled structure without cross linking. In this model sperm penetration need not occur through channels (as in Odebland's theory) but in a random fashion through a loosely entwined network. The orientation of sperm penetration is thus a consequence of stress induced orientation of the mucus imposed during the standard form of sperm penetration test (Kremer 1965)   not progressive movement through channels in a miscelle. The cyclic changes in rheology may thus arise from changes in the fluid content of the glycoprotein miscelle rather than variation in its cross linking. These predictions from the random coil model are in agreement with several others experimental findings:

1.   The water content of cervical mucus decreases by 10% during the luteal phase as a result of gestagen dominance (Kopito et al 1973).

2.   Mucus spontaneously goes into solution indicating the absence of covalent bonds (i.e. cross linking) in the maintenance of the gel structure (Mayer & Silberberg 1980).

3.   Adjustment of the degree of hydration of mucus can give the range of viscoelasticity seen during the menstrual cycle (Wolf et al 1977a,b,c).

The penetration of mucus produced by the cervix under oestrogen dominance consequently appears to be dependent simply upon their entry of the enclosed low viscosity phase. The action of the pioneering spermatozoa is thought to change the local environment of the mucus glycoprotein strands facilitating the entry of other sperm. The active digestion of the mucus by sperm acrosin to provide channels through which migration can occur is probably not significant as sperm penetration is insensitive to the presence of acrosin inhibitors (Bayler & Zaneveld 1979). When sperm have moved from the seminal plasma into the mucus gel they show a different flagellar beat pattern and frequency indicating an interraction with the high viscosity phase of the mucus (Katz, Mills & Pritchett 1978).

CHEMICAL COMPOSITION

Cervical mucus glycoprotein is highly polydisperse and has a molecular weight reported in the range of $5 \times 10^5 - 4 \times 10^6$ with

70-80% consisting of oligosaccharide side chains containing 7-10
monosaccharide residues.

Amino Acid composition

The protein core in a human mucus glycoprotein sub-unit obtained
by reduction or proteolysis and having a molecular weight of about
5×10^5 contains about 1000 amino acid residues. A discreet region
exist in the polypeptide chain which is not glycosylated (the "naked
peptide") which consists of about 200 amino acid residues and contains
about 10% cysteine; this area is susceptible to trypsin digestion.
In contrast, the core region containing about 800 amino acid residues,
is highly glycosylated and has about 200 oligosaccharide side chains
containing about 10 sugar residues which are O-glycosidically linked
to serine or threonine (Mayer et al 1980).

Only a single NH_2 terminal threonine residue is present in the
reduced glycoprotein of the human (Wolf et al 1980) which is in
contrast to the cow which has an NH_2 terminal alanine (Bhushano Rao
& Masson 1977) and Maccaca radiata which has arginine (Hatcher et
al 1977). Examination of the number of NH_2 terminal amino acids
suggests the existence of a very small sub-unit; 37,000 in the human
(Wolf, Sokoloski and Litt 1980) and 27,000 in the cow (Bhushano Rao
& Masson 1977). Further supportive evidence for this putative
sub-unit in the cow is indicated by the presence of a repeating unit
of 62 amino acids. However, the use of detergents, reduction or
limited proteolysis only produces fragments of 5×10^5 in the
analytical ultracentrifuge and the small sub-unit has never been
demonstrated in free solution.

In the human, the amino acid composition of the mucus glycoprotein core does not appear to change in response to the different
hormonal conditions of the menstrual cycle (Wolf, Sokoloski and
Litt 1980) and the same is probably true in the cow, as bovine
oestrus mucus is homogenous (Bhushano Rao and Masson 1977). There
is some disagreement in the case of Maccaca radiata which has been
reported to have a constant amino acid sequence during the menstrual
cycle (Hatcher et al 1977) or to show several glycoproteins with
different NH_2 terminal amino acids (Nasir-Ud-Din et al 1977c).

Carbohydrate composition

The oligosaccharide chains present in purified human cervical
mucus glycoprotein are 9-10 monosaccharide units in length (Hatcher
et al 1977) and contain five residues: L fucose, galactose, N acetyl
glucosamine, N acetyl galactosamine and N acetyl neuraminic acid.
N acetyl galactosamine is at the reducing end of the oligosaccharide
chain and is O-glycosidically linked to either serine or threonine
in the polypeptide core. The molar ratio of the monosaccharide
units is shown in Table 1. Good agreement exists for the content

TABLE 1

Molar ratio of the constituent monosaccharide residues of the oligosaccharide chains of human cervical mucus glycoprotein. The ratios have been adjusted to unit concentration of galactose for comparison

Reference		Fuc	Gal	GlcNAc	GalNAc	NeuAc
Van Kooij et al (1980)	Blood Group A	0.4	1	0.5	0.45	0.25
	Blood Group B	0.6	1	0.5	0.45	0.35
	Blood Group O	0.45	1	0.5	0.45	0.25
Wolf et al (1980)		0.54	1	0.62	0.46	0.39
Yurewicz & Moghissi (1979 Private communication)		0.47	1	0.65	0.59	0.35

of the monosaccharide and their ratio appears to be independent of
both the ABO blood group of the woman (Van Kooij et al 1980) and
the time during their menstrual cycle (Van Kooij et al 1980;
Wolf, Sokoloski and Litt 1980).

The sequence of the oligosaccharide chains in the human mucus
glycoprotein still remains to be elucidated though alkaline
borohydride reduction of bovine (Yurewicz and Moghissi 1980) and
human (Yurewicz and Moghissi, unpublished data) yields both
neutral and acidic fragments,consequently it is possible that more
than one type of chain sequence exists. Several fragments have
been sequenced from Maccaca radiata cervical mucus glycoprotein
which indicate a variation in the carbohydrate structure dependent
upon the time in the menstrual cycle. In the periovulatory mucus
galNAc 1→6(fucα1→2)gal and gal 1→3(fucα-1→2)gal have been isolated
with a core sequence of glcNAc1→3(NeuNAcα2→6)galNAc-Oser/thr.
In contrast luteal phase mucus yields a peripheral fragment NeuNAc
α2→3(1→6/2fuc, gal,galNAc)gal with a core sequence glcNAc1→6(1→3fuc,
gal,galNAc) galNAc- Oser/thr (Nasir-Ud-Din et al 1977,b). It is
possible, however, that the mucus from Maccaca radiata is radically
different to human mucus as, unlike the human,the luteal phase mucus
from Maccaca radiata is immunologically distinct from that secreted
during the periovulatory phase (Nasir-Ud-Din et al 1979a) indicating a
completely different core structure depending on the ovarian hormonal
environment.

INFLUENCE OF STRUCTURE ON PHYSICAL PROPERTIES

The possible relationship between cyclical variation in the
mucus composition in the menstrual cycle and its rheological properties has been reviewed previously (Chantler & Debruyne 1977).
Briefly, three theories have been advanced to explain the molecular
changes causing the normal variation in the sperm penetration associated with mucus rheology. Firstly, based on the observed cyclic
variation in the soluble protein content of the whole mucus, variable
interaction between these proteins and the mucus glycoprotein has been
suggested (Gibbons 1969). A strong association between the mucus
glycoprotein and several of the soluble proteins present in the
cervical mucus does exist e.g. albumin, IgA and most rigorous
dissassociating conditions e.g. 0.22M sodium thiocyanate are required
for their removal (Scudder 1980). In addition it is known that the
rheological properties of the mucin from pig stomach are affected
by varying the albumin concentration (List et al 1978), therefore
this type of interaction probably occurs in vivo. However, the
artificial addition of serum proteins to ovulatory mucus (as occurs
naturally during the luteal phase of the menstrual cycle, Elstein
1970) does not reduce the sperm penetration (Masson 1973).

Another indication of the involvement of the protein fraction
of the mucus in its physical properties is the effect of limited

pepsin digestion on bovine pregnancy mucus. This treatment which
does not affect the carbohydrate moieties, converts the firm
impenetrable plug to a thin gel which is visibly similar to oestrus
mucus and allows sperm penetration (Gibbons & Sellwood 1973).

Consequently, whilst there is indirect evidence for the involvement of the protein fraction of the mucus in its ability to show a
range of viscoelastic properties, no direct evidence for this type
of interaction in vivo exists.

Secondly, changes in the charge present as a result of sulphate
and carboxyl groups on the oligosaccharide chains of the glycoprotein
may affect mucus rheology. There have been a number of reports of a
variation in the sialic acid content of the mucus during the menstrual
cycle but there is disagreement as to the time of the cycle at which
the levels are maximal (see Table II). The most probable reason for
these conflicting results is the varying degree to which the mucus
glycoprotein has been purified and none of the determinations reported
in Table II have been performed on a mucus glycoprotein which has
been prepared under conditions which would remove the strongly
bound serum proteins/glycoproteins. More recent determinations of
the carbohydrate content of mucus glycoprotein (Wolf, Sokoloski, Litt
1980; Van Kooij et al 1980) find no variation in the sialic acid
content during the menstrual cycle. There is some evidence however
(Wolf et al 1980) which does implicate the ratio of the two alternative
terminal monosaccharides L fucose or sialic acid in the viscoelasticity. Whilst no correlation between sialic acid level and the
storage modulus G' was found there was a strong correlation with the
fucose:sialic acid ratio. This observation requires further investigation to relate this fact to the apparent independence of mucus
rheology to the sialic acid level alone.

Further evidence for the non-involvement of sialic acid in the
mucus rheology is indicated by the observation that the rheological
properties of bovine oestrus mucus are not affected by the almost
complete removal of sialic acid by neuraminidase treatment (Mayer
et al 1975). Indeed, the fact that the viscoelasticity of mucus is
not affected by variation in the ionic strength or pH of the medium
indicates that there is no electrostatic interaction between any
charged groups within the mucus gel which influence the rheology
(Litt, Khan, Shih and Wolf 1977).

Finally, variation in the degree of hydration of the mucus
glycoprotein has been shown to have a considerable effect on its
viscoelasticity. In a detailed study of the relationship between
cervical mucus rheology and sperm penetration a direct linear
relationship was established between the rate of sperm penetration
and the storage modulus (G') of the mucus (Wolf et al 1977a,b,c).
Furthermore, it was shown that mucus taken at any time during the
menstrual cycle could be manipulated by lyophillisation and

TABLE II

Sialic acid content of cervical mucus glycoprotein during the normal human menstrual cycle

Species	Preovulatory Phase	Ovulatory Phase	Postovulatory Phase	Reference
Human	-	20-25mg/g dry whole mucus	50mg/g dry whole mucus	Carlborg et al (1968)
Human	1.5mg/ml whole mucus	0.46mg/ml whole mucus	1.65mg/ml whole mucus	Moghissi et al (1976)
Human	30mg/g mucus glycoprotein	-	46mg/g mucus glycoprotein	Iacobelli et al (1971)
Human	-	90mg/g total carbohydrate	260mg/g total carbohydrate	Bhushano Rao et al (1976)
Human	4mg/g dry whole mucus	12mg/g dry whole mucus	2mg/g dry whole mucus	Sobrero et al (1973)
Maccaca radiata	60mg/g mucus glycoprotein	80-90mg/g mucus glycoprotein	50mg/g mucus glycoprotein	Hatcher et al (1977)
Human	46mg/g mucus glycoprotein	102mg/g mucus glycoprotein	43mg/g mucus glycoprotein	Chantler et al (1977)
Human	29mg/g dry washed mucus	-	12mg/g dry washed mucus	Daunter (1978)

rehydration to give an identical G' value, showing that it is the
concentration of mucus glycoprotein which affects the viscoelasticity.
In vivo the production of mucus glycoprotein does not vary significantly during the menstrual cycle consequently to achieve the observed
concentration dependent variation in G' the degree of hydration
must vary. This variation results in a six-fold decrease in the
mucus non-dialysable solid concentration during the ovulatory
phase of the menstrual cycle compared to the luteal phase resulting
in the lowest levels of storage modulus and enhanced sperm
penetration (Wolf et al 1977c).

     The variation in the water content of cervical mucus during
the menstrual cycle is well recognised and these data indicate
that this may be a key factor in the modulation of the viscoelasticity. However, the question still remains what is the mechanism
for altering the hydration of the mucus gel?

CONTROL OF GLYCOSYLATION

     The oligosaccharide chains of cervical mucus have either L-fucose
or sialic acid (normally N acetyl neuraminic acid in the human) as
their terminal residue. These monosaccharides are universal to
epithelial glycoproteins and their levels have been found to be in
inverse proportion in a great many cases (Dische 1963). As discussed
in the previous section, this relationship was suggested to be an
hormonally-dependent one in the cervical mucus and the different
chemical properties of the two monosaccharides being the basis of
the observed hormonally dependent variation in its physical properties. Whilst this simple relationship between structure and
function is now seen to be incompatible with the more critical
structural and physical analysis reviewed here there still remains
the question of why there are these alternative termination
sequences at the non-reducing end of the oligosaccharide chains
and what their significance is.

     The purification to homogeneity of a number of the glycosyl
transferases responsible for glycoprotein biosynthesis from a number
of mammalian tissues has revealed one reason for the alternative
termination sequences (Bayer et al 1979). It was shown that whilst
oligosaccharide chains containing both fucose and sialic acid
could be achieved in vitro this was not a usual biosynthetic route.
In nine cases where a sialyated acceptor was tested against a
fucosyl transferase or a fucosylated acceptor was tested against a
sialyl transferase only the fucα1-2gal β 1-3(NeuAcα2-6)galNAc
sequence could be synthesised which contains both terminal fucose and
sialic acid. In the other eight cases, six were totally blocked and
two were strongly inhibited. However, as the acceptor specificities
of sialyl and fucosyl transferases frequently overlap the inhibitory
action one residue has on the addition of the second terminator
monosaccharide does not absolutely define which of the two will

appear at the non-reducing end of the oligosaccharide chain. In addition, sequences have been found in the cervical mucus glyco-protein from Maccaca radiata (Nasir-Ud-Din et al 1979a,b) which contain both fucose and sialic acid on the same oligosaccharide fragment e.g. NeuNAcα(2-3)(1-2Fuc)gal. This structure could not be synthesised using porcine sub-maxillary gland β galactoside α2 3 sialyl transferase and β galactoside α1-2 fucosyl transferase (Bayer et al 1979a,b, Sadler et al 1979) indicating a different specificity for these enzymes when present in the cervical epithelium. In the human cervical epithelium several fucosyl transferases and one sialyl transferase have been described (Scudder & Chantler 1981a,b). There are two β galactoside α1-2 fucosyl transferases and evidence exists for a β N acetyl glucosamine α1-3 fucosyl transferase and an N acetyl galactosamine fucosyl transferase. The presence of both a neutral and an acid pH optimum for the β galactoside α1-2 fucosyl transferase activity considered in conjunction with the decreased sensitivity to inhibition by p chloromercuribenzoate at neutral pH and the inability to utilise asialofetuin as an acceptor at that pH indicates the presence of two separate forms of β galactoside α1-2 fucosyl transfer-ase.

An interesting feature of both human cervical β galactoside α1-2 fucosyl transferase and sialyl transferase is their sensitivity to GDP and CMP respectively. In both cases the nucleotides act as competitive inhibitors; GDP having a Ki = 50μM and CMP a Ki = 180μM. This property may be a further control mechanism in the choice of either of the terminal monosaccharides. Additionally, the transfer-ases responsible for these glycosylation steps are sensitive to the free nucleotide triphosphates. Thus the fucosyl transferase reaction is inhibited by 20μM GTP but lower levels have a stimulatory effect. Similarly CTP is a powerful inhibitor of sialyl transferase in the cervix but unlike the dual action shown by GTP, it inhibits the sialyl transferase reaction at all concentrations.

In summary, there are three mechanisms which could affect the ratio of sialic acid to fucose at the non-reducing end of the cervical mucus glycoprotein.

1.   In many instances the addition of one of these residues prevents the later addition of the alternative one by the formation of an unfavourable substrate.

2.   The continued addition of fucose or sialic acid within golgi secretory vesicles may be prevented by feed-back inhibition from the nucleotide products of the transferase reaction. Thus if there is an overlap between the substrate specificity for the two terminal monosaccharide transferase reactions, this feedback inhibition may allow both transferases to be expressed in the following way. As the more favoured transferase reaction becomes progressively inhibited by

feed back inhibition by the nucleotide product,the less active transferase becomes more active in adding the terminal residue.

3. The sensitivity of both terminal transferases to the level of their associated nucleotide triphosphates indicates the possibility of modulation of the termination step by the nucleotide potential, however, this nucleotide ratio is known to be relatively constant in at least one secretory tissue (the submandibular gland, Phelps & Stevens 1978). There may be some potential for control by this method however as fucosyl transferase can be stimulated if concentration of GTP falls to low levels, an ability not shared by sialyl transferase which is always inhibited by CTP.

REFERENCES

Beyer, T.A., Prieels, J.P., and Hill, R.L. In eds. Gregory, J.D., and Jeanloz, R.W. Glycoconjugate Research : Proceedings of the Fourth International Symposium on Glycoconjugates Vol II Academic Press, New York pp 641-643.

Beyer, T.A., Rearick, J.I., Paulson, J.C., Prieels, J.P., Sadler, J.E. and Hill, R.L. (1979) J. Biol. Chem 254: 12531-12541.

Beyler, S.A., and Zaneveld, L.J. (1979) Fertil Steril 32: 671-5.

Bhushano Rao, K.S.P., Barbier, B., Masson, P.L., Heremans, J.F. and Ferin, J. (1976) In ed. Hafez, E.S.E. Human semen and fertility regulation in men, pp 237-242. Mosby. St Louis, Missouri.

Bhushano Rao, K.S.P., Masson, P.L. (1977) J. Biol Chem 252: 7788-7795.

Carlborg, L., McCormick, W., and Gemzell, C (1968) Acta Endocrinol (Kbh) 59: 636-643.

Chantler, E.N., and Debruyne E. (1977) In eds. Insler, V., and Bettendorf, G. The Uterine cervix in Reproduction Georg Thieme, Stuttgart pp 77-82.

Daunter, B. (1978) Contraception 17: 27-35.

Dische, Z. (1963) Ann. N.Y. Acad. Sci 106: 259-270.

Elstein, M. (1970) J. Obstet Gynaecol. Brit. Cmwlth 77: 443-456.

Gibbons, R.A. (1969) In: Peeters H, ed. Protides of the biological fluids: Proceedings of the Sixteenth Colloquim, Bruges, 1968 pp 299-305. Permagon Press, Oxford.

Gibbons, R.A. & Sellwood, R. (1973) In: Blandau, R.J. and Moghissi, K.S. The Biology of the cervix pp 251-265, The University of Chicago Press, Chicago, Ill.

Hatcher, V.B., Schwarzmann, G.O.H., Jeanloz, R.W., McArthur, J.W. (1977) Fertil Steril 28: 682-688.

Iacobelli, S., Garcea, N., and Angeloni, C. (1971) Fertil Steril, 22: 727-734.

Katz, D.F., Mills, R.N., Pritchett, T.R. (1978) J. Reprod. Fert.
     53: 259:265.
Kesseru, E. (1973) In Eds. Elstein, M., Moghissi, K.S., and Borth,
     R. Cervical Mucus in Human Reproduction Scriptor, Copenhagen
pp 45-57.
Kopito, L.E., Kosasky, H.J., Sturgis, S.H., Lieberman, B.L. and
     Shwachman, H. (1973) Fertil Steril 24: 499:511.
Kremer, J. (1965) Int J Fertil 10: 209-215.
Lee, W.L., Verdugo, P., Blandau, R.J., Gaddum-Rosse, P. (1977)
     Gynecol. Invest. 8: 254-266.
List, S.J., Findlay, B.P., Forstner, G.G., Forstner, J.F. (1978)
     Biochem J 175: 565-571.
Litt, M., Khan, M.A., Shih, C.K. & Wolf, D.P. (1977) Biorheology
     14: 127-133.
Masson, P.L. (1973) In Eds. Elstein, M., Moghissi, K.S. and Borth, R.
     Cervical Mucus in Reproduction Scriptor, Copenhagen pp 82-92.
Meyer, F.A., King, M.A., Gelman, R.A. (1975) Biochim Biophys Acta
     392: 223-232.
Meyer, F.A. and Silberberg, A. (1980) Biorheology 17: 163-8.
Moghissi, K.S., and Syner, F.N. (1976) Int. J. Fertil 21: 246-250.
Nasir-Ud-Din, Jeanloz, R.W., Nash, T.E., McArthur, J.W. (1979,a)
     in Glycoconjugates Eds. Schauer R, Boer, P., Buddecke, E.,
     Kramer, M.F., Vliegenthart, J.F.G., Wiegandt, H., Georg
     Thieme, Stuttgart, pp 548-9.
Nasir-Ud-Din, Jeanloz, R.W., Reinhold, V.N., and McArthur, J.W.
     (1979,b) Carbohydr. Res. 75: 349-356.
Nasir-Ud-Din, McArthur, J.W. and Jeanloz, R.W. (1979,c) in N.
     Alexander (Ed.) Animal Models for Research on Contraception
     and Fertility, Harper and Row, Hagerstown, Maryland.
Odeblad, E. (1973) In Eds. Elstein, M., Moghissi, K.S., and Borth,
     R. Cervical Mucus in Human Reproduction, Scriptor Copenhagen
     pp 58-74.
Phelps, C.F., and Stevens, A.M. (1978) In: Respiratory Tract Mucus
     (Ciba Foundation Symposium 54, Elsevier; Excerpta Medica;
     North Holland, Amsterdam pp 91.
Sadler, J.E., Rearick, J.I., Paulson, J.C., and Hill, R.L. (1979)
     J. Biol Chem 254: 4434-4443.
Scudder, P.R. (1980) PhD Thesis Manchester University.
Scudder, P.R. and Chantler, E.N. (1981a) Biochim Biophys Acta
     660; 128-135.
Scudder, P.R. and Chantler, E.N. (1981b), Biochim Biophys Acta
     660: 136-141.
Sobrero, A.J., Szlachter-Aisemberg, B.N. Musacchio, I. and
     Epstein, J.A. (1973) In: Blandau, R.J., and Moghissi,
     K.S. The Biology of the Cervix pp 357-366 The University
     of Chicago, Chicago, III.
Van Kooij, R.J., Roelofs, H.J.M., Kathmann, G.A.M., Kramer, M.F.
     (1980) Fertil Steril 34: 226-233.

Wolf, D.P., Blasco, L., Khan, M.A., Litt M (1977a) Fertil Steril
    28: 41-46.
Wolf, D.P., Blasco, L., Khan, M.A., Litt, M (1977b) Fertil Steril
    28: 46-52.
Wolf, D.P., Sokoloski, J., Khan, M.A., Litt, M. (1977c) Fertil Steril
    28: 53-58.
Yurewicz, E.C. and Moghissi, K.S. (1978) Fed Proc. 37: 1440

CONTROL OF HUMAN CERVICAL MUCIN GLYCOSYLATION BY ENDOGENOUS

FUCOSYL AND SIALYLTRANSFERASES

Peter R. Scudder[*] and Eric N. Chantler

Department of Obstetrics & Gynaecology
University Hospital of South Manchester
Manchester, M20 8LR

[*] Division of Communicable Diseases
Clinical Research Centre
Harrow, Middlesex HA1 3UJ

In many epithelial glycoproteins a reciprocal relation
exists between levels of L-fucose and sialic acid which could be
the result of their mutually exclusive incorporation during
oligosaccharide synthesis[1]. Changes in the visco-elastic
properties of human cervical mucus occurring during the
menstrual cycle may be related to alterations in the structure[2]
or composition[3] of cervical mucin involving terminal L-fucosyl
and sialyl residues. In the present study the possible role of
endogenous glycosyltransferases in controlling levels of
L-fucose and N-acetylneuraminic acid in cervical mucin was
investigated by measuring the variation in activity of β-galactoside fucosyl and sialyltransferases in the cervical epithelium
throughout the menstrual cycle.

Using phenyl β-D-galactoside and asialofetuin as
acceptor-substrates parallel measurements of GDP-L-fucose:
β-D-galactoside-α-2-L-fucosyltransferase and CMP-N-acetylneuraminic acid:asialofetuin(NeuNAc → Gal) sialyltransferase
activity were made in 45 individual biopsies of cytologically
normal cervical epithelium taken from women of reproductive age
at hysterectomy. For the purpose of statistical analysis the 45
biopsies were divided into 18 proliferative phase (day 1-12 of
the menstrual cycle), 5 peri-ovulatory phase (day 13-15) and 22
luteal phase (day 16-28). A one-way analysis of variance was
performed on the grouped data.

Optimum conditions for the assay of β-galactoside α-2-L-fucosyltransferase and β-galactoside sialyltransferase in cervical tissue have been described [4,5]. The activity of both enzymes is stimulated by Triton X-100. However, for maximum activity, α-2-L-fucosyltransferase also requires the presence of a divalent metal cation, either manganese or magnesium. The assay products, phenyl-L-[^{14}C]fucosyl-β-D-galactoside and N-acetyl[^{14}C]neuraminic acid:fetuin, were isolated by ion exchange chromatography (Dowex AG-1X2) and high voltage electrophoresis respectively. In neither assay could incorporation of labelled sugar into endogenous acceptor be demonstrated. In each assay the reaction was zero order with respect to the acceptor-substrate and the rate of incorporation of ^{14}C-sugar into the exogenous acceptor was linear with respect to time and enzyme concentration.

Both enzymes were present in the cervical epithelium throughout the menstrual cycle although the specific activity of β-galactoside sialyltransferase was invariably 20-30 times higher than that of α-2-L-fucosyltransferase. This difference does not appear to be a reflection of the reduced affinity of α-2-L-fucosyltransferase for low molecular weight acceptors since Michaelis constants for phenyl β-D-galactoside and asialofetuin are similar (Scudder, unpublished observations). Table 1 illustrates the variation in fucosyl and sialyl-transferase activity during the menstrual cycle; results were

Table 1   β-Galactoside Sialyltransferase and β-Galactoside α-2-L-Fucosyltransferase Activity in the Human Cervical Epithelium.

Phase of Cycle	Sialyltransferase Activity	Fucosyltransferase Activity
	mUnits. g^{-1} tissue[a]	
Proliferative n = 18[b]	0.70 ± 0.25	0.033 ± 0.022
Peri-ovulatory n = 5	0.57 ± 0.21	0.027 ± 0.018
Luteal n = 22	0.68 ± 0.35	0.020 ± 0.018

[a] One unit of enzyme activity will catalyse the incorporation 1 μmole sugar into the appropriate acceptor per min at pH 6.0 at 37°C.

[b] n = the number of individual biopsies in each group.

similar whether enzyme activity was expressed relative to tissue
wet weight, protein or DNA concentration. Fucosyltransferase
activity was significantly increased (p <0.05) in the proliferative phase when compared with the luteal phase of the menstrual
cycle. Mean levels of sialyltransferase activity were essentially the same in the proliferative and luteal phase groups but
were reduced, although not significantly, in the peri-ovulatory
phase group.

Maximal β-galactoside α-2-L-fucosyltransferase activity in
the proliferative phase and minimum activity in the luteal phase
correlates with the observed variation in content of cervical
mucus glycoprotein L-fucose, which decreases progressively
throughout the menstrual cycle, and suggests that changes in the
activity of this enzyme may be important in controlling levels
of cervical mucin L-fucose. Also of significance may be the
elevated levels of endogenous α-L-fucosidase present in late
cycle cervical mucus[3]. The constant high ratio of β-galactoside
sialyltransferase activity to β-galactoside α-2-L-fucosyltrans-
ferase activity indicates that observed cyclic changes in
cervical mucin L-fucose and N-acetylneuraminic acid content
cannot be accounted for by simple competition between these two
glycosyltransferases for a possible common β-galactoside
acceptor residue.

REFERENCES

1. T.A. Beyer, J.I. Rearick, J.C. Paulson, J-P. Prieels,
    J.E. Sadler and R.L. Hill, Biosynthesis of mammalian
    glycoproteins. Glycosylation pathways in the synthesis of the
    nonreducing terminal sequences, J. Biol. Chem. 254:12531 (1979).
2. Nasir-Ud-Din, R.W. Jeanloz, V.N. Reinhold, J.W. McArthur,
    Changes in the glycoprotein structure of the cervical mucus
    of the bonnet monkey during the menstrual cycle. Study on
    the pre-menstrual phase mucus, Carbohydr. Res. 75: 349 (1979).
3. E.N. Chantler and E. Debruyne, Factors regulating the
    changes in cervical mucus in different hormonal states, Adv.
    Exp. Med. Biol. 89: 131 (1976).
4. P.R. Scudder and E.N. Chantler, Glycosyltransferases of the
    human cervical epithelium. I. Characterisation of a
    β-galactoside α-2-L-fucosyltransferase and the identification
    of a β-N-acetylglucosaminide α-3-L-fucosyltransferase,
    Biochim. Biophys. Acta (1981) in press.
5. P.R. Scudder and E.N. Chantler, Glycosyltransferases of the
    human cervical epithelium. II. Characterisation of a CMP-N-
    acetylneuraminic acid:glycoprotein sialyltransferase,
    Biochim. Biophys. Acta (1981) in press.

CYCLIC CHANGES IN GLYCOPROTEIN SYNTHESIS AND SECRETION

BY THE HUMAN ENDOCERVIX

R.J. van Kooij and M.R. Kramer

Department of Histology and Cell Biology
Medical School, State University Utrecht
Nicolaas Beetsstraat 22
3511 HG Utrecht, The Netherlands

INTRODUCTION

The aim of our study was to investigate whether the well-known variations in properties of cervical mucus during the menstrual cycle possibly have their basis in synthetic and secretory behaviour of the mucous glycoprotein (M.G.) producing columnar cells, lining the endocervix.

RESULTS

The secretory activity of the endocervix cannot be estimated directly from the amount of the mucus produced, because of the unknown amount left behind in the crypts during suction of the mucus from the cervical canal, especially in the luteal phase of the cycle. We therefore applied an indirect method determining the concentrations of M.G. and of blood plasma (transudated) proteins in mucus samples serially taken during the cycle of 6 women. The day of ovulation in those series was estimated from the plasma estradiol and progesterone level and the basal body temperature. The concentration of blood derived proteins in the mucus is very low during the ovulatory phase, when much water and ions are secreted. The plasma protein concentration rises rapidly during the luteal phase when the volume of mucus becomes very small. We assumed that the transudation of plasma proteins into the mucus is less variable than the secretion of M.G. and water. Under that assumption the ratio of M.G. to plasmaprotein concentration in the mucus gives an indication of the rate of secretion of M.G. This ratio reaches a peak level in the peri-ovulatory phase of the cycle and decreases rapidly after ovulation (Van Kooij et al., 1980).

269

To obtain information about the synthetic activity of the epithelial cells, we took small biopsies of the endocervical wall, incubated them with ^{3}H-galactose and processed sections for autoradiography. The number of sivergrains per unit area of cytoplasm of cervical columnar cells gives a value for the rate of incorporation of ^{3}H-galactose. That value is used as a measure of the amount of M.G. synthesized by the cells.
We found a clear correlation between changes in galactose incorporation and plasma estradiol level, strongly suggesting an influence of the hormone on the rate of M.G. synthesis by the cells.
The estradiol peak just preceding ovulation and the second, post- ovulatory rise in estradiol level coincide with two peaks of galactose incorporation. There was no evidence for an effect of progesterone on incorporation. Changes in plasma progesterone level are not paralleled by changes in galactose incorporation nor did the mean rate of incorporation change during the postovulatory rise of progesteron.

CONCLUSION

The data strongly suggest a positive effect of the plasma estradiol level on the rate of glycoproteinsynthesis in the colmumnar epithelial cells of the human cervix.
The first peak of galactose incorporation fits remarkably the ovulatory peak in M.G. secretion described before. Apart from its positive effect on M.G.-synthesis estradiol might stimulate exocytosis. The second peak of galactose incorporation is not related to a known peak in glycoprotein secretion. Perhaps the increasing level of progesterone inhibits the assumed stimulatio n of exocytosis by estradiol. Such an antisecretory effect of progesterone might also explain why the large flow of fluid, produced in the ovulatory phase of the cycle, is absent during the second postovulatory peak.
In 2 series of biopsies taken from 2 women with normal cycle, we could not observe substantial differences in ultrastructure of the endocervical epithelium, though some quantitative variations can not be excluded. The rather constant ultrastructural aspects of formation, storage and secretion of the mucous glycoproteins and of the ciliated cells shall be shown.

REFERENCES

Kooij, R.J. van, Roelofs, H.J.M., Kathman. G.A.M. and Kramer. M.F., 1980, Human cervical mucus and its mucous glycoprotein during the menstrual cycle, Fertil. and Steril. 34:226.

CARBOHYDRATE CHAINS OF HUMAN PRE- AND POSTOVULATORY CERVICAL MUCOUS

GLYCOPROTEIN

R.J. van Kooij and M.F. Kramer

Department of Histology and Cell Biology
Medical School, State University Utrecht
Nicolaas Beetsstraat 22
3511 HG Utrecht, The Netherlands

INTRODUCTION

Mucous glycoproteins (M.G.) of cervical mucus are characterized by their great number of O-glycosidically linked carbohydrate chains. For cervical mucus of Macaca radiata differences in these chains have been reported between the various phases of the menstrual cycle, apart from variations in bound neuraminic acid (AcNeu) content (Nasir ud Din et al, 1979).Our study of human cervical M.G. (Van Kooij et al., 1980) and that of Wolf et al. (1980) did not reveal systematical variations with cycle phase, in AcNeu content nor in the relative amount of the other sugars. In the present study we report data on size and nature of the oligosaccharide chains of human cervical M.G.

RESULTS

The oligosaccharide chains were isolated by reductive β-elimination (2-3 mg/ml M.G., 0,05N KOH, 2M KBH_4, 16 hr 45^0C.). M.G. was purified by CsCl density gradient centrifugation at 150.000 g, after solubilization of the cervical mucus by sonication in water. After β-elimination the unreacted material was precipitated with alcohol after neutralization and removal of borate. The supernatant, containing more than 80% of the hexoses of M.G. was chromatographed on P2 and P6 (Biogel) columns after passing over Dowex X8 (H^+). Part of the purified M.G. was treated with (^3H)-borohydride to get radioactive chains, allowing the analyses of smaller aliquots. M.G. from postovulatory mucus was studied by the latter method only. The eluted fractions from preovulatory M.G. gave 2 hexose peaks on P2, one in the void

271

volume of the column and one retarded. Elution of the excluded
material on P6 gave 2 peaks. The larger peak contained more
AcNeu than the smaller one. Removal of AcNeu by acid hydrolysis
before β-elimination resulted in the appearance of one peak on
P6, which is also included on P2. Elution of radioactive
carbohydrate chains, originating from postovulatory and
preovulatory M.G. gave the same P2 and P6 patterns.

CONCLUSION

    M.G. from pre- and postovulatory mucus has at least 3 types
of carbohydrate chains. Two larger types differ in their AcNeu
content, the smaller type has little or no AcNeu.

REFERENCES

Kooij, R.J. van, Roelofs, H.J.M., Kathman, G.A.M. and
        Kramer, M.F., 1980, Human cervical mucus and its mucous
        glycoprotein during the menstrual cycle, Fertil. and
        Steril. 34:226.
Nasir ur Din and Jeanloz, R.W., 1979, Changes in glyco-
        proteinstructure in the bonnet monkey, Carbohydrate
        Res. 75:349.
Wolf, D.P., Sokoloski, J.E. and Litt, M., 1980, Composition and
        function of cervical mucus, Biophys. Biochem. Acta
        630:545.

ISOLATION AND PURIFICATION OF

THE MUCIN COMPONENT OF HUMAN CERVICAL MUCUS

I. Carlstedt, J. Sheehan, U. Ulmsten and L. Wingerup

Department of Physiological Chemistry 2
University of Lund
P.O.Box 750, S-220 07 Lund 7, Sweden

INTRODUCTION

Cervical mucus plays a vital role in the protection of the uterine cavity and in the regulation of sperm survival and penetrability. The mucus is described as a hydrogel consisting of a network of high-molecular weight glycoproteins, the so-called mucins. The purpose of this work was to prepare mucins from human cervical mucus for chemical and physical studies. Great care was exercised to avoid the risk of degradations during handling.

PREPARATION OF MUCINS

Specimens of human cervical mucus plugs formed during pregnancy and released just before delivery were pooled according to the blood-group status of the donor. Diisopropyl fluorophosphate (DFP) was added and the mucus gel was gently agitated for 10 minutes. Ice-cooled 6 M GuHCl[*]/10mM Na-phosphate/5mM EDTA/5mM NEM[*], pH 6.5 was added and the sample was carefully dispersed in a Dounce homogenizer and stirred over-night at 4°C. After high-speed centrifugation, solid CsCl and buffer were added to the supernatant to reach a final density of 1.39 g/ml in 4M GuHCl. Isopycnic density-gradient centrifugation was carried out in an angle rotor at 36k rpm for approximately 65 h at 15°C. Tubes were emptied from the bottom and fractions were analysed for sialic acid, absorbance at 280 nm and density. Fractions containing mucin (density 1.40-1.48 g/ml) were pooled and the density-gradient centrifugation step was repeated. The mucin preparation thus obtained was finally subjected to density-gradient

---

[*]GuHCL: guanidine hydrochloride; NEM: N-ethyl-maleimide

centrifugation in CsCl/0.2M GuHCl (starting density 1.52 g/ml) performed under the same conditions as described above.

RESULTS AND DISCUSSION

        The mucus gel readily dissolved in 6M GuHCl leaving some 5% as
an insoluble residue. DFP and EDTA were added to inhibit the possible
activity of serin- and metallo-proteases, respectively. NEM served
the dual function to inhibit thiol-proteases and to block any free
thiols on the mucins, thus minimizing thiol-disulfide exchange reactions.

        To assess the efficiency of the purification procedure, analyti-
cal isopycnic density-gradient centrifugation was carried out on the
starting material(Fig. 1a) and on the mucin fractions after each purification step. A small amount of residual protein is evident after
the first centrifugation in 4M GuHCl as is a sharp band corresponding
to the position of DNA(Fig.1b). The proteins were removed by the
second centrifugation step(Fig.1c) and DNA by the third one in 0.2M
GuHCl(Fig.1d). Removal of proteins was performed under dissociative

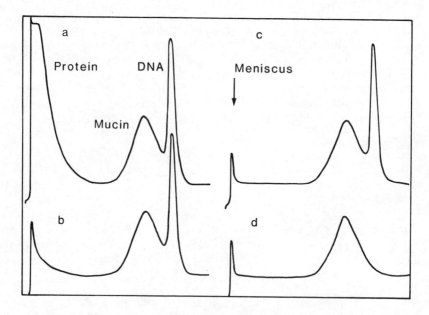

Fig. 1.   Analytical density-gradient equilibrium patterns of (a) the
          starting material and (b-d) of mucins recovered after the
          first, second and third purification steps, respectively.
          The experiment was carried out in an MSE Centriscan 75 ana-
          lytical ultracentrifuge at 55k rpm; 40 h; 25°C on samples
          (about 0.5 mg of mucin/ml) dialysed against 4M GuHCl/CsCl
          ρ_0 = 1.42 g/ml. Absorption optics at 280 nm was used.

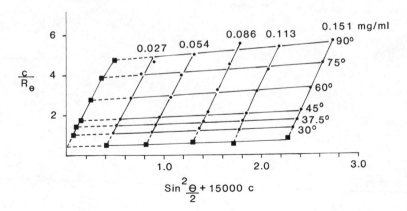

Fig. 2.   Zimm plot of the light-scattering data obtained from mucins
          dissolved in 6M GuHCl. Measurements were performed using a
          Sophica light-scattering photometer. The dn/dc was estimated
          to 0.138 ml/g. The molecular weight and the R_G were extra-
          polated to 6.4×10^6 daltons and 214 nm, respectively.

conditions in order to avoid binding of proteins to the mucins and
to minimize enzymatic modifications.

     The purified mucins chromatographed in the void volume of a
Sepharose 2B column and gave rise to a unimodal schlieren peak in the
analytical ultracentrifuge. Molecular weights were assessed by light-
scattering measurements and a representative Zimm plot is shown in
Fig. 2. The result varied with the solvent conditions. In 6M GuHCl
molecular weights around 6×10^6 were obtained, while measurements
in 0.2M NaCl and 4M GuHCl gave 11×10^6. The corresponding radii of
gyration suggested that an increasing expansion of the macromolecules
took place from 0.2M NaCl over 4M GuHCl to 6M GuHCl.

CONCLUSIONS

     The mucins prepared by this procedure are pure as shown by ana-
lytical density gradient centrifugation. It should be pointed out
that the absorption optics favour proteins and especially DNA since
these molecules have a much higher molar absorptivity than do the
mucins.

     The high molecular weight and apparent homogeniety of the mucins
suggest that extensive degradation has not occurred during handling.
It is noted that dissociation is incomplete in 4M GuHCl and 0.2M NaCl,
a fact that should be considered when molecular weight measurements
are carried out.

SPEED-DEPENDENT SEDIMENTATION-VELOCITY

OF HUMAN CERVICAL MUCINS IN THE ANALYTICAL ULTRACENTRIFUGE

John Sheehan and Ingemar Carlstedt

Department of Physiological Chemistry 2
University of Lund
P.O.Box 750, S-220 07 Lund 7, Sweden

INTRODUCTION

Highly purified human cervical mucins were prepared by isopycnic density-gradient centrifugation (Carlstedt et al., 1981). The macro-molecules elute in the void volume of a Sepharose CL 2B column and five separate preparations gave molecular weights between 4.8 and 6.5 x 10^6 by light-scattering in 6M guanidine hydrochloride (GuHCl). To study the size heterogeneity and polydispersity, sedimentation-velocity experiments were carried out in the analytical ultracentri-fuge. In the course of this work new observations were made on the sedimentation behaviour of large mucin molecules.

EXPERIMENTAL

Sedimentation-velocity experiments were carried out in an MSE Centriscan 75 analytical ultracentrifuge. Approximately 350 ul of sample were used in cells (1 cm path length) fitted with quartz windows. Samples (0.8-1.1 mg/ml) were extensively dialysed to the appropriate starting conditions. The sedimenting boundary was moni-tored by schlieren optics with a knife edge angle of 75°. Experiments were performed at 20^oC in 0.2M NaCl, 4M, 6M and 8M GuHCl.

RESULTS

A typical sedimentation experiment in 4M GuHCl carried out at a rotor speed of 30k rpm is shown in Fig.1a. The material sedimented as a single peak with an apparent sedimentation rate of 13.3 S. Three features of the experimental result arouse suspicion (i) boundary broadening occurred to such a degree that the boundary could not be monitored for more than 80-90 minutes, (ii) a perturbed baseline

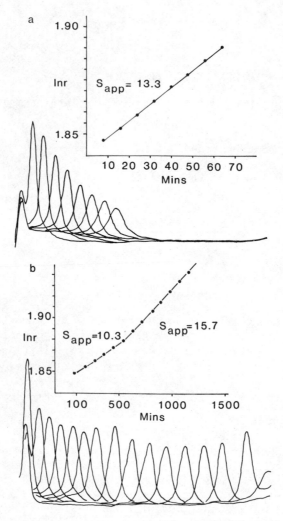

Fig. 1.   Sedimentation-velocity patterns obtained with human cervical mucins at (a) 30k rpm (scans every 8 minutes) and (b) at 10k rpm (scans every 90 minutes). Sedimentation is from left to right. Samples (approximately 1 mg/ml) were extensively dialysed against 4M GuHCl. Schlieren optics at 550 nm was used.

behind the sedimenting boundary, and (iii) a loss of material that could not be accounted for by radial dilution.

The boundary broadening did not appear consistent with the high molecular weight of the molecules. An examination of the DNA literature (See e.g. Aten and Cohen, 1965; Schumaker and Zimm, 1973) revealed that similar observations have been made on very large DNA molecules and this prompted an investigation of the sedimentation-rate as a function of rotor speed. Fig. 1b shows the same sample run at 10k rpm and virtually no boundary broadening is evident over a period of 1 500 minutes. The sedimentation-rate is complex, there being a change from a slow-moving state (S_{app} = 10.3) to a fast-moving one (S_{app} = 15.7). At 15k rpm the apparent sedimentation-rate was the same as at 30k rpm. However, the boundary was more stable at 15k rpm than at 30k rpm even though there was a smear of trailing material behind the boundary.

CONCLUSIONS

Our preliminary studies of these phenomena can be summarized as follows: (i) the effects are independent of the solvent conditions, e.g. 0.2M NaCl, 4M, 6M or 8M GuHCl, (ii) the effects are highly reproducible (a high-speed experiment may be followed by a low-speed one after shaking the cell and vice versa) and (iii) experiments with mineral oil layered over the samples suggest that there is no pressure dependence. Rather the transition from a slow to a fast sedimentation-rate is dependent on the velocity of the boundary.

The phenomena may be explained by speed-dependent aggregation as noted for DNA (i.e. clusters of molecules sediment ahead of the boundary) and a rate-dependent sticking of the mucin molecules to the surface of the centrifuge cell.

REFERENCES

Aten, J.B.T. and Cohen, J.A.,1965, J. Mol. Biol.,12:537.
Carlstedt, I., Sheehan, J. and Wingerup, L.,1981, Proceedings of the
      Second Symposium on Mucus in Health and Disease.
Schumaker, V.N. and Zimm, B.H.,1973, Biopolymers, 12:869, 877.

ENDOCERVICAL CHANGES IN THE GUINEA PIG AFTER TREATMENT WITH 17-β -

OESTRADIOL AND 9,10 DIMETHYLBENZANTHRACENE

John A. Thomas   and   Frederick D. Dallenbach

Department of Pathology   Institute of Experimental
St John's Medical College Pathology
Bangalore 560 034, India   German Cancer Research Centre
                           D-6900 Heidelberg, W. Germany

INTRODUCTION

Revadi and Thomas (1979) have established the existence of normal cyclic variations in the human uterine endocervical mucosa. Their further studies (1980) on endogenously-produced oestrogen states and the human uterine endocervix showed that oestrogen produces a progressive increase in endocervical mucosal cell heights which is unrivalled by the endocervix, even in the pregnant state. On the basis of these studies, they hypothesized, that though oestrogen, was probably responsible for the secretion of excessive quantities of endocervical mucus, the release of this mucus was probably mediated by progesterone. In their studies basal cell proliferation was never seen.

Controversy exists as to the occurrence of "Atypical" changes in the cervix in "pill-users", the "pill" containing variable amounts of oestrogen and progesterone which are obviously of exogenous origin.

The present study was instituted to observe the possible synergestic effect of exogenous oestrogen and a potent carcinogenic hydrocarbon on guinea-pig-endocervices. Particularly looked for, were any changes in the basement membrane, the basal cells, the mucus-producing cells and the mode of release of mucus from these latter cells, if any.

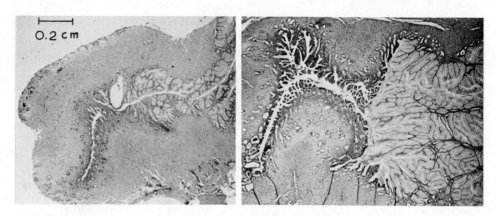

Fig.1. Observe the marked mucosal hyperplasia of the treated
        guinea pig endocervix. (right).
        The control guinea pig endocervix is on the left.

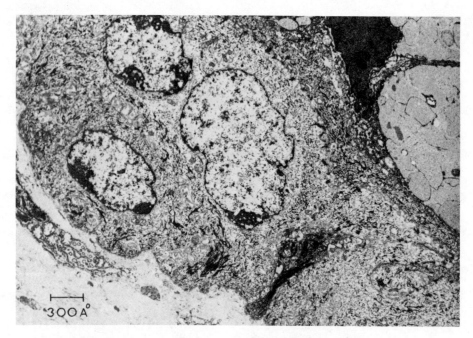

Fig.2. Treated guinea pig. Basal cells showing immature
        metaplasia and mild dysplasia.

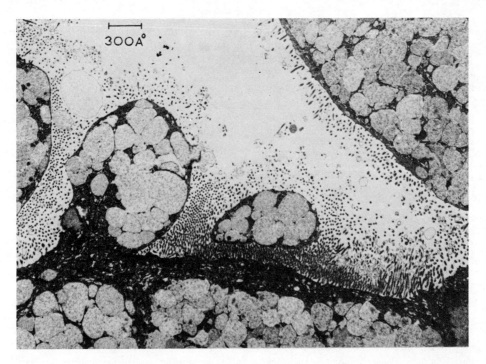

Fig. 3.   Treated guinea pig. Domed shaped luminal surface of mucus
producing cells with nodularity.

RESULTS

One hundred, four-month old female Duncan-Hartley-Pirbright guinea pigs with suitable controls, were used for this study. 17-β-oestradiol (Depot) was injected into the pectoral region of these animals and at laparotomy 9,10 Dimethylbenzanthracene in sesame oil was injected into one uterine horn. Representative animals were killed weekly and the endocervices examined grossly with the light and the electron microscope.

The endocervices of the treated animals showed marked endocervical mucosal hyperplasia both grossly and with the light microscope (Fig.1). With the electron microscope, basement membrane thickening, was a notable feature, with proliferation of basal cells showing early immature squamous differentiation, mild dysplasia (Fig.2), mucus cell differentiation and combined squamous and mucus cell differentiation (Fig.3). The mucus cells showed excessive membrane-bound mucus-globule formation with the mucus contained within these globules showing electron-density and electron-lucency. The luminal surfaces of these cells showed a generally dome-shaped, microvillus-covered appearance with many irregular surface nodularities. Some of the microvilli showed evidence of branching. The release of mucus was seldom seen but when it occurred was by a process of single mucus-globule extrusion or by the release of partially or wholly membrane-bound globules. No cilia were observed in this study. The stroma showed increased amounts of ground substance with increased amounts of ground substance with increased activity in the myofibroblastic cells.

CONCLUSION

Exogenous oestrogen with 9,10 Dimethylbenzanthracene produced, basement membrane thickening, proliferation of basal cells, immature squamous metaplasia, mild dysplasia, mucus cell differentiation and mucus and squamous differentiation. Excessive quantities of mucus seemed to be present in the well differentiated mucus cells with little of it being released. When such release did occur, it was observed as mucus-globule extrusion, or release of partial or wholly-membrane-bound mucus-globule packets.

REFERENCES

Revadi, P.S. and Thomas, John A. 1979. Uterine endocervical morphology in different phases of the menstrual cycle.
Indian J. Med. Res. 70, 49-54.
Revadi, P.S. and Thomas, John A. 1980. Uterine endocervical
Indian J. Med. Res. 71, 884-889.

HISTOCHEMICAL ALTERATIONS IN THE UTERINE ENDOCERVICAL MUCOSA IN
DIFFERENT PHASES OF THE NORMAL MENSTRUAL CYCLE AND IN THE ALTERED
CYCLE

V. Nirmala   and   John A. Thomas

Department of Pathology
St. John's Medical College
Bangalore 560 034
India

INTRODUCTION

Cervical mucus contributes towards the hostile vaginal environment and interferes with the entry of the spermatozoa into the uterus
at all times except during the periovulatory period of the normal
menstrual cycle. During this period the normally acidic, thick,
viscid cervical mucus becomes less acid, thin and watery. The
resultant changes in pH and consistency is believed to protect the
spermatozoa from the hostile vaginal environment and also help in
sperm transport. (Elstein, 1978; Kroeks, Kremer, 1977; Moghissi,
1973).

The present studies were made to provide and elucidate a histochemical basis for the changes that take place in human endocervical
mucus cells in different phases of the normal menstrual cycle and in
human endocervices under the influence of prolonged endogenous
oestrogen and prolonged endogenous progesterone stimulation. The
latter portion of these studies, was done to delineate histochemically the effect, if any, of each of these hormones on the components
of cervical mucus and its fractions.

RESULTS

The cervices used in these studies required to fulfil two
criteria - firstly in having adequate amounts of endocervical mucosa
and secondly in having sufficient relevant endometrium for phasing
or diagnosis.

285

Seventy five specimens from hysterectomy material and four from biopsy material were used in the study on endo-cervices of the normal menstrual cycle. These tissues had been removed from women who were in the reproductive period of life, for histologically confirmed non-cervical and non-endometrial conditions. Using the phased endometria, the cervices were made into five groups. Twenty-five belonged to the early oestrogen phase, 17 to the mid-oestrogen phase, 14 to the late oestrogen phase, 10 to the early progesterone phase and 13 to the late progesterone phase.

Thirty-one cervices were chosen for the study on the altered menstrual cycle. Ten were from post-menopausal women with atrophic endometrium, 10 from patients with cystic hyperplasia of the endometrium and nine from endometrial carcinoma. All these patients were grouped under the head "prolonged endogenous oestrogen action". To demonstrate "prolonged endogenous progesterone action", only uteri with histologically viable chorionic villi and a pronounced decidual endometrial reaction were included for study. One two uteri satisfied these criteria.

By the use of standard histochemical methods the changes in the quantity of acid mucins and its tainable fractions and neutral mucins occurring in the endocervical mucosal epithelial cells, were assessed.

In the normal menstrual cycle a marked decrease in both acid and neutral mucin production was observed at the periovulatory period though the decrease in neutral mucin production was to a much lesser extent. The fractions of acid mucin showed a decrease in sialomucins and a concomitant increase in sulfomucins during this period of the normal cycle.

The human endocervices subjected to increased amounts of endogenously-produced oestrogen and progesterone action, showed a statistically significant increase in the production of total acid and neutral mucins. The ratio of total acid to neutral mucins, however, remained unaltered, when comparisons were made to the oestrogen and progesterone phases of the normal cycle, respectively. In conditions where sustained oestrogen action was present the sialomucin fraction showed a decrease and the sulfo-mucin fraction an increase. With sustained prolonged progesterone action the quantities of these substances showed an increased of sialomucin and a decrease in sulfomucins.

CONCLUSION

The increase in released cervical mucus pH and decrease in its viscosity around the time of ovulation in the normal menstrual cycle can be accounted for histochemically by the decreased production of total acid mucin and a decrease in sialomucin levels.

The prolonged unopposed action of endogenous oestrogen or progesterone resulted in the overproduction of mucus with the resultant increase in production of both acid and neutral mucins. There was, however, no alteration in the ratio of these fractions to each other when comparisons were made with levels in the normal cycle. The decreased sialomucin fraction of acid mucin and increased sulfomucin fraction production in the endogenously produced oestrogen-stimulated state resulted in thin watery mucus being produced within the endocervical epithelial cells. The reverse changes in the levels of these fractions in the endogenously produced progesterone state resulted in thick viscous mucus.

REFERENCES

Elstein, M., 1978. Functions and physical properties of mucus in the female genital tract.
Br. Med. Bull. 34, 83-88.
Kroeks, M.V.A.M. and Kremer, J., 1977. The pH in the lower third of the genital tract. IN: The Uterine cervix in Reproduction.
Ed. V. Insler and G. Buttendorf Georg Thieme Publishers, Stuttgart.
Moghissi, K.S., 1973. IN: Cervical mucus in human reproduction.
M. Elstein; K.S. Moghissi, R. Borth., Scriptor., Copenhagen.

A PRELIMINARY REPORT ON THE INTRACERVICAL CONTRACEPTIVE DEVICE -

ITS EFFECT ON CERVICAL MUCUS

H.A. Pattinson, I.D. Nutall and M. Elstein

University Hospital of South Manchester
West Didsbury
Manchester M20 8LR

INTRODUCTION

A progestogen releasing intracervical contraceptive device
(I.C.D.) has been developed, designed to release 20-25 mcg of
norgestrel per day. This study was designed to ascertain whether
the device had any adverse effect on uterine histology or
bacteriology, and to study the effect on cervical mucus of this
rate of local progestogen release.

METHODS

Previously sterilised women, about to undergo hysterectomy for
conditions unlikely to effect the physical characteristics of the
cervix or cervical mucus, were chosen for the study.

The physical properties, i.e. the amount, clarity, spinnbarkeit,
cellularity, viscosity and ferning of the cervical mucus were
assessed in the cycle prior to insertion of the device. Post coital
tests were performed[1]. An I.C.D. was then fitted. Half of the
devices contained a norgestrel releasing capsule, the remainder
being inert. The cervical mucus was then assessed in two successive
test cycles prior to hysterectomy.

The bacteriology of the cervix and uterine cavity was assessed
as described by Sparks et al[2]. The cervix was carefully examined
histologically particularly with reference to the area in contact
with the device.

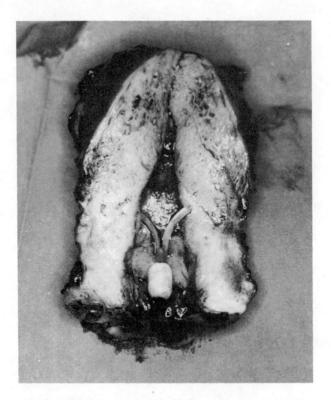

Fig. 1   The intracervical device in situ in an opened uterus.

RESULTS

    To date, eight patients have been fitted with an active device,
and seven with an inert device.   Insertion of the device was easy
in nine cases and difficult in six, usually due to a tight internal
cervical os.   One further patient withdrew from the trial because
of discomfort at attempted insertion, and in two further cases the
device fell out at some time after insertion.

Cervical Mucus

    Of the seven patients fitted with an inert device, one developed
a trichomonal infection, with heavy vaginal discharge and irregular
bleeding, necessitating removal of the device.   Two patients had
apparently anovular mucus in the control cycle, with a small volume
of viscous mucus which demonstrated only minimal spinnbarkeit and
ferning.   Both these patients developed ovular mucus in their test
cycles, with 2-3 mls of clear watery mucus, spinnbarkeit of over
6 cms and 50 to 60 per cent ferning.   The remaining four patients
had satisfactory mucus in the control cycle.   Of the eight test
cycles, four showed anovular appearances, as described above, and

four showed continued satisfactory ovular type mucus. Post coital
tests were performed in five control and eight test cycles. In the
control cycles, 1 - 10 sperm were seen per high power field, though
in one case all were non motile. In the test cycles, 1 - 6 sperm
per high power field were seen with non motile sperm only in four
cases.

Only one of the eight patients fitted with an active device had
anovular mucus in the control cycle. Her mucus became ovulatory in
one of the test cycles. Of the remaining seven control cycles, all
showed thin clear ovular mucus, ferning of 60 to 90 per cent and
spinnbarkeit of 5 cms or more. In all, fifteen test cycles were
studied. In two, no mucus at all could be obtained for testing.
Of the remainder, all but one demonstrated progestational mucus, with
a small volume of thick viscous, cellular mucus with no more than
10 per cent ferning and spinnbarkeit of 2 cms. In only one test
cycle were motile sperm present at post coital testing.

Bacteriology

Bacterial colonisation of the uterine cavity was detected in
nine of the fifteen subjects studied to date. Streptococcus,
Staphylococcus and Bacteroides being the commonest invading organisms.

Histology

The only significant abnormality detected to date has been some
local acute inflammatory changes, and small areas of focal ulceration
at the site of the device in three cases.

CONCLUSIONS

Though the numbers in this study to date are small and data
incomplete, certain trends are apparent. The progestogen releasing
I.C.D. does appear to have an effect on cervical mucus, rendering
it inpenetrable to sperm. The incidence of bacterial colonisation
of the uterus is high, though this may be a transient phenomenon
following insertion. The focal ulceration and local inflammation
is of some concern, but earlier reports of dysplasia have not been
confirmed.

These results would seem to suggest that the I.C.D. might be
an effective contraceptive method, and warrant further investigation.

REFERENCES

1. J. Kremer and S. Jager, Fertil. Steril. 27: 335 (1976).
2. R.A. Sparks, B.G.A. Purrier, P.J. Watt and M. Elstein,
    Br. Med. J. 282: 1189 (1981).

A SCANNING ELECTRON MICROSCOPIC STUDY OF HUMAN CERVICAL MUCUS

Malcolm F. Beeson, Greg R. Parish*, Stuart L. James[+]
and Christopher Marriott[+]

Beecham Pharmaceuticals Research Division, Great Burgh
Epsom, KT18 5XQ and *Brockham Park, Betchworth, RH3 7AJ
and [+]Department of Pharmacy, Brighton Polytechnic
Brighton, BN2 4GJ; U.K.

Scanning electron microscopy (SEM) of cervical mucus has
suggested a honeycomb structure, with interconnecting channels
separated by thin membranous walls (Chretien et al, 1973), which
may represent glycoprotein bundles or sheets. The channels were
considered to exist in the native gel and provide routes along
which spermatozoa could swim on their journey up the cervical canal.
The apparent size of such channels was shown to change at differing
stages in the ovarian cycle (Chretien and David, 1978), and was
accompanied by changes in the viscoelasticity of the secretion
(Wolf et al, 1977); maximum channel diameter and minimum viscoelasticity occurring around the time of ovulation, facilitating
fertilization. Forstner et al. (1977) have shown that intestinal
goblet cell mucin presents a similar structure in the SEM. The
purpose of this work was to reassess the changes reported to occur
in cervical mucus, taking into account the findings of Parish et al.
(1981) concerning the artefactual nature of the porous structure
demonstrable in respiratory mucus.

Specimens of cervical mucus were obtained by gentle suction
from women attending a gynaecological out-patients' clinic for a
variety of reasons. The samples were frozen immediately, as
described by Parish et al. (1981), using either boiling liquid
nitrogen (LN) or solidifying dichlorodifluoromethane (Freon) as
the coolant.

Scanning electron microscopy of these samples confirmed the
previous finding, established for purified human respiratory mucin,
that the size of the observed 'pores' (crystal voids) is inversely

293

proportional to the freezing rate. No variation was seen between samples of cervical mucus taken from women during pregnancy, or early or late in the ovulatory cycle. The only specimen to show a differing response to freezing was from a woman taking an oral contraceptive preparation (see Fig. 1). In this instance the appearance of the frozen mucus resembled that reported for calcium-treated mucin (Forstner et al., 1977; Beeson et al., 1981).

Further specimens of cervical mucus were studied by fluorescence microscopy, using fluorescein isothiocyanate conjugated lectin as the fluorescent probe. Midcycle specimens frozen in LN yielded a network structure analogous to that observed using the SEM. In contrast, identical wet specimens showed no such structure.

These findings support the conclusion of Parish et al., (1981) that the porous structure seen in frozen specimens of mucin is a product of the freezing procedure used prior to examination in the SEM, and does not represent a structure which is present in the natural state. Nevertheless, the observed change in the artefact in one specimen, and its similarity to that which can be induced by calcium ions, suggests that alterations in the SEM appearance may still parallel in vivo modifications in mucus composition, which may in turn be relevant to fertility.

REFERENCES

Beeson, M. F., Parish, G. R., Brown, D. T. and Marriott, C., 1981, The effects of low molecular weight mucus modifying substances on the appearance of mucin gels in the scanning electron microscope, this publication.

Chretien, F. C., Gernigon, C., David, G. and Psychoyos, A., 1973, The ultrastructure of human cervical mucus under scanning electron microscopy, Fertil. Steril., 10:746.

Chretien, F. C. and David, G., 1978, Temporary obstructive effect of human cervical mucus on spermatozoa throughout reproductive life: a scanning electron microscopic study, Europ. J. Obstet. Gynec. Reprod. Biol., 8:307.

Forstner, G., Sturgess, J. and Forstner, J., 1977, Malfunction of intestinal mucus and mucus production, in: Mucus in Health and Disease, M. Elstein and D. V. Parke, eds., Plenum Publishing Corporation, N.Y.

Parish, G. R., Beeson, M. F., Brown, D. T. and Marriott, C., 1981, A freezing artefact associated with the preparation of mucin for examination using the scanning electron microscope, this publication.

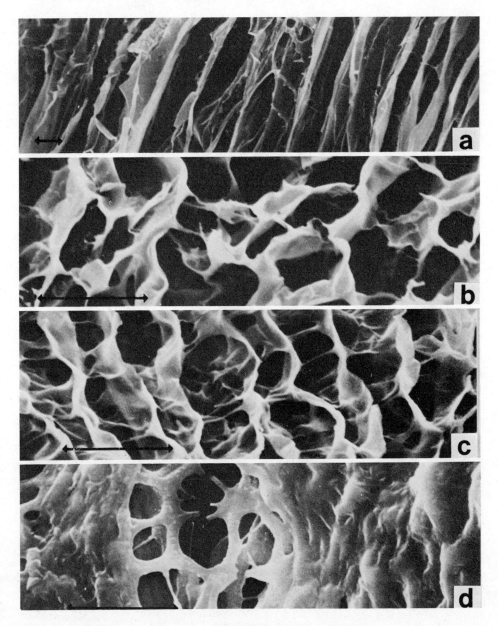

Fig. 1   SEM of human cervical mucus, frozen in LN.
          (a) pre-ovulatory (+8 days)
          (b) post-ovulatory (+22 days)
          (c) pregnant (+108 days)
          (d) patient taking oral contraceptive agent (no periods)
                    (Scale marks = 10 mµ)

Wolf, D. P., Blasco, L., Khan, M. A. and Litt, M., 1977, Human
cervical mucus II, Changes in Viscoelasticity during the
ovulatory menstrual cycle, Fertil, Steril., 28:47.

A FREEZING ARTEFACT ASSOCIATED WITH THE PREPARATION OF MUCIN FOR EXAMINATION USING THE SCANNING ELECTRON MICROSCOPE

Greg R. Parish, Malcolm F. Beeson*, David T. Brown[+] and Christopher Marriott[‡]
Beecham Pharmaceuticals Research Division, Brockham Park, Betchworth, RH3 7AJ, and *Great Burgh, Epsom, KT18 5XQ; [+]Department of Pharmacy, University of Nottingham, Nottingham, NG7 2RD, and [‡]Department of Pharmacy, Brighton Polytechnic, Brighton, BN2 4GJ; U.K.

Scanning electron microscopy (SEM) of freeze-dried cervical mucus has been used to demonstrate that a filamentous or honeycomb structure exists (Chretien et al., 1973), with interconnecting channels separated by thin, membranous walls. This is interpreted as support for the channel theory of sperm migration (Odeblad, 1968). Forstner et al. (1977) have demonstrated a similar structure for intestinal goblet cell mucin. The purpose of the present study was to investigate the behaviour of human respiratory mucin in this respect, and to investigate parameters which might modify the observed appearance, with a view to understanding its biological significance.

Human respiratory mucus obtained from bronchitic subjects was purified by the method of Marriott et al.(1979). The concentration of the resulting mucin gel could be adjusted during the final ultrafiltration stage of the procedure. Samples (50 μl) of the gel were placed in the end of hollow rivets of 2 mm internal diameter and quench frozen in either boiling liquid nitrogen (LN) or solidifying dichlorodifluoromethane (Freon). These provided slow or fast rates of freezing respectively (Costello and Corless, 1978). Samples were freeze-dried at a vacuum of 3×10^{-4} Pa at 193°K for 18 hours, sputter-coated with gold, and studied in a Philips 501 SEM. Some samples were also fixed by immersion in 1% glutaraldehyde prior to freezing, but this was found to have no effect on the structures visualised.

The internal structure of the samples was examined by viewing a natural fissure running perpendicular from the surface towards the

297

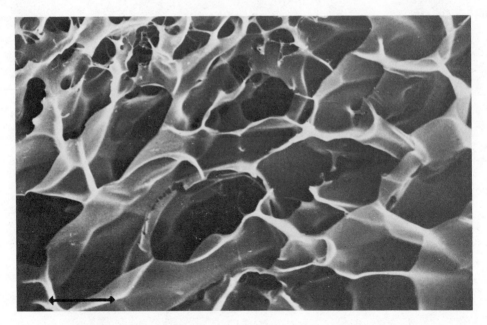

Fig. 1. Scanning electron micrograph of human respiratory mucin,
         prepared by freezing in liquid nitrogen. Scale bar = 10 μm.

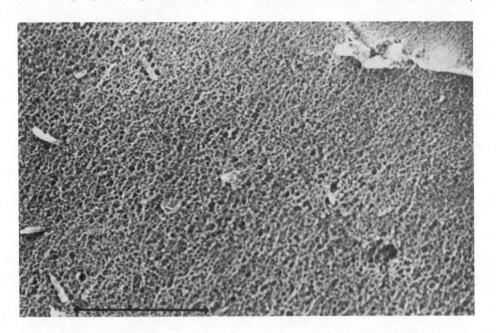

Fig. 2. Scanning electron micrograph of human respiratory mucin,
         prepared by freezing in Freon. Scale bar = 10 μm.

centre of the sample. A porous structure was evident in all
samples, with the pore size generally increasing from the surface
towards the centre of the sample. Consequently, the sample surface
was taken as the point of reference, all further work being
restricted to an examination from the surface down to approximately
100 μm depth into the sample. The size of the pores differed
markedly according to the freezing process used. Samples frozen in
LN showed irregularly shaped pores, 3-15 μm diameter (Figure 1),
whereas those frozen in Freon had more regular pores, up to 1 μm
in diameter (Figure 2).

In view of the relationship between pore size and freezing
rate it is clear that the pores are an artefact and are 'crystal
voids'; that is, the spaces left by ice crystals which grew within
the gel during freezing. During growth, the ice crystals displace
the mucin glycoprotein into bundles or sheets. Ice crystal size
in bulk hydrated samples varies inversely with freezing rate
(Nei, 1978; Moor, 1973) and the freezing rate reduces from the
sample surface to the interior (Moor, 1973; Bald and Crowley, 1979).
Thus, in our experiments the lower freezing rate (LN) gave larger
pores than the higher freezing rate (Freon), and the pore size
increased towards the sample centre. When the freezing conditions
were rigorously controlled the results were highly reproducible.

Ice crystal size is also related to water content (Moor, 1973;
Franks, 1978; Nei, 1978) and this factor was investigated using
hog gastric mucin. As expected, a decrease in water content
produced an overall reduction in crystal size. In a 0.6% (w/v) gel
the pores were clearly resolvable at the sample surface when
examined at 6,000x magnification, whereas in a 3.3% gel there was
a surface layer, approximately 2 μm thick, in which the pores were
unresolvable.

The differences in pore size reported by some workers
(Chretien and David, 1978) could be due to less rigorously
controlled freezing, storage and freeze-drying conditions, and also
to variations in water content. Our findings cast serious doubt on
the in vivo existence of pores or channels in mucus gels.

REFERENCES

Bald, W. B. and Crowley, A. B., 1979, On defining the thermal
history of cells during the freezing of biological specimens,
J. Microscopy, 117:395.

Chretien, F. C., Gernigon, C., David, G. and Psychoyos, A., 1973,
The ultrastructure of human cervical mucus under scanning electron
microscopy, Fertil. Steril., 10:746.

Chretien, F. C. and David, G., 1978, Temporary obstructive effect of
human cervical mucus on spermatozoa throughout reproductive life: a

scanning electron microscopic study, Europ. J. Obstet. Gynec.
Reprod. Biol., 8:307.

Costello, M. J. and Corless, J. M., 1978, The direct measurement of
temperature changes within freeze-fracture specimens during rapid
quenching in liquid coolants, J. Microscopy, 112:17.

Forstner, G., Sturgess, J. and Forstner, J., 1977, Malfunction of
intestinal mucus and mucus production, in: Mucus in Health and
Disease, M. Elstein and D. V. Parke, eds., Plenum Publishing
Corporation, N.Y.

Franks, F., 1978, Biological freezing and cryofixation, J.
Microscopy, 111:3.

Marriott, C., Barrett-Bee, K. and Brown, D. T., 1979, A comparative
evaluation of mucus from different sources, in: Glycoconjugates,
R. Schauer et al. eds., Georg Thieme, Stuttgart.

Moor, H., 1973, Cryotechnology for the structural analysis of
biological material, in: Freeze-etching techniques and applications,
F. L. Benedetti and P. Favard, eds., Soc. Francaise de Microscopie
Electronique, Paris.

Nei, T., 1978, Structure and function of frozen cells: freezing
patterns and post-thaw survival, J. Microscopy, 112:197.

Odeblad, E., 1968, The functional structure of human cervical mucus,
Acta Obstet. Gynec. Scand., 47 (Suppl. 1):57.

CERVICAL MUCUS: ITS PHYSIOLOGICAL ROLE AND CLINICAL SIGNIFICANCE

Max Elstein
Professor of Obstetrics & Gynaecology
University Hospital of South Manchester
Manchester   M20 8OR

INTRODUCTION

The functions of the cervix are mediated primarily by its mucus which undergoes unique changes dependant upon the endocrinal influences acting upon it.

Its situation at the entrance to the upper genital tract ensures its major role in the reproductive processes of fertilisation, implantation and the development of pregnancy.   It is the site through which spermatozoal transport occurs which is an extremely rapid process, given healthy motile spermatozoa and penetrable cervical mucus.   It plays a most important role in protecting the upper genital tract from invasion by micro-biological organisms.   Its unique physical properties are related to these functions.   Our understanding of the molecular biology of this mucus is based on a number of concepts which explain the mechanisms of spermatozoal movement through the mucus when desired for reproduction and their prevention at all other times.   Because of its position in the genital tract, abnormalities of the cervix might be invoked as possible mechanical factors in the aetiology of infertility.   Indeed, abnormalities of the mucus itself can cause infertility and the management of these abnormalities and their recognition is ill-understood.   In view of its strategic position in the genital tract, locally acting contraceptives exerting their action at the level of the cervix, would be ideal.

The present state of knowledge of cervical mucus in relation to its important role in reproduction in the female is reviewed.

FUNCTIONS OF THE CERVIX AND ITS MUCUS

The uterine cervix may be likened to a biological valve, allowing variable entry of spermatozoa to the uterus and thereby playing a passive and active role in respect of spermatozoal migration. Most of the functions of the cervix are affected by the secretion of the cervical epithelium. This secretion has the following functions:

1. Transmission of spermatozoa at or near the time of ovulation and interference of entry at all other times;

2. Protection of spermatozoa from the hostile environment of the vagina and from being phagocytosed;

3. Supplementing the energy requirements of spermatozoa;

4. Filtering of abnormal and poorly motile spermatozoa and only allowing actively motile spermatozoa of normal shape into the upper genital tract when appropriate for reproduction;

5. To serve as a reservoir of spermatozoa;

6. A site where the process of maturation of spermatozoa, i.e. capacitation or similar such process, takes place;

7. To protect the upper genital tract from invasion of microorganisms from the vagina.

In the human, cervical mucus has a prominent role in spermatozoal migration and is similar to that of the bovine and ovine species, in which the ejaculate has a concentrated number of sperm and is of low volume. During intercourse, up to a half-a-billion sperm are ejaculated into the upper vagina, and undergo specific movements to ensure their migration through the cervical canal.

PRODUCTION OF THE CERVICAL MUCUS

While there are other locations for mucus production in the human female genital tract, such as the endometrium, fallopian tubes and Bartholin's glands, the major site of mucus production is within the cervical canal. Little is known of the mucus production from the other sites because of paucity of the secreted material. It is well-known that there is a considerable increase in the amount of fluid content of the fallopian tube at the time of ovulation and this may well be related to the increase in peritoneal fluid at that time. Studies into the nature and function of the fluid in the peritoneal cavity in the pelvis and the fallopian tube is currently being under-

taken but their character does not come within the definition of a
mucus secretion.   In the case of Bartholin's glands, however, a
copious clear secretion is produced following sexual stimulation
which lubricates the vaginal introitus.   No biochemical studies or
researches into its physical properties have thus far been reported.
Although it is possible that the contents of the endometrial cavity
and the fallopian tube may contribute to the cervical secretion, this
would likely be of minor significance since observations on the mucus
secretion from a cervix in a woman who had undergone sub-total
hysterectomy do not differ to any marked extent from that of a woman
whose upper genital tract is intact.   These observations are based
on anecdotal reports and no detailed comparative investigations have
been reported.   In the descriptions to follow the cervical mucus has
been extracted from a woman with an intact upper genital tract and
the contributions, if any, from the endometrial cavity and fallopian
tube are included.

Structure of the Cervical Canal

The cervix is situated in the upper dilated portion of the vagina
at the gateway to the upper genital tract in which the reproductive
processes take place.   It points backwards and slightly downwards to
the posterior fornix and following ejaculation, its external os is
immersed in the seminal pool which collects therein.

Cyclic alterations occur in the diameter of the external os and
the cervical canal which favour spermatozoal transport during the
pre-ovulatory phase of the menstrual cycle.   A progressive widening
of the external os occurs with the increasing oestrogen influence of
the proliferative phase, reaching its maximum at ovulation (Mann,
McLarn & Hayt,1961).   Indeed, at ovulation the external os can
measure over 3mm in diameter compared with that of 1mm post-menstrually
and following ovulation, when there is increase in tone of the
internal os and to a lesser extent of the external os which persists
until the onset of the next period.   Pari passu with this morphological change there is an increase in quantity of cervical mucus
which becomes profuse and watery at the time of ovulation with a
cascade dipping from the cervix into the posterior vaginal fornix.
Following ovulation and during the luteal phase, the mucus becomes
scanty, thick opaque and viscid.   This increase in the amount of
alkaline cervical secretions buffers the acidity of the vagina and
provides a favourable milieu for the spermatozoa in the ejaculate and
facilitates their movement into the cervical canal.

The cervix comprises a canal which is spindle-shaped with a
narrowing at both ends described as the internal and external os,
with the sphincter being more evident at its junction with the
endometrial cavity.   The cervix itself is a fibro-muscular organ
with collagen tissue predominating in its lower half with muscular

tissue being more evident in its upper half.   The collagenous
material undergoes interesting changes in pregnancy in relation to
its preparation for child bearing and the relationship of these
to its mucus production remains to be established.

The endocervical mucus membrane is arranged into an intricate
system of grooves and out-pocketings of the epithelium, referred to
as crypts macroscopically described as the "arbor vitae."   The
structure of these crypts varies with the age and parity of the
woman.   Studies on hysterectomy specimens on the number and size
and shape of these crypts during the menstrual cycle indicate that
there is a massive increase in the surface area of these crypts in
the late proliferative phase (Insler, Glezerman, Bernstein et al,
1981).   Indeed, these workers have shown that the largest surface
area of the crypts was at the lower end of the cervical canal and
these decrease in extent in the upper reaches of the canal nearer the
internal os.   There was also a gradient of spermatozoa present in
the cervical canal which moved upwards with an increasing duration of
time following their deposition at the external os.

The cervical epithelium comprises different types of non-
ciliated secretory and ciliated cells.   The number of these cells,
their physiological and histo-chemical characteristics and the nature
and abundance of the secretory granules vary in different parts of
the cervix (Jordan, 1976).   Non-ciliated cells, presumably the
secretory cells, are covered by microvilli and contain massive amounts
of cytoplasmic granules.   The release of the secretory granules is
associated with disruption of the cell membrane.   During the phase
when storage of secretion takes place, the cell membrane is covered
over its luminal surface by microvilli.

The luminal surface of the ciliated cells is covered by well-
developed cilia.   The cytoplasm of these cells contains few granules,
Golgi vesicles and endoplasmic reticulum.   The kinocilia have been
observed to beat towards the vagina.   The active beating of these
kinocilia may facilitate the orientation and flow of mucus from the
surface of the secretory cells into the upper vagina.   This down-
ward flow of mucus would appear to result in an impedence to the
ascent of the spermatozoa but in fact this effect occurs primarily
in the periphery of the canal and indeed encourages the rapidly
moving spermatozoa to remain within its centre.

The daily mucus production varies from 600 mg during
mid-cycle to 20-60 mg during other periods of the cycle.   The
cervical mucus is not homogenous but is a heterogenous mixture of
secretions.   Although originally it was thought that each crypt
produced a single type of mucus, this view is not thought likely
since single crypts have been observed to produce mixed mucus on the
basis of their staining properties.   Biochemical studies on mucus

production from single crypts are not possible with present day
technology.

Physical Properties of the Cervical Mucus

Cervical mucus reveals a number of rheological properties which
include viscosity, spinnbarkeit, flow elasticity or retraction,
plasticity and tack.   Rheology is the study of the flow and
deformation of matter.   In the case of cervical mucus, however, this
term is sometimes extended to include various biophysical properties
and also some characteristics of molecular physics.

Viscosity.   With regard to viscosity, cervical mucus cannot be said
to be viscous or to have simple viscosity, because it is not truly a
fluid in the physical sense.   In reality it has a changing viscosity,
or shows anomalous viscous behaviour.   Viscometers are not generally
clinically available and therefore this is not measured in practice.
At the time of ovulation, the viscosity is minimal and spermatozoal
penetrability is favoured.

Spinnbarkeit pertains to the capacity of a liquid to be drawn into
threads.   From the end of menstruation and until ovulation there is
a progressive increase in spinnbarkeit and this physical property is
maximal at the time of ovulation.   It is easily measured by simply
drawing mucus away from the slide with a cover slip and measuring the
length when fracture of the thread occurs.   It is performed usually
at the time of a post-coital test.   This phenomen is attributed to
the presence of the long macromolecules in the mucus and depends upon
them and the strong inter-molecular forces between them.   Following
ovulation spinnbarkeit decreases with the advent of the progestogen
effect of the luteal phase.   It is useful to determine the fertile
time and has an application for artificial insemination.   It forms
the basis of the sympto-thermal method of periodic abstinence for
contraceptive practice.

Flow elasticity or retraction refers to the tendency of cervical mucus
to assume its original shape following deformation produced by stress
or pressure.   This characteristic parallels spinnbarkeit.

Plasticity and tack are physical features of mucus noted during
pregnancy.   Plasticity pertains to that property of the mucus which
permits it to be deformed without rupture, while tack refers to the
stickiness of the mucus demonstrated by pulling a cover slip away
from a sample of mucus on a slide.   These properties have little
application in clinical practice excepting when the tack is present

abnormally at mid-cycle in a woman with a cervical factor in
infertility.

Ferning.   Of all the properties of the cervical mucus, its arborization phenomenon is one of the more sensitive to the changes of the
levels of sex hormones.   Like spinnbarkeit, ferning is used in the
detection of ovulation and for the presence of a progestational
effect.   It is particularly useful in the induction of ovulation
with gonadotrophins since its occurrence serves as an indicator for
measuring systemic levels of oestrogen when a particular regimen of
administration of gonadotrophins is used.   This arborization
phenomenon is not specific for cervical mucus and can be seen in any
organic material containing proteins and electrolytes.   Although non-
specific, ferning of cervical mucus is distinctly dependant upon the
action of oestrogen and bears a close relationship to the physical
properties of viscosity and spinnbarkeit.   The extent of mucus
crystallisation bears a direct relationship to spermatozoal recepti-
vity.

     Several precautions must be observed in performing the fern test
in order to make the appropriate inferences from it.   Blood will
inhibit the phenomenon and the speculum and instruments used in
obtaining the mucus specimen must not be stored or rinsed in a saline
solution.   In practice any semi-quantitative criteria with a grading
based on the amount or extent of crystal formation have been super-
ceded by measurements of oestrogen in urine and blood.   Nevertheless
a qualitative assessment of the degree of ferning has a correlation
with the amount of oestrogenic stimulation of the endocervical mucosa,
but does not increase in amount once a threshold level of oestrogenic
stimulation has occurred.   Maximal ferning is associated with the
crystal formation at 4 right-angles to each other and this occurs when
the level of oestrogen approximately coincides with that found during
the late proliferative phase.

     Ferning and spinnbarkeit can be of considerable value in assess-
ing a patient with infertility.   It is a useful means of indicating
the relative degree of endogenous oestrogen activity.   A hypo-
oestrogenic state can be suspected if the cervical os is small with
very little mucus present and if the fern is atypical or absent at
mid-cycle.   Similarly abundant, watery, spinnbar mucus which ferns
readily is obtained from women who have an optimal or even excessive
output of oestrogens.

                    ODEBLAD CONCEPT OF CERVICAL MUCUS

     Early investigations by Odeblad (1973) using nuclear resonance
(NMR) and other physical techniques enabled him to propose an hypothesis which explained the unique rheological properties of cervical

mucus, which is composed of a high and a low viscosity component.
In this concept, he stated that the high viscosity component of the
cervical mucus was made up of long, flexible macromolecules (fig.1)
which were joined up in bundles which he called "micelles" which
averaged about 0.5 mu in diameter, which under oestrogenic stimulus
increase in size and under the influence of progestogens become very
much smaller.   Thus, when there is a progestational influence the
bonding between the glycoprotein macromolecules alters to form a
finer mesh group which prevents entrance of the spermatozoa.   However,
when the bundles are larger and the "micelles" are forming, there are
free channels between the high viscosity component which facilitate
the transmission of spermatozoa through the aqueous phase.

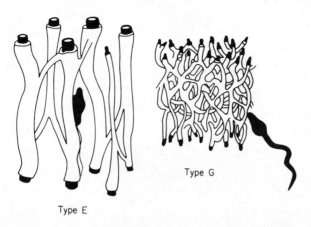

Type G

Type E

Fig.1.   Schematic three-dimensional view of the structure of mucus
type E & type G.   The macromolecular cores (consisting of several
long molecules side by side) are shown in black together with the
surrounding hydration cells (white).   A sperm moving in the
cervical plasma between the micelles of type E, and a non-
invading sperm outside type G are also shown.   From Odeblad (1972).
Reproduced with the permission of the editors of Cervical Mucus in
Human Reproduction, World Health Organisation publication.

       Odeblad and his group in 1978 have now proposed "a dynamic mosaic
model" of the human ovulatory cervical mucus (fig.2) demonstrating
the heterogenous nature of the cervical mucus.   The more viscid
parts, which they call E_L mucus lie as "loafs" in a more fluid
material.   After drying, the "loafs" are easily recognised in the
ferning material by their large crystals with a flower-like arrangement.   Also, the distribution of spermatozoa in the cervical mucus
at post-coital testing, shows a heterogenous distribution.   Indeed,
progressively moving spermatozoa are usually confined to narrow high-
ways of more fluid mucus which they call "strings" consisting of E_s
mucus.   They propose a 3-dimensional mosaic pattern, with a high
degree of order inside the cervical canal comprising components which

are being continuously altered to form such a dynamic mosaic. This
arrangement is thought to be important for effective upward spermatozoal progression, since only complete "strings" of E_s material have
the capacity to convey the spermatozoa rapidly upwards in the cervical
canal. This E_s material is identical with that described previously
as type E mucus. At mid-cycle the crypts responsible for the
production of "strings" are said to be most abundant in the upper
part of the cervical canal which is not in keeping with the morphological findings of Insler and his co-workers (see above). The whole
system is not static, but dynamic and undergoes continuous changes
with time. After invading a "string" the spermatozoa rapidly
progress to some crypts within a few minutes. These crypts serve as
a sperm reservoir with an average storage time of 15 hours. However,
some spermatozoa rapidly enter into the uterine cavity probably
utilising "stair-cases" made up of these fused E_s fragments.

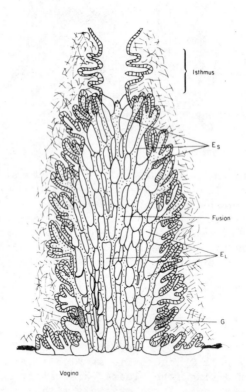

Fig. 2. Dynamic mosaic model of the cervical canal (Odeblad 1978).
E_S "strings of mucus threads."
Fusion of "strings" of mucus containing columns of motile
spermatozoa.
E_L "loaf" of mucus of more dense material
G thick tacky mucus. (Reproduced with permission of the author),

ULTRA STRUCTURE OF CERVICAL MUCUS

A variety of electron microscopic techniques have been explored to explain the hypothesis based on the studies of Odeblad which have attempted to explain the unique biological properties of cervical mucus.   All electron microscopic techniques involve the production of artifacts and preparative processes which are liable to degrade the mucus and have resulted in cautionary comments made about inter-pretation based on electron microscopic data.   Most transmission electron microscopic techniques involve extensive degradative processes in order to render the mucus soluble so that it might be applied to the grid.   In this way some changes in the orientation of the organic material had been demonstrated which only partly explain Odeblad's concept.   Scanning electron microscopy usually involves less severe preparative procedures and produces reproduci-ble artifacts and would therefore give a better indication of the three-dimensional structure of the high viscosity component of the mucus.   A number of studies have been reviewed which gave varying findings which are probably related to differences in techniques of preparing specimens.   Those techniques involving less dehydration demonstrated clear channels with the glycoproteins and their water hydration forming a macromolecular meshwork through which there were wide aqueous channels at the time of oestrogenic influence through which the spermatozoa could readily travel.   When the specimens were more intensely dehydrated, a meshwork was demonstrated consisting of filaments of organic material which coalesced to form fibres of different size.   At times of progestational effects there was a dense meshwork impeding entrance of spermatozoa to the upper genital tract (Elstein & Daunter, 1977).   The underlying mechanisms to explain the changing physical properties of cervical mucus in rela-tion to our current state of knowledge of the biochemistry of this hydrogel will be dealt with in the symposium by Dr Chantler.

MOLECULAR ARRANGEMENT OF CERVICAL MUCUS:

BASED ON LASER LIGHT-SCATTERING SPECTROSCOPY

The concept of linear channels formed by micelles as defined by Odeblad (see above) has been refuted by the work of Lee, Verdugo, Blandau and Gaddum-Rosse (1977).   In their view, when a cross-linked network is subjected to sheer (as between two planes moving parallel in opposite directions), it will deform as long as the sheer is maintained.   However, cervical mucus will begin to flow after the initial deformation if the sheer stress is maintained.   By using the monochromatic coherent light produced by lasers, they were able to study the molecular dynamics of cervical mucus by performing time correlation of light-scattering by cervical mucus.   In this way they were able to study cervical mucus and yet cause negligible perturbation of the system which they regard as critical in the study

of molecular systems such as cervical mucus,which is sensitive to
external stress.   By using laser light-scattering spectroscopy they
found that the molecular arrangement of mid-cycle human cervical mucus
should be more accurately described as an ensemble of entangled,
random-coiled macromolecules similar to raw or unvulcanised rubber
rather than the orderly micellar arrangements described by Odeblad
(fig.3 a + b).

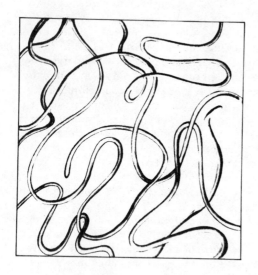

Fig.3A. A model of
entangled macromolecules
representative of human
mid-cycle cervical mucus.

Fig.3B.   A model of
entangled macromolecules
after "stress" has been
induced.   The randomly
coiled and entangled macromolecules align themselves
along the direction of
stress.

(Reproduced with permission
of authors & editors of
Journal Gynecol.Invest.
8:254.(1977)).

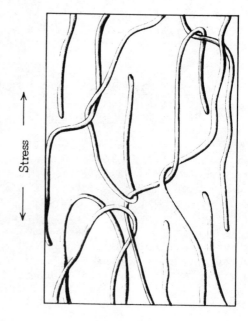

They suggest that changes in the rheological properties of cervical mucus may be due to variations in the fluid content of the entangled ensemble rather than to changes in the degree of chemical cross-linking. Although luteal phase cervical mucus contains 10% less water than mid-cycle mucus, they are of the view that these changes can be attributed to differences in water content. The chemical basis for this discussion will be dealt with by Dr Chantler but it should be borne in mind that mucus does not swell when brought into contact with saline (Silberberg 1977). Their view is that the molecular arrangements are similar for luteal phase mucus as for mid-cycle mucus, only that the spaces between these random-coiled chains were reduced. The degree of entanglement was thought to increase owing to ionic interactions since these become more pronounced when intersegmental distances are reduced. This would result in the mucus becoming more compact and develop this dense network structure which makes it impossible for spermatozoa to penetrate the mucus. The cyclic changes in the molecular structure would also affect the decay time of the auto-correlation function of the laser scattered light.

This concept of the molecular arrangement of the cervical mucus is such that spermatozoal penetration would be an entirely mechanical phenomenon not requiring any enzymic lysis of chemical cross-links of these molecules. When stress is applied to the cervical mucus entanglement of macromolecules this would induce linear orientation of these random-coiled molecules. This orientation would facilitate the penetration of the spermatozoa into an axis parallel to the orientated macromolecules. Indeed, it is a well-known observation that spermatozoa penetrate more readily into the axis of such aligned mucus and perpendicular to it. Such "stress-induced orientation" of the mucus probably occurs in vivo as the mucus is secreted and drains into the vagina through the cervical canal. In this aligned state the sperm readily would travel up the cervical canal. Lee and his co-workers have demonstrated a particular pattern of spermatozoal movement through cervical mucus by high speed cinematography which supports their concept of entangled random-coiled macromolecules of cervical mucus which resembles the molecular configuration of raw rubber supporting the view that cervical mucus is a "super elastic material."

MECHANISMS OF SPERM MOVEMENT THROUGH CERVICAL MUCUS

Current data indicate that ejaculated spermatozoa rapidly enter the mid-cycle cervical mucus. There is a rapid early transport of spermatozoa into the cervical canal so that they are seen at the internal os in about 1.5 to 3 minutes following ejaculation during coitus (Sobrero & McLeod 1962). This rapid migration is thought to be accomplished principally by the instrinsic mobility of the spermatozoa. It might also be influenced by the proteolytic activity

of the seminal plasma and the spermatozoa and their particular
orientation into the cervical mucus as demonstrated by the phalanx
formation.   However, the primary factor is probably the particular
orientation of the organic material of the cervical mucus as discussed
above.   This molecular arrangement is likely to facilitate their
storage in the crypts and the gradual release of these spermatozoa
over an extended period of time into the uterus and hence on towards
the fallopian tubes.   During their sojourn in the cervical crypts
the spermatozoa obtain oxygen and substrates for metabolism by
diffusion through the cervical plasma.   This retention in the crypts
also protects them from parasites within the female genital tract
since pre-ovulatory cervical mucus contains very few, if any,
leucocytes.   Furthermore, only motile spermatozoa of optimal shape
will be able to travel through the cervical mucus into the uterus and
only normally shaped spermatozoa are seen above the endocervix with
the abnormal forms being trapped within the cervical canal.   This
mechanism ensures a constant release of motile spermatozoa into the
upper genital tract so that there is an even distribution of the
spermatozoa into the ampulla to meet the ovum with its short life span.

     There are thus three phases of spermatozoal transport;   first,
the rapid initial passage through the genital tract immediately
following intercourse, followed by the colonisation of the cervical
mucus reservoir in the crypts of the cervical canal and a subsequent
prolonged release of spermatozoa allowing them to travel through the
genital tract in cohorts following their release from the storage in
the cervical crypts.

     Experimental data support this concept of the rapid transport
through the genital tract since spermatozoa have been found at the
ampullary end of the tube 5 minutes after deposition of the external
os with a constant level for 15 to 45 minutes (the duration of the
study) after insemination (Settlage, Motoshima and Tredway, 1973).

     These data are particularly intriguing since the in-vitro sperm
velocity in pre-ovulatory mucus varies between 0.1 to 3mm a minute.
Their arrival at the ampulla in such a short time suggests that their
travel is facilitated by some transit mechanisms in the female tract
which still require to be elucidated.   Such a factor has been
suggested by Hoglund and Odeblad (1977), which they have called
Axreveillian, a low molecular weight substance, which they allege
is present in the isthmic region of the uterus.   They have suggested
that spermatozoa which colonise the crypts are immotile or slowly
motile, but are re-activated by this factor which may play a role in
their rapid transit and activation through the genital tract.   The
colonisation of the crypts is such that shortly after intercourse
there is a greater number of spermatozoa in the crypts in the lower
regions of the canal as compared with the upper part of the canal;
with the passage of time there is a gradient as the spermatozoa
ascend the genital tract (Jaszczak & Hafez, 1973).   This is

consistent with the changes in the surface area of the endocervical canal described above and the fact that the crypts are more abundant in the lower part of the canal and the surface area of the endo-cervical mucosa is greatest at mid-cycle.

The linear movement of spermatozoa through the cervical mucus was originally demonstrated by Tampion & Gibbons (1962). They showed that spermatozoa orient themselves in a linear way, particularly when the mucus was placed on the stretch. When mucus and spermatozoa are placed in contact with each other in an in vitro slide test there is initially a phalanx formation and then invasion of the mucus with spread throughout the material. This phalanx formation was thought to be a function of proteolytic invasion of the cervical mucus. However, studies at high speed cinematography by Gaddum-Rosse,Blandau and Lee (1980) indicate that phalanx formation is not due to active deformation or penetration of the mucus by spermatozoa, but rather to a movement of spermatozoa into spaces or clefts existing between folds at the edge of the mucus. They showed that the spermatozoa in each phalanx were orientated with the head towards the interior of the mucus. Despite the active flagellation, the forward movement was slow. Once they left the phalanx and penetrated the mucus they lost their uni-directional orientation and moved in a random way. The initial move-ment was a figure-of-eight motion.However, when the mucus was aligned in a linear fashion the spermatozoa showed a distinct orientation, swimming actively along an axis parallel to the axis of stretch of the mucus. The initial pattern of flagellation of the spermatozoa during the penetration into the cervical mucus was that of a figure-of-eight, which appeared to be the extra driving force required by the "pioneering" spermatozoa. These "front-line" spermatozoa made their way through the mucus, forged a pathway when the mucus was randomly orientated and loosely intertwined. These workers consi-dered that the pioneering spermatozoa changed the local molecular arrangement and allowed the following spermatozoa to enter into a more orientated environment and move in an undirectional way. After the initial figure-of-eight movement, spermatozoa developed a longitudinal rotation with rapid unidirectional movement along the axis of stretch of the mucus.

These patterns of spermatozoal movement support their concept of the randomly coiled entangled macromolecules of the cervical mucus, resembling raw rubber as described above.

PROTECTIVE ROLE OF THE CERVICAL MUCUS

The uterine cervix appears to be the major site of secretory immunological activity in the female genital tract. The cervix is liberally equipped with plasma cells and cervical mucus contains immunoglobulins part of which is locally secreted. This can be demonstrated by finding secretory immunoglobulin A as serum

immunoglobulin A, and immunoglobulin G in the mucus.   These
immunoglobulins are formed locally at the mucus surfaces with the
secretory component coming from the columnar epithelium.   Large
amounts of these immunoglobulins have been found in the mucus, in the
lumen of the cervical crypts and on the surface of the epithelium.
The combination between secretory component and IgA to give
secretory IgA acts as a barrier to the passage of bacteria by
bacterial lysis with complement and lysozyme and by agglutination
following coating of the bacteria (Rebello, Green & Fox, 1975).
Increases in immunoglobulins have been described in the cervices
of women with a variety of cervical infections or following local
immunisation (Chipperfield & Evans, 1972; Jones, 1976). Sparks,
Purrier, Watt & Elstein (1981) have demonstrated that a tailed
Intra-uterine-Device interferes with these protective mechanisms and
allows bacteria to ascend into the uterus.   This ascent of bacteria
through cervical mucus and into mucus attached to a cervical
appendage above the level of mucus has been demonstrated in an in-
vitro model resembling conditions in the cervical canal (Purrier,
Sparks, Watt & Elstein, 1979).

     It is clear, therefore, that the cervix not only prevents sperm
entering the upper genital tract at times other than when required
for fertilisation, but also prevents the ascent of micro-organisms
into the upper genital tract.   Pelvic infections tend to occur
following mentruation when these protective mechanisms can be
interfered with.   The impairment of these protective mechanisms
during menstruation forms the basis of the restrictions on intercourse at the time of menstrual bleeding in many of the religious
tenents.   Clearly these protective mechanisms need to be operational
at mid-cycle when spermatozoal penetration is optimal.   The presence
of a cervical appendage to an IUD has been shown to interfere with
these mechanisms.

                    THE CERVICAL FACTOR IN INFERTILITY

     Abnormalities of the cervix and its secretions are considered
by some authors to be responsible for over 15% of cases of infertility, whereas others are of the opinion that the true incidence is
probably not greater than 5-10% (Moghissi, 1976).   Nevertheless,
within the cervix there might be a significant factor in a couple's
infertility per se or the examination of its mucus might be the means
of recognising other important causes of the failure to reproduce.
It has long been suggested that the proteolytic enzymes of spermatozoa play a role in their passage through the mucus and in
fertilisation.   Seminal plasma, spermatozoa and cervical mucus
contain inhibitors to these enzymes.   An inter-action between these
enzymes and the inhibitors is probably involved in spermatozoal transport, capacitation and fertilisation.   Indeed, the role of acrosin
in the acrosomal cap of the sperm remains to be firmly established

in this regard.   However, Moghissi (1981)and his group have recently
found that   oligospermic males manifesting infertility have diminished
acrosin available and there are degenerative changes in their acrosomal cap.   Whether there is a capacitation-decapacitation mechanism
in the human cervical canal remains to be established, although the
likelihood of changes in the acrosome and its inhibitors as part of
this mechanism seems probable.

In all cases of infertility there are multifactorial aspects and
of these the importance of the cervical factor needs to be appreciated by clinicians.   Due attention to the significance of this factor
is likely to improve the management of the infertile couple.   The
assessment of the physical properties of cervical mucus gives an
excellent indication of the biological effect of oestrogen on the
genital tract.   The post-coital test is of considerable importance
because it provides a wealth of information concerning the couple's
reproductive potential.   It provides information on the couple's
coital technique; the quality and mobility of the sperm within the
mucus as well as the quality and physical properties of the cervical
mucus (Elstein 1978).   The finding of a post-coital test with
impaired motility of sperm is an indication for in vitro sperm-
mucus interaction tests which have been described in detail previously
(Elstein 1974).

Treatment of the cervical factor where cervical mucus hostility
per se has been demonstrated is difficult.   Artificial insemination
within the cervical canal offers the best chance of success.   This
method of management is also appropriate with a couple where the
male is oligospermic using the first part of a split-ejaculate where
the optimal spermatozoa are usually obtained.   This method of treatment is only resorted to when the mechanical and endocrinal factors
have been corrected.   In the case where ovarian function is abnormal
this needs to be optimised by use of drugs such as anti-oestrogens,
e.g. clomiphene and tamoxiphen.   An end-organ inadequate response to
circulating oestrogens by the cervical mucosa can be treated by
supplementary oestrogens at mid-cycle and this is often effective.
Difficulty in evaluating data on the treatment of the cervical
factor is that there are often multiple factors in the couple's
infertility and the possibility of controlled studies is remote.

CONTRACEPTIVE MECHANISMS MEDIATED BY CERVICAL MUCUS

In view of its strategic position in the genital tract, the
cervix is an obvious site at which contraceptives might act.   Indeed,
mechanical occlusive diaphragms are placed in close proximity to the
cervix, but the more effective hormonal methods involve changes
in the mucus per se.   The possibility of a specifically designed
contraceptive modality which would alter the physical properties of
the mucus itself without any systemic effects, would be ideal, but

continues to be hypothetical.   The combined oestrogen formulations, even the low-dose preparations and the tri-phasic ones, all have changes in the cervical mucus as part of their ancillary contraceptive mechanisms.   Indeed, the tri-phasic formulations do try to mimic the rise and ebb of ovarian steroids during the menstrual cycle, but throughout their course of administration some progestogen is present in order to ensure the cervical mucus blocking effect.   The low-dose progestogens have changes in cervical mucus as their primary mode of action.   A recent development is the release of progestogens, levonorgestrel in particular, from vaginal rings retained continuously in the upper vagina.   Effectiveness and acceptability studies currently underway indicate that this is a valuable method of contraception in caucasian women.   A multi-centre trial currently underway has demonstrated that other than a small group of women in Southeast Asia this is a highly effective and most acceptable method of contraception with its primary action by altering the cervical mucus mainly by its systemic effect, although the possibility of a local diffusion into the mucus cannot be excluded.   The notion of a local diffusion of steroids into the cervical mucus from an intra-cervical device continues to be explored and a detailed study on women prior to undergoing hysterectomy will be reported at this meeting.   The problem is the risk of local irritative changes in the cervical epithelium particularly at the squamo-columnar junction because of the presence of a foreign body inside the cervical canal.   Indeed, changes have been demonstrated in the cervical epithelium owing to the presence of the foreign body but these are also present in women wearing an intra-uterine device with a cervical appendage.   Ascent of organisms into the lower part of the uterus have also been demonstrated with an intra-cervical device.   Intra-cervical devices loaded with levonorgestrel do result in changes in the cervical mucus which impair sperm migration.   However, in view of the morphological changes in the cervical epithelium, prolonged toxicological studies will be required to ensure that the minimal metaplastic changes demonstrated in the cervical epithelium are not sinister.   In view of this, the WHO Programme with intra-cervical devices has been curtailed.

Developments aimed at improvements in occlusive diaphragms being placed in adjacent positions to and covering the cervix are urgently required since this is likely to yield a safer and more acceptable contraceptive modality.   Such an occlusive diaphragm would need to release substances aimed at acting either directly on the spermatozoa, either by inhibiting their motility or their fertilizing capacity, or directed towards altering the physical properties of the cervical mucus.

REFERENCES

Chantler, E.N.C. (1981)
    This proceeding

2   Chipperfield, E.J. & Evans, B.A.
        Clin Exp Immunol II, 219
3   Elstein, M. & Daunter, B (1977)
        in The Uterine Cervix in Reproduction pp 52-61
        eds. Insler, V. & Bettendorf, G. Georg Thieme
        Publishers, Stuttgart.
4   Elstein, M. (1978)
        Br Med Bull 34 83
5   Elstein, M (1974)
        Clin Obstet Gynaecol 1 345
6   Gaddum-Rosse, P. Blandau, R.J. & Lee, W.I. (1980)
        Fertil Steril 33 636
7   Hoglund, A. & Odeblad, E. (1977)
        in The Uterine Cervix in Reproduction pp 129-134
        eds. V Insler & Bettendorf, G.   Georg Thieme
        Publishers, Stuttgart
8   Insler, V. & Glezerman, M. Bernstein, D. Zajdel, L. &
        Misgav, M (1981)
        in Advances in the diagnosis and treatment of infertility.
        eds. Insler, V. & Bettendorf, G. Proceedings of a
        meeting held at Badreichenhall (1980) in Press.
9   Jasczcsak, S. & Hafez, E.S.E. (1973)
        in Cervical Mucus in Human Reproduction pp 34-44
        eds. M. Elstein, K.S. Moghissi, & R. Borth, Scriptor
        openhagen
10   Jones, W.R. (1976)
        In: Jordan, J.A. & Singer, A. (eds.)
        The Cervix. pp 184-192 W.B. Saunders, London
11   Jordan, J.A. (1976)
        In: Jordan, J.A. & Singer, A. (eds.)
        The Cervix pp 44-50 Saunders, London.
12   Lee, W.I. Verdugo, P. Blandau, R.J. & Gaddum-Rosse, P.(1977)
        Gynecol Obstet Invest 8,254
13   Mann, E.C. McLarn, W.D. & Hayt, D.B. (1961)
        Am J Obstet Gynecol 81 209
14   Moghissi, K.S. (1976)
        Fertil Steril 27 117
15   Moghissi, K.S. (1981)
        Personal communication
16   Nuttall, I.D. & Elstein, M. (1981)
        in Progress in Obstetrics & Gynaecology 2 1982
        ed. J Studd, Churchill, Livingstone, London (in Press)
17   Odeblad, E. (1973)
        in Cervical mucus in human reproduction pp 58-74
        eds. Elstein, M. Moghissi, K.S. & Borth, R
        Scriptor, Copenhagen
18   Odeblad, E. (1978)
        Personal communication
19   Purrier, B.G.A. Sparks, R.A. Watt, P.J. & Elstein, M.(197●)
        Br J Obstet Gynaecol 86 374

20   Rebello, R.   Green, F.H.Y. & Fox, H. (1975)
         Br J Obstet Gynaecol 82 812
21   Settlage, D.S.F.   Motoshima, M. & Tredway, D.R. (1973)
         Fertil Steril 24 655
22   Silberberg, A. (1977)
         in Mucus in health and disease
         eds.   M. Elstein & D.V. Park, pp 181   Plenum Press
         New York & London
23   Sobrero, A.J. & MacLeod, J. (1962)
         Fertil Steril 13 184
24   Sparks R.A.   Purrier, B.G.A.   Watt, P.J.   & Elstein, M.(1981)
         Br Med J 282 1189

THE MECHANISMS AND ANALYSIS OF SPERM MIGRATION THROUGH CERVICAL
MUCUS

David F. Katz[1] and James W. Overstreet[1,2]

Departments of Obstetrics and Gynecology[1] and
Human Anatomy[2]
School of Medicine, University of California
Davis, California 95616 USA

INTRODUCTION

The migration of spermatozoa through cervical mucus has been
studied extensively by physiologists and clinicians. Yet our
understanding of this biological phenomenon remains superficial.
This has been due at least in part to a lack of scientific objec-
tivity in the experimental procedures employed, as well as a tendency to rely upon phenomenological thinking rather than fundamental principles of chemistry and physics. In this chapter our
attention will be directed initially to the biophysical context
within which the mechanisms of sperm transport in cervical mucus
should be considered. Our current knowledge of those mechanisms
will be summarized, and some of the applicable experimental techniques and mathematical models of analysis will be discussed.
Finally, some new strategies for future investigation will be
suggested. We will bring the reader's attention to a number of
scientifically sound and biologically relevent studies, but we will
not attempt a comprehensive survey of the research in this area.
For this information the reader is referred to a number of recent
reviews on sperm transport[1] and cervical mucus.[2,3,4,5] In beginning this chapter it should be emphasized that among mammals it is
primarily the ruminants and primates which produce substantial
amounts of cervical mucus, and it is to these species that our
comments will apply. The human is by far the best studied of these
species and most of the specifics of our discussion will refer to
human sperm-cervical mucus interaction.

A BIOPHYSICAL PERSPECTIVE OF SPERM TRANSPORT IN CERVICAL MUCUS

In thinking about the mechanisms of sperm transport through cervical mucus we can make a distinction between the kinetics and the mechanics of the process. Kinetics refers to the time-course of sperm accumulation and the variation in the distribution of sperm numbers in the mucus. Thus, investigation of the kinetics of sperm transport requires simultaneous attention to sperm numbers and migratory velocities in relation to time and location within the mucus. Kinetic analysis can provide information on where individual spermatozoa go over a given time interval, i.e. a knowledge of the trajectories of a sperm population that has entered the mucus. Such studies are mathematical in design and require formulation of various transport models. Examples of such models are the use of mathematical group theory by Odeblad,[6] and a Boltzmann-like transport equation system which we are developing.[7]

We use the term mechanics to refer to the production and action of forces which propel spermatozoa through the mucus. Any investigation of the "mechanisms" of sperm migration must take into account the generation of propulsive forces, resistances and/or migratory velocities of spermatozoa. Sperm transport through cervical mucus is a complex process that results from interacting physical, electrical, and chemical events. The expression of these interactions is often mathematical and requires careful use of statistics to distinguish between biological variability and biophysical causality.

Our understanding of the mechanisms of sperm migration through mucus has been developed from studies of spermatozoa, from information on the chemistry and physics of the cervical mucus, and from direct observations of interaction between these factors and the female genital tract. It should be appreciated that the mechanisms of mucous secretion, and thus the properties of the mucus itself exhibit cyclic, endocrine-controlled variation. It is only during a period of a very few days at the time of ovulation that cervical mucus can be penetrated by spermatozoa. During the remainder of the reproductive cycle the mucous microstructure is altered such that permeation by spermatozoa does not occur. In evaluating the current data on cervical mucous composition and structure we must also bear in mind that microscopic properties of the mucus experienced by the individual spermatozoa may be very different from those assessed in the laboratory with larger mucous volumes. The homogenization and/or dehydration of the mucus which are necessary for many physico-chemical assays alter its native gel state, and may obscure many physical properties which are important in its structure-function relationship with spermatozoa.

THE CURRENT UNDERSTANDING OF TRANSCERVICAL SPERM MIGRATION

A. Kinetics

   Substantial numbers of human spermatozoa have been observed
in midcycle cervical mucus recovered within 90 seconds after
coitus,[8] and sperm may continue to migrate into the cervix for
several additional hours before being immobilized by the acidic
vaginal environment.[9] Sperm numbers in the human cervix appear to
reach a maximum by 15 minutes after artificial vaginal insemination,
and within 20 minutes 95 percent of the cervical sperm population
has migrated from the vagina.[10] Surprisingly, there are no comparable data available for any other species.

   Although the evidence is not conclusive, experiments in a
number of mammals have demonstrated a rapid sperm transport phenomenon in which spermatozoa are catapulted by female visceral
contractions through the cervix and uterus to the upper oviduct
and peritoneal cavity.[1] Our work with rabbits suggests that these
spermatozoa probably do not engage in fertilization, and that their
transit through the female is lethal to most of the sperm cells
which reach the oviduct.[11] The biologically important result of
these early transport events may be rapid sperm colonization of
the cervical mucus. Rapid sperm transport lasts only a few seconds
and spermatozoa which subsequently reach the mucus must migrate
upward from the vagina. Following artificial insemination, sperm
numbers in human cervical mucus remain constant for approximately
4 hours and then begin to decline gradually.[10] In some women,
actively motile sperm can be recovered from the mucus for two or
more days after coitus.[2,3,4,5] It has been hypothesized that the
cervix is one of several anatomical regions in the female tract
which act as "reservoirs" for spermatozoa, and through gradual
release of the stored cells a continuing presence of spermatozoa at
the site of fertilization is insured.[2,3,4,5]

B. Mechanics

   Sperm passage into and through the cervix results from mechanical interaction between the mucus, the spermatozoa, and the female
tract. The mucus is a heterogenous, polydisperse secretion with
biophysical characteristics which result in part from its glyco-
protein-dominated microstructure.[12] The spermatozoa must pass within
the interstices of this microstructure. These spaces measure only
a few micrometers,[13,14] and thus spermatozoa must travel in very
close proximity to the mucous macromolecules. Direct physical as
well as hydrodynamic interactions are therefore substantial.[13] In
a rheological sense, the mucus behaves as a viscoelastic fluid of
some non-linearity.[15,16,17,18] Considerable external pressures are
applied to the mucus during coitus and the deformations that the
mucus undergoes in response to these applied pressures are probably

not uniform. Thus, the kinetics of sperm migration in vivo is mod-
ulated by the interplay among mucous heterogeneity, its strong physical interaction with spermatozoa, and the externally applied
stresses associated with coitus.

When mucus is "stretched" in vitro, its macromolecular backbone is aligned producing a physical arrangement which results in
directed sperm swimming along the local lines of strain. Observations of unidirectional sperm migration within stretched mucus in
vitro led some investigators to hypothesize that similar phenomena
may occur in vivo, spermatozoa being "guided" to the site of mucous
secretion in the cervical mucusa.[19,20] Others suggested that
the naturally occurring mucous heterogeneity may result in organized
"strings" of low viscosity mucus, along which spermatozoa preferentially migrate to particular cervical "crypts".[6] The alignment of
mucus in vivo would depend on the rate of mucous secretion, the
effects of external forces and the natural tendency of mucus to
"heal" itself after shearing. Unfortunately, we have no direct
observations on the organization of cervical mucus in vivo. It has
been presumed that the cervix can "store" spermatozoa in mucosal
crypts, although the actual mechanisms of sperm retention and
release are not known. Mucous properties at or close to the sites
of secretion may be substantially different from those of the whole
secretion as studied in vitro, and these differences could play a
role in the hypothetical mechanism of sperm retention. In the
rabbit, a species with little cervical mucus, sperm may be stored
in the cervix and lower isthmus of the oviduct via a combination of
inhibited sperm motility plus reversible adherence to epithelial
surfaces.[21,22] It is possible that similar mechanisms are also at
work in the primate and ruminant cervix. For example, Odeblad has
stated, without providing evidence, that spermatozoa are rendered
temporarily immotile or weakly motile in the crypts of the human
cervix.[6]

With our current understanding of the hydrodynamics of sperm
swimming, it is easy to appreciate that differences in local
properties of mucus can modulate the migration of spermatozoa.[13,23]
A number of factors may contribute to such mucous variability in
vivo. The naturally occurring polydispersity and heterogeneity of
mucus may result, in part, from the secretion of different "types"
of mucus by different glands within the mucosa.[24,25] The transition
of individual mucous secretory granules into a unified network undoubtedly creates zones of varying mucous properties. The shear
forces experienced by mucus will vary depending on its position in
the cervical canal and these will also affect local mucous characteristics. Experiments in vitro have shown that the mechanics of
the interaction between bull spermatozoa and bovine mucus differ in
relation to their proximity to a solid surface.[26,27] This distinction is probably due to an alteration of mucous properties near the
solid surface as well as a direct effect of the surface on sperm
movement.[27]

Both sperm motility and the contractility of the female viscera are important contributors to the mechanism of sperm migration through the mucus in vivo. These factors may act in two ways: by generating propulsive forces, e.g., the rapid transport phenomenon, and by altering local mucous properties. The latter possibility is well illustrated by recent experiments in our laboratory in which the "vanguard" spermatozoa (human, bovine) which first penetrated aligned mucus were observed to swim faster than the "following" spermatozoa which later arrived at the same location.[27,28] This could not be explained by the flagellar beats, which were not different in the two sperm populations. Rather, it seems likely that the following spermatozoa develop less propulsive thrust and/or encountere more resistence because of alterations in local mucous properties. It is possible that an increase in mucous microelasticity had occurred, and/or that the adherent forces between the sperm surface and mucous microstructure were altered as a result of the earlier penetration by the vanguard sperm. As yet, little is understood about these forces and their variation. Electrostatic repulsions between the negatively charged sperm plasma membrane and mucous glycoprotein are probably involved. In this context, it is interesting to note that most morphologically abnormal human spermatozoa are unable to penetrate the mucus in vivo or in vitro.[29,30,31] Although the motility of the abnormal cells in semen is not as vigorous as that of the normal ones,[32,33] it is our impression that factors additional to motility and morphology are involved, i.e. surface interactions with the mucus.

It seems likely that proteases endogenous to human seminal plasma assist in the passage of spermatozoa across the semen-mucus interface, but there is no conclusive evidence that the proteolytic enzymes contained within the sperm acrosome facilitate mucus penetration. Preincubation of mucus in vitro with seminal proteases has been shown to alter its physical properties, resulting in improved sperm progression.[34] Simple 1:1 dilution of human semen with Tyrode's solution reduces the relative number of spermatozoa that can initially enter the mucus, as compared with a 1:1 dilution of the semen with its own plasma.[35] In these experiments both groups of spermatozoa swam with equal velocities after having entered into the mucus proper. Thus, sperm ability to penetrate the semen-mucus interface should be distinguished from the ability to subsequently permeate the mucus.

The mechanical contractions of the female viscera and the pressures that result appear to be important mediators of initial sperm entry into the cervix. Because of the spatial heterogeneity in the organization of the mucous microstructure in vivo, because of its relatively small interstices, and because of its unique electro-chemico-physical properties, experiments which examine the

importance of external pressure gradients in transcervical sperm
migration must be very carefully designed. The intravaginal and
intrauterine pressures which have been recorded during human
coitus[36] produce forces which are hydrodynamically substantial
in comparison with the thrust developed by a sperm flagellum.
Small microparticles[37] as well as non-motile spermatozoa[38] have
been shown to enter the mucus after intravaginal insemination in
vivo. Under controlled conditions, we have demonstrated similar
phenomena in vitro. In contrast, radioopaque liquids do not readily
permeate the mucus after introduction into the vagina.[39,40] The
results of experiments with the latter materials have been cited
in arguments against pressure-mediated sperm transport into the
cervix.[41] Given the nature of cervical mucus, negative results
with liquids or large microspheres (e.g. 50 μm in diameter) are
not surprising from a biophysical point of view; and they may have
little biological relevance to the migration of spermatozoa.

METHODS OF MEASUREMENT AND ANALYSIS

A. Sperm Movement

     To study the kinetics of sperm migration, we must be able to
measure the swimming speeds of spermatozoa. Studies of the mechanics of sperm-mucus interaction may require considerably more information, including the shapes of the swimming trajectories, and the
time-dependent geometrical details of the flagellar beat. The
suitability of techniques for measuring net swimming speed alone
and/or the details of sperm movement is a primary consideration in
judging their application to a specific study. A second major
technical distinction between various methods concerns their
relative abilities to determine a mean value or central tendency
of a movement characteristic for a sperm population, to determine
the distribution of that characteristic, or preferably to provide
the per-sperm distribution of a set of movement characteristics.
Direct photomicrographic techniques (cinemicrography, videomicro-
graphy, time-exposure micrography) are the only methods capable of
eliciting this latter information, and we use these exclusively in
our laboratory for study of sperm-mucus mechanics. Assessment of
the kinetics of sperm penetration of mucus in vitro is a valued
clinical diagnostic procedure. Such tests are often performed
with the mucus contained in capillary tubes of circular[42,43] or,
more recently, rectangular[44] cross-section. Attention is generally
paid to the movement of spermatozoa within the mucus and the
numbers of spermatozoa that succeed in mucous entry. The classical
capillary tube penetration test measures by direct visual observation,
the distrance traversed by the forwardmost vanguard spermatozoa.[42]
More recent methods define a "percentage of successful collisions"
of those spermatozoa which are exposed to the mucous interface and
which succeed in crossing it.[45] This latter computation focuses,
thus, upon all spermatozoa within the mucus. Since it is now

evident that vanguard and following spermatozoa behave differently
in mucus in vitro, it should be appreciated that these two approaches
measure different functions of the sperm population.

Direct applications of the principles of mechanics and hydrodynamics to data on sperm movement in mucus have been scarce,
although our laboratory has begun to conduct such studies.[27,28]
Our current approach is to use the spermatozoon as a biophysical
probe of the mucus, determining from its mechanical interaction
with the mucus microstructure some of the physical characteristics
of that structure. In such studies, the relationships between
swimming velocity and flagellar beat kinematics are examined in
relation to time and space. High speed cine films are analyzed
frame by frame, and the flagellar beat is characterized paramet-
rically. Analysis of covariance and stepwise multiple regression
techniques are then used to apply the data to mathematical models
of sperm-mucus hydrodynamics. Thus, the swimming speed is
treated as the dependent variable, and the beat parameters
(frequency, amplitude, wavelength, etc.) are taken as independent
variables and covariates. The analyses of covariance provide a
means of distinguishing between biological and biophysical variation,
and the stepwise multiple regression methods provide the best
quantitative fit of the experimental data to the qualitative
hydrodynamic models. This combination of experimentation, hydrodynamic theory, and applied statistics has enabled us, for example,
to elicit the distinctions between "vanguard" and "following"
spermatozoa mentioned previously.

Indirect methods for measuring swimming speed can also be
applied to sperm movement in mucus. Automatic techniques, such
as image analyzing computers[46] and laser light scattering spectro-
meters[47] could be used in some experimental designs. However,
the interpretation of such measurements requires detailed understanding of their physical and mathematical basis, as well as
application of a set of qualitative assumptions about the nature
of sperm movement, e.g. its geometry and distribution within the
suspension. These assumptions must be tested experimentally. Thus
any application of a new instrument that assesses sperm movement
indirectly should commence with a careful evaluation of its accuracy
and precision as compared with direct measurement on the same sperm
suspensions.

B. Mucous Properties

A variety of objective biophysical and physico-chemical techniques have been used to study the properties of cervical mucus.
These include flow birefringence for study of the shape of the
mucous glycoprotein;[48] rheometry for assessment of viscoelastic
properties;[15,16,17,18] NMR for measurement of diffusion and relaxation
times;[24,25,49] laser light scattering for assessing intermolecular

associations;[50] and flow permeation experiments for assessing the geometry of the microstructure.[14] These methods have provided the variety of information on the bulk structure and macroscopic behavior of the mucus. As such, they have contributed to the characterization of the mucus in classical physico-chemical terms,[12,20] However, even with micro-miniaturization, these techniques must measure bulk properties on a scale of several hundred micrometers. A spermatozoon is primarily influenced by the mucus within one body length of it.[23,50] Thus, these methods do not perceive the mucus on the scale that a sperm does, and we cannot be sure how the properties determined compare with those experienced by the sperm cell in vivo.

FUTURE DIRECTIONS

In spite of the significant amount of previous research on cervical mucus, we have little specific information on the structure-function relationship between mucus and spermatozoa or on the mechanisms of sperm transport into and through the cervix. Future work must emphasize departures from the traditional approaches of the past, and will necessarily be multi-disciplinary. Cervical mucus must be studied on several scales: the scale of individual spermatozoa, that of groups of spermatozoa, and still larger regions up to and including the entire cervical content. Account must be taken of changes in mucous properties due to cyclic variation in mucous synthesis, rates of secretion and efflux, as well as those resulting from external forces and the presence of sperm cells themselves. Careful attention must be paid to the forces which propel spermatozoa and the resistances they must overcome. Particularly important are the surface-surface interactions between spermatozoa and the mucous microstructure, as well as those involved with sperm-epithelium interaction.

Future work must involve more studies in vivo. Much of our current conception of sperm transport is based on extrapolation from in vitro work, and the properties of the mucus within the cervix itself must now be examined directly. Future in vitro studies must be designed to overcome the substantial variability in mucous properties that hinders current work. A significant advance toward this goal will be the standardization of methods for mucous collection and storage, concerning which considerable controversy and contradiction still exist. The first important steps in attaining these goals must include the uniform application of experimental methodology which is objective, quantitative, accurate and precise. Measurements and analysis must be founded on fundamental scientific principles. We can then shift our attention away from the limitations of previous experiments and focus our concern on the true biological variability, composition and function of the cervical mucus.

REFERENCES

1. J. W. Overstreet and D. F. Katz, Sperm Transport and Selection
    in the Female Genital Tract, in: "Development in Mammals II,"
    M. Johnson, ed., North Holland Publishing Company, New York
    (1977).
2. V. Davajan, R. M. Nakamura, K. Kharma, Spermatozoan transport
    in Cervical Mucus, Obstet. Gynec. Survey 25:1 (1970).
3. R. J. Blandau, and K. Moghissi, "The Biology of the Cervix,"
    University of Chicago Press, Chicago (1973).
4. M. Elstein, K. Moghissi, and R. Berth, "Cervical Mucus in
    Human Reproduction," Scriptor, Copenhagen (1973).
5. V. Insler and G. Bettendorf, "The Uterine Cervix in Reproduc-
    tion," Geo. Thieme Publ., Stuttgart (1977).
6. A. Hoglund and E. Odeblad, Sperm Penetration in Cervical
    Mucus, a Biophysical and Group-Theoretical Approach, in:
    "The Uterine Cervix in Reproduction," V. Insler and G.
    Bettendorf, eds., Geo. Thieme Publ., Stuttgart (1977).
7. D. F. Katz, and S. Paveri Fontana, A Stochastic Model of
    Transcervical Sperm Migration, (submitted for publication)
    (1981).
8. A. J. Sobrero and J. MacLeod, The Immediate Postcoital Test,
    Fert. Ster. 13:184 (1962).
9. W. H. Masters and V. E. Johnson, The Physiology of Vaginal
    Reproductive Function, West. J. Surg. Obstet Gynec. 69:105
    (1961).
10. D. R. Tredway, D.S. Fordney Settlage, R. M. Nakamura, M.
    Motoshima, C. U. Umezaki, and D. R. Mishell, Significance of
    Timing for the Postcoital Evaluation of Cervical Mucus,
    Amer. J. Obstet. Gynec. 121:387 (1975).
11. J. W. Overstreet, and G. W. Cooper, Sperm Transport in the
    Reproductive Tract of the Female Rabbit. I. The Rapid
    Transit Phase of Transport, Biol. Reprod. 19:101 (1978).
12. R. A. Gibbons and R. A. Sellwood, The Macromolecular
    Biochemistry of Cervical Secretions, in "The Biology of the
    Cervix," R. J. Blandau and K. S. Moghissi, eds., Univ. Chicago
    Press, Chicago (1973).
13. D. F. Katz and S. A. Berger, Flagellar Propulsion of Human
    Sperm in Cervical Mucus, Biorheology 17:169 (1980).
14. P. Y. Tam, D. F. Katz, S. A. Berger, and G. F. Sensabaugh,
    Flow Permeation Analysis of Bovine Cervical Mucus, Biophys. J.
    (1981).
15. S. W. Hwang, M. Litt and W. C. Forsman, Rheological Properties
    of Mucus, Rheolog. Acta. 8:438 (1969).
16. N. Eliezer, Viscoelastic Properties of Mucus, Biorheol.
    11:61 (1974).
17. D. P. Wolf, L. Blasco, M. A. Khan and M. Litt, Human Cervical
    Mucus. I. Rheological Characteristics, Fert. Ster. 28:41
    (1977).
18. P. Y. Tam, D. F. Katz and S. A. Berger, Non-linear Visco-

elastic Properties of Cervical Mucus, Biorheology (1981).

19 P. E. Mattner, Formation and Retention of the Spermatozoan
    Reservoir in the Cervix of the Ruminant, Nature, Lond. 212:1479
    (1966).

20. R. A. Gibbons, and P. E. Mattner, The Chemical and Physical
    Characteristics of the Cervical Secretion and Its Role in
    Reproductive Physiology, in: "Pathways to Conception",
    A. I. Sherman, ed., Charles C. Thomas, Publ., Springfield, Ill.,
    (1971).

21. J. W. Overstreet, G. W. Cooper and D. F. Katz, Sperm Transport
    in the Reproductive Tract of the Female Rabbit. II. The
    Sustained Phase of Transport, Biol. Reprod. 19:115 (1978).

22. G. N. Cooper, J. W. Overstreet and D. F. Katz, Sperm Transport
    in the Reproductive Tract of the Female Rabbit. Sperm Motility
    and Patterns of Movement at Different Levels of the Tract,
    Gamete Res. (1979).

23. R. D. Dresdner and D. F. Katz, Relationships of Mammalian Sperm
    Motility and Morphology to Hydrodynamic Aspects of Cell Func-
    tion, Biol. Reprod. Suppl 1, 73A (1980).

24. E. Odeblad, The Functional Structure of Human Cervical Mucus,
    Acta. Obstet. Gynec. Scand. (Suppl. 1) 47:59 (1968).

25. E. Odeblad and C. Rudolfsson, Types of Cervical Secretions:
    Biophysical Characteristics, in: "The Biology of the Cervix,"
    R. J. Blandau and K. Moghissi, eds., Univ. Chicago Press
    (1973).

26. D. F. Katz and J. W. Overstreet, Mammalian Sperm Movement in
    the Secretions of the Male and Female Genital Tracts, in:
    "Testicular Development, Structure, and Function," E. Stein-
    berger, and A. Steinberger, eds., Raven Press, New York,
    (1980).

27. D. F. Katz, T. Bloom, and R. BonDurant, The Movement of Bull
    Spermatozoa in Cervical Mucus, (Submitted for publication)
    (1981).

28. D. F. Katz, B. T. Brofeldt, J. W. Overstreet and F. W. Hanson,
    Vanguard Spermatozoa Alter the Properties of Cervical Mucus,
    (Submitted for publication) (1981).

29. B. Fredricsson and G. Bjork, Morphology of Postcoital
    Spermatozoa in the Cervical Secretion and Its Clinical
    Significance, Fert. Ster. 28:841 (1977).

30. G. Perry, M. Glezerman, and M. Insler, Selective Filtration of
    Abnormal Spermatozoa by the Cervical Mucus in Vitro, in:
    "The Uterine Cervix in Reproduction," V. Insler and G. Betten-
    dorf, eds., Geo. Thieme, Publ., Stuttgart (1977).

31. F. W. Hanson and J. W. Overstreet, The Interaction of Human
    Spermatozoa with Cervical Mucus In Vivo, Amer. J. Obstet.
    Gynec. (In press) (1981).

32. J. W. Overstreet, M. J. Price, W. F. Blazak, E. L. Lewis and
    D. F. Katz, Simultaneous Assessment of Human Sperm Motility
    and Morphology by Videomicrography, J. Urol. (In press) (1981).

33. D. F. Katz, L. Diehl and J. W. Overstreet, Differences in the Movement of Morphologically Normal and Abnormal Human Seminal Spermatozoa, Fert. Ster. 35:256 (1981).

34. K. S. Moghissi and F. N. Syner, The Effect of Seminal Protease on Sperm Migration Through Cervical Mucus, Int. J. Fert. 15:43 (1970).

35. J. W. Overstreet, C. Coats, D. F. Katz and F. W. Hanson, The Importance of Seminal Plasma for Human Sperm Penetration of Cervical Mucus, Fert. Ster. 34:569 (1980).

36. C. A. Fox, H. S. Wolff and J. A. Baker, Measurement of Intra-Vaginal and Intra-Uterine Pressures During Human Coitus by Radio-Telemetry, J. Reprod Fert. 22:243 (1970).

37. G. E. Egli and M. Newton, The Transport of Carbon Particles in the Human Female Reproductive System, Fert Ster. 12:151 (1961).

38. P. E. Mattner, and A. W. H. Braden, Comparison of the Distribution of Motile and Immotile Spermatozoa in the Ovine Cervix, Aust. J. Biol. Sci. 22:1069 (1969).

39. N. H. Masters and V. E. Johnson, "Human Sexual Response," Little Brown and Company, Boston (1966).

40. A. J. Sobrero, Sperm Migration in the Human Female, In: "Proc. 5th World Congr. Fert. Ster." B. Weston and N. Wiquist, eds., No. 133 Exerpta Medica, Amsterdam (1967).

41. K. S. Moghissi, Sperm Migration Through the Human Cervix, in: "The Uterine Cervix in Reproduction," V. Insler and G. Bettendorf, eds., Geo. Thieme Publ., Stuttgart (1977).

42. J. Kremer, A Simple Sperm Penetration Test, Int. J. Fert. 10:209 (1965).

43. M. Ulstein, Evaluation of a Capillary Tube Sperm Penetration Method for Fertility Investigations, Acta Obstet. Gynec. Scand. 51:287 (1972).

44. R. N. Mills, D.F. Katz, A Flat Capillary Tube System for Assessment of Sperm Movement in Cervical Mucus, Fert. Ster. 29:43 (1978).

45. D. F. Katz, J. W. Overstreet and F. W. Hanson, A New Quantitative Test for Sperm Penetration into Cervical Mucus, Fert. Ster. 33:179 (1980).

46. D. F. Katz and H. M. Dott, Methods of Measuring Swimming Speed of Spermatozoa, J. Reprod. Fert. 45:263 (1975).

47. P. Jouannet, B. Volochine, P. Deguent, C. Series and G. David, Light Scattering Determination of Various Characteristic Parameters of Spermatozoa Motility in a Serie of Human Sperm, Andrologia 9:36 (1977).

48. R. A. Gibbons and F. A. Glover, The Flow Birefringence of Purified Bovine Cervical Mucoid, in: "Flow Properties of Blood and Other Biological Systems," A. L. Copley and G. Stainsby, eds., Pergamon Press, New York (1960).

49.   D. F. Katz and J. R. Singer, Water Mobility in Bovine Cervical
      Mucus, Biol. Reprod. 17:843 (1978).
50.   W. I. Lee, P. Verdugo, R. J. Blandau and P. Gaddum-Rosse,
      Molecular Arrangement of Cervical Mucus: A Reevaluation Based
      on Laser Light-Scattering Spectroscopy, Gynecol. Invest.
      8:254 (1977).

ACKNOWLEDGEMENTS

    The unpublished studies reported in this chapter were supported
by NIH grant 12971 (DFK), and Research Career Development Award
HD00224 (JWO). We thank Dr. Frederick W. Hanson for his participation in many of the clinical studies reported in this chapter.
The assistance of Ms. Sherry Stroud in preparing the manuscript
is gratefully acknowledged.

SPERMATOZOAL VELOCITY IN HUMAN CERVICAL MUCUS MEASURED BY

LASER DOPPLER VELOCIMETRY

Catherine Serres and Pierre Jouannet

Laboratoire d'Histologie et Embryologie
Centre Hospitalier
Kremlin Bicetre, 94270 France

Sperm penetration into and migration within the cervical mucus are dependent upon two factors: the intrinsic motility of the spermatozoa and the macromolecular structure of the cervical secretions.

A quantitative and objective evaluation of the migration of spermatozoa into cervical mucus can be made using Laser Doppler Velocimetry (LDV) (see Dubois et al., 1974). When an object moving at speed v receives monochromatic light having a frequency ν_0, it scatters the light which undergoes a frequency change ν according to the Doppler effect. When several objects (e.g. spermatozoa) move at different speeds, but in the same direction, the frequency change gives rise to a frequency spectrum $S(\nu)$ directly representative of $N(v)$, N beeing the number of objects moving at the speed v.

Preovulatory endocervical mucus was gently aspirated using a syringe attached to a special pipette made of optical quality glass. This pipette has a 1mm square lumen along its central axis and acts as a scattering cell. Pipettes were placed in test tubes containing 0.3 ml of the semen to be studied and incubated at 37° C.

Analysis of the light scattered by the spermatozoa moving in the cervical mucus gave $S(\nu)$ spectra having a definite peak at a frequency value shifted with respect to $\nu =0$ in 79% (74/94) of cases. Such a spectrum is clearly indicative of an orientated unidirectional movement of the spermatozoa. After one hour of incubation, the spermatozoal mean progressive velocity was $28 \mu s^{-1} \pm 1.12$ (SD, n=74) at 2.5-3 cm from the sperm-mucus interface. There was a good positive correlation ($p<0.01$) between the values of this parameter and mean

331

instantaneous velocity of spermatozoa in seminal plasma, $V = 56.4 \mu s^{-1}$ ± 3.1 (SD, n=63).

There was a good agreement between the results obtained by LDV and the progressive velocities measured by microcinematographic frame-by-frame analysis of movement in cervical mucus.

Measurement made at various times after the beginning of contact and at different distances from the interface showed an initial selection of the fastest cells, followed by an equilibration of the two populations by one hour. At this latter time there was no significant difference between velocities measured at 2.5-3 cm and 5- 6 cm from the interface.

Studies of the migration of spermatozoa from different ejacula- tes into the same mucus sample and of sperm from the same ejaculate into different mucus samples indicate that spermatozoa velocity is not the only factor determining the rate of penetration into the mucus.

REFERENCES

Dubois, M., Jouannet, P., Berge, P. and David, G., Nature, 252: 711, 1974

RESPIRATORY MUCUS

RESPIRATORY MUCUS: STRUCTURE, METABOLISM

AND CONTROL OF SECRETION

John T. Gallagher* and Paul S. Richardson**

*Cancer Research Campaign Dept. of Medical Oncology
Manchester Univ. and Christie Hospital and Holt Radium
Institute, Manchester M20 9BX, England
**Department of Physiology, St. George's Hospital
London SW17 ORE, England

Mucus has at least three important roles in the protection
of the airways:

(i) It is essential for the transport of dust, debris,
irritants and bacteria from the lungs. Two transport mechanisms
clean the airways. Ciliated cells line all the conducting
airways (i.e. those concerned with distributing air to the gas
exchanging parts of the lung rather than the process of gas
exchange itself) and these beat continuously to propel mucus
from the smaller to the larger airways and eventually out of the
lung altogether. In the absence of mucus or something with
equivalent physical properties, cilia are powerless to move dust
etc. (King et al, 1974). The other cleaning mechanism is cough
which is only called upon when the burden of irritation or debris
threatens to overwhelm the cilia. As with mucociliary transport,
mucus is essential for effective coughing. A dry cough fails to
remove dust from the lungs (Yeates et al, 1975).

(ii) Mucus also has a role in diluting irritants which
enter the airway and so it renders them less harmful.

(iii) There is growing evidence that mucus has antibacterial
and antiviral properties. It contains immunoglobulins, principally secretory IgA (Kaltreider, 1976)., lactoferrin which chelates
iron necessary for the growth of some bacteria (Masson & Heremans
1966) and lyzozyme which destroys some bacteria (Lorenz et al,
1957).

335

To the biochemist, however, mucus is a dreadful substance
(refs. not needed!):   it sticks to everything, disappears on
columns, clogs up filters, won't dissolve or dissociate, invariably it behaves anomalously and may end up being hurled through
the nearest window together with the most recent addition of
'Mucus in Health & Disease'!

In this account, we will deal principally with mucus produced by the cat. We will try to give some indication of the
structure of the epithelial glycoprotein components (i.e. the
mucins), their cellular source and the physiological control of
their secretion. We will attempt to relate our findings to
respiratory secretions of other mammalian species and give some
indication of the more compelling avenues for future investigation.

The Cat Tracheal Perfusion System

This has been described in detail elsewhere (Gallagher et
al, 1975). Briefly, the trachea of an anaesthetised cat is cut
beneath the larynx and above the mainstem bronchi. The exposed
trachea is flushed periodically with warm (37°C) Krebs-Henseleit
medium and the animal breathes through a cannula inserted
caudally to the lungs. The principal advantages of the system
are that the mucus is derived exclusively from a healthy trachea,
the blood and nerve supply is intact and the mucins can be radiolabelled biosynthetically by introducing appropriate radio-
isotopic precursors into the tracheal lumen. In most experiments
described here, the mucins were labelled by introducing $Na_2{}^{35}SO_4$
and ^3H-glucose (more specific oligosaccharide labels, e.g.gluco-
samine, were poorly incorporated) simultaneously into the lumen
in 2 ml of Krebs-Henseleit. After 1 hour the non-incorporated
isotopes were flushed out and the lumen was then washed through
at intervals with medium to collect radioactive mucins.

The influence of various drugs, nervous stimulation and
other possible 'modulators' of mucous secretion can be easily
evaluated and structural changes in mucins which follow repeated
stimulation can be monitored.

Nervous Control of Secretion.

Parasympathetic nerves. Kokin (1896), in Pavlov's laboratory,
described how electrical stimulation of the vagus nerves resulted
in blebs of mucus appearing on the lining of the dog trachea and
then spreading out to cover the surface. Florey et al (1932)
confirmed this and showed that atropine would block the effect.
Gallagher et al (1975) were able to show that the vagal stimulation enhanced the secretion of mucus glycoproteins as well as
the volume of secretions. There has been no direct demonstration

that man shares this parasympathetic control of airway secretion,
but there is quite persuasive indirect evidence that this is so.
Cholinergic drugs cause secretion from human bronchial tissue
(Sturgess & Reid, 1972: Boat & Kleinerman, 1975). The submucosal
glands receive nerve fibres, many of which appear to be cholinergic on ultrastructural grounds (Bensch et al, 1965: Meyrick
& Reid, 1970). It is probable then, that the parasympathetic
nervous system controls airway secretion even in man.

Sympathetic nerves. Electrical stimulation of the sympathetic
nerve supply to the cat trachea increases the output of mucous
glycoproteins (Gallagher et al, 1975). There is good anatomical
evidence that submucosal glands in the cat receive a sympathetic
nerve supply (Murlas et al, 1980).

     Until recently it seemed unlikely that sympathetic nerves
play any part in controlling airway secretion in man. Earlier
studies failed to show that sympathomimetic drugs enhance mucus
secretion from human bronchi in vitro (Sturgess & Reid, 1972:
Boat & Kleinerman, 1975) but recently we have found that both
α- and β- adrenoceptor agonists stimulate the secretion of
radiolabelled mucins from human bronchial tissue in vitro
(Williams et al, 1981).   The method involves opening a bronchus
removed at operation and mounting it between the two halves of an
Ussing chamber. Mucins are radiolabelled with ^{35}S and secreted
mucus is collected by draining the fluid from the luminal chamber.
These experiments have shown clearly that, as ion the cat, both
α- and β-adrenoceptor stimulating drugs cause secretion and that
the appropriate adrenoceptor blocking drugs can prevent these
effects. It is now necessary to search for sympathetic nerves
which may innervate the submucosal glands in man.

Reflexes which control airway secretion

     If autonomic nerves influence secretion it is almost certain
that there must be reflexes which act through these nerves. In
fact this is so in the cat. Irritation of the airway at different
levels augments mucus secretion and this acts through the parasympathetic and sympathetic nerve pathways described above.
Gentle rubbing of either the laryngeal epithelium or the nasal
and nasopharyngeal epithelium increase secretion of mucins from
the trachea (Phipps & Richardson, 1976). Only after both
sympathetic and parasympathetic nerves have been cut does the
effect vanish completely. Administration of ammonia vapour into
the lower airway at concentrations sufficient to cause coughing
also causes secretion from the trachea in the neck. Atropine is
sufficient to block this, so this effect depends upon parasympathetic nerve fibres (Phipps & Richardson, 1976).

Two of these stimuli (rubbing the larynx and the presence of
ammonia vapour in the lungs) caused coughing. We and others
have argued that a cough would only be effective at removing
irritating material from the airway wall if this were embedded
in a thick layer of mucus (Phipps & Richardson, 1976: Camner et
al, 1973a: Yeates et al, 1975). Perhaps an increasesd production of mucus should be regarded as an essential part of the
cough reflex.

     Little is known about other reflexes which augment respir-
atory tract mucus secretion or even whether the same influences
apply in man. Camner et al (1973b) have shown that the inhalation of dust-laden air increases the rate at which mucociliary
transport clears marker particles from the lungs. They suggest
that inhaled dust sets up vagal reflexes which encourage mucus
secretion and that this, in turn, assists mucociliary transport.

Local influences on airway secretion.

     Autonomic nerves are not the only influence on secretion:
many local mechanisms have now been described which increase the
rate of secretion even when extrinsic nerves have been cut or
their influences blocked with pharmacological agents. We shall
describe these in three groups: irritants, serum and mediators.

Irritants. Most of those chemicals that feel irritant to breathe
also seem to provoke mucus secretion when administered locally
inmto a segment of trachea. That studied most has been ammonia
vapour which provokes copious mucus secretion even in the trachea
in vitro. There is some evidence that ammonia acts via local
reflexes (Richardson et al, 1978) but this has not been tested
properly. It would be easy to do this by seeing whether tetrodotoxin, which prevents nerve conduction without blocking
secretion, abolishes the effect in vitro. To the extent that
tetrodotoxin fails, ammonia must act independently of nerves.

     Sulphur dioxide is another irritant with a local action.
At a concentration of 200 parts per million it enhances mucus
secretion and in two preliminary experiments the effect has
survived cutting of both sympathetic and parasympathetic nerve
supplies. Similar results have been found with cigarette smoke
(Richardson et al, 1978).

     The riot control agent CS gas (O-chlorobenzilidene malononi-
trile) which causes intense pain at all levels in the respiratory
tract, failed to influence secretion in the cat trachea. Despite
its inactivity in the cat, it was a particularly strong secretagogue in the goose trachea. All the evidence supported the idea
that this was a direct effect with no reflex component (Richardson et al, 1978).

The direct action of a variety of irritants in promoting secretion immediately suggests that the response would dilute the irritant and hasten its removal by ciliary action. Dahlamn(1956) has pointed to a possible flaw in this argument: he found that this sort of irritation quickly paralysed the cilia and so the respiratory tract would be left loaded with copious mucus which it was incapable of clearing. The hypothesis might still be true: sufficiently dilute irritants stimulate mucus secretion without causing ciliary paralysis. Even if the irritant is present transiently at high concentrations, the cilia may recover their function quickly as the irritant is diluted.

Serum. Whenever the airway becomes inflamed serum proteins leak from local capilliaries, through the respiratory epithelium and into the airway lumen. We have described recently (Hall et al, 1978) that when dilute serum comes into contact with the airway epithelium it provokes an outpouring of mucus into the cat trachea. This is a direct effect: atropine and propranolol, at concentrations which block the effects of parasympathetic and sympathetic nerve stimulation, do not inhibit the secretagogue action of serum. The concentration of serum required to evoke this effect in the cat airway are well within those often found in human sputum, so it is likely that this mechanism contributes to the flow of mucus into the inflamed airway.

It is still uncertain how serum stimulates secretion, but it is the large molecules which seem to induce the effect. When serum was fractionated by dialysis it was only the molecules held within the dialysis bags (i.e. >12,000 Daltons) which promoted secretion. When dialysed serum was fractionated into three broad molecular weight bands by gel filtration on Sephadex G-200, all fractions promoted secretion. We think, therefore, that it is some general property of proteins to elicit secretion rather than some specific component of serum. In keeping with this, we have found that certain foreign proteins (bovine serum albumin, horse-radish peroxidase) also stimulate secretion into the airway.

It is likely that this is an adaptive response: invasion of the airway by serum, bacteria or foreign proteins encourage mucus secretion and, we suggest, enhances the rate of mucociliary clearance. The latter part of the hypothesis is speculative and needs to be tested directly.

Mediators. Some of the chemical mediators released in inflamm-ation promote mucus secretion into the airway. These include several of the prostaglandins and histamine (Lopez-Vidriero et al, 1977: Richardson et al, 1978). It is probable that irritants and infections have at least part of their action on the airway epithelium by releasing such mediators which then cause release of mucus, dilution of the irritant and its removal.

This account has stressed the variety of control systems
which lead to the release of mucus into the airway and suggested
ways in which the output of mucus may assist in the protection
of the lungs.   What has been less emphasised is the variety of
cells producing the airway secretions and the histochemical and
immunocytochemical evidence that that contribute different components to mucus (Jones, 1978).   It is now almost certain that
different control systems cause release of different types of
mucus, though the study of the differences between different
sorts of mucus and the different ways in which they protect the
airway has scarcely reached its cradle (King & Viires, 1979).

Biochemical Studies of Secreted Mucins.

We have mainly analysed cat mucins produced in experiments
involving sequential use of pilocarpine and ammonia.   Our major
objectives were to compare mucins produced under resting conditions with those released in response to the administration
of first pilocarpine (submucosal gland cell mucins) followed by
ammonia vapour.

General properties of tracheal mucin.

Some of our findings have been previously described (Gallagher et al, 1977).   In general, secreted cat tracheal mucin
could be resolved by gel filtration on Sepharose CL-4B, under
dissociating conditions (6M urea or 4M guanidine chloride) into
two peaks: the major component eluted in the void volume and the
smaller, minor component was partially retarded (Fig.1a).   The
large component also eluted in the void volume of Sepharose CL-2B
and was not dissociated in the presence of reducing agents
(e.g. DTT).   In control or resting level secretions, the smaller
component had a higher ^{35}S-^{3}H ratio than the larger one.
Pilocarpine-stimulated mucins showed a considerable increase in
^{35}S:^{3}H ratio, particularly in the small component but this ratio
was also elevated in the excluded fraction (Fig.1b).   Repeated
stimulation with pilocarpine still resulted in the release of
high and low molecular weight fractions and no 'unusual peaks',
possibly indicative of the release of 'immature' mucin components were observed.   We might conclude from these findings that
the integrity of native mucin structure is maintained despite
the continuous demand for secretion.   This conclusion is
supported by monosaccharide analyses of the secreted material
which showed that in the crude mucin fraction repeated pilocarpine caused no striking changes in carbohydrate composition.
Galactose, N-acetylglucosamine and fucose were the main sugars,
with smaller amounts of sialic acid and N-acetylgalactosamine
also present (Gallagher et al, 1977).   Interestingly, the overall
carbohydrate content of the separated large and small mucin
components were very similar to the unfractionated material.

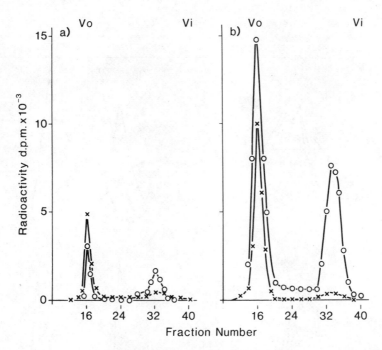

Fig.1. Gel filtration on Sepharose CL-4B of cat tracheal mucin.

An isolated, perfused cat trachea was incubated for one hour
with ^{3}H-glucose and Na_2 $^{35}SO_4$ introduced into the tracheal lumen.
The radiolabels were then washed out and biosynthetically-labelled
mucins collected at various time intervals.   The mucins were
dialysed, concentrated and applied to the Sepharose CL-4B column
(80 cm x 2 cm) which was eluted with 6M urea, 1mM dithithreitol
in 20mM sodium phosphate buffer, pH 6.8.   The samples were dialysed
against this buffer before fractionation.   4 ml fractions were
collected and the flow rate was 6 ml/hr. (x-x-x = ^{3}H, o-o-o = ^{35}S).

The graphs indicate the amount of radiolabelled mucin
secreted during a 15 min. 'control' period (panel a) and after a
15 min.stimulation with pilocarpine (panel b) introduced into the
tracheal lumen at a concentration of 0.025 mM.   The graphs illustrate the stimulatory effect of pilocarpine on the secretion of
both high and low molecular weight mucins but the stimulation is
particularly notable for the ^{35}S-sulphate content of the smaller
component.

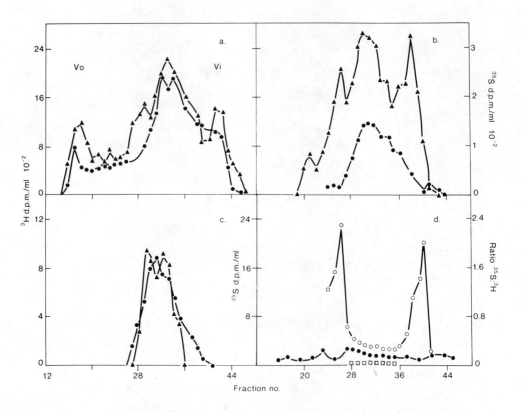

Fig.2.   Glycopeptides in cat tracheal mucin.

      The high molecular weight mucin from Sepharose CL–4B chroma-
tography of control, pilocarpine- and ammonia-stimulated secretions
was pooled, dialysed against distilled water and concentrated.
Each preparation was then digested with papain (1 mg/ml of enzyme
at 60°C for 18 hours in 50 mM sodium phosphate buffer, pH 7.0,
containing 5 mM cysteine and 1 mM EDTA).   The digest was cooled
to 4°C and fractionated on a Sepharose CL–4B column as described
in the legend to Fig.1.

       Panel a - control mucin ($\bullet$–$\bullet$–$\bullet$ = ^{3}H, $\blacktriangle$–$\blacktriangle$–$\blacktriangle$ = ^{35}S)

       Panel b - pilocarpine mucin   "

       Panel c - ammonia mucin   "

       Panel d - ratio of ^{35}S : ^{3}H in control ($\bullet$–$\bullet$–$\bullet$),
                 pilocarpine (o-o-o) and ammonia (□□□□□)mucins

Ammonia-stimulated mucin was considerably depleted in the small component and the ^{35}S:^{3}H ratio of the excluded material was much lower than either pilocarpine or control mucins.   The carbohydrate composition of this material was enriched in sialic acid and reduced in fucose (Gallagher et al, 1977: Hall, R.L., Richardson, P.S. and Gallagher, J.T. - unpublished observations)

When the high molecular weight component from Sepharose CL-4B was digested with papain (essentially reduction and proteolysis since the buffer contained 5mM cysteine) and re-chromatographed on the column most of the radioactivity was retarded by the gel (Fig.2).   The resolution observed on the degraded material was variable but usually two or three partially resolved peaks were observed.   Each peak had a similar ^{35}S:^{3}H ratio in the control mucins (Fig.2a).   However, in pilocarpine-stimulated material, the higher and lower molecular weight components were enriched in ^{35}S-sulphate and most of the ^{3}H-label was in the intermediate peak (Fig.2b).   Only one component, low in ^{35}S, was observed after proteolysis of the main ammonia-stimulated mucin: it corresponded in size to the intermediate component from pilocarpine and control secretions (Fig.2c).

These findings indicate that the high molecular weight mucin consists of a complex of two or three populations of glycoprotein or mucin subunits which contain protease-sensitive regions – these may correspond to the naked peptides described by several workers (Roberts, 1976: Creeth et al, 1977: Allen, 1977). These naked peptides are presumably in some way involved in the formation of an aggregated high molecular weight complex. Our inability to induce any measurable dissociation with DTT suggests that disulphide bonds linking subunits, of the type described for gastric mucus (Allen, 1977) and human sputum (Roberts, 1976) are not involved in maintaining the complex. Alternatively, disulphide bonds may be protected from reduction in the native mucin (i.e. the aggregate of subunits) or possibly the glyco-sylated domains may be part of the same core protein, separated by a 'continuous' naked peptide of the type described by Roberts (1974). Disulphide bonds may link such complex subunits, but the subunit itself may be sufficiently large to be excluded from Sepharose CL-4B and CL-2B. Clearly, we need more information on the molecular structure of this high molecular weight material. One important point that does emerge, however, is that the glyco-peptide domains within it are heterogeneous. Pilocarpine appears to elicit the secretion of glycopeptides which, although rich in ^{35}S-sulphate, are low in ^{3}H-label. If we accept the general notion that ^{3}H-glucose will label mainly carbohydrate residues, in order to explain the production of a structure of high ^{35}S-content but very low ^{3}H, it is necessary to conclude that for certain glycopeptides there is a considerable delay in time between glycosylation and sulphation or that an endogenous pool

of glycosylated mucin exists which is only sulphated at the time
of secretion.   As will be indicated below, our electrophoretic
and ion-exchange data suggest an increase in charge density for
pilocarpine stimulated mucins and are thus in agreement with the
conclusion that these mucins have a higher sulphate content.
There was no increase in the proportion of sialic acid in pilocarpine mucin.   However, the rate of secretion of mucin-associated
monosaccharides was increased about fourfold by pilocarpine (Hall,
R.L., Richardson, P. & Gallagher, J.T.. - unpublished results).
This establishes that the enhanced relase of radiolabelled mucin
is a real increase in chemical terms and not simply the secretion
of products of high specific radioactivity.   Some of the
important problems raised by these experiments concern the
metabolic coupling of glycosylation and sulphation in submucosal
gland cells (the source of pilocarpine mucin) and the influence
of secretory stimuli on the sulphation process. The possibility
should be considered that certain pools of intracellular mucins
may be secreted only in response to an appropriate stimulus -
resting level secretions may thus have a distinctive character
from stimulated secretions, features strongly evident from our
histochemical and biochemical findings.

Electrophoretic Analyses.

     We have studied the properties of crude and fractionated
mucins by zone electrophoresis on cellulose acetate paper.   In
this system, the crude mucin fractions (i.e. concentrated
dialysed tracheal washings) move as broad bands or smears towards
the anode. (Fig.3).   Mucins were identified by staining with
Alcian Blue (AB) or PAS and although each stain was reactive over
an extensive area of the mucin band, they were not entirely coin-
cident.   Mucins of intermediate mobility were reactive with both
stains but slower components stained only with PAS and faster
components only with AB.   The faster moving AB positive zone
was noticeably enhanced in pilocarpine mucin and absent in
ammonia mucin.   Both mucins gave a strong PAS reaction.

     Analysis of mucin fractionated on Sepharose CL-4B showed
that the larger mucin fraction contained the slower moving
components and the smaller fraction was faster moving and stained
with particular intensity with AB.   These findings are consistent
with the strong AB reactivity and high ^{35}S-sulphate content of
unfractionated pilocarpine mucin (NB. this material was enriched
in the smaller mucin component) and with the absence of a highly
^{35}S-sulphated small mucin in ammonia secretions.

     When the distribution of radioactivity along the electro-
phoretic strip was determined, both the glucose and sulphate
radiolabelled precursors were localised mainly over the PAS and
AB reactive regions, although some counts were always found at

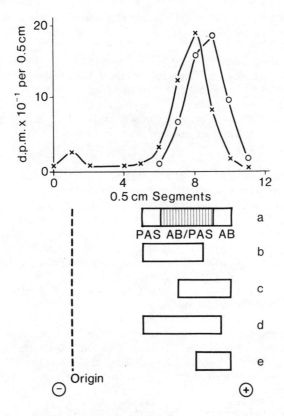

Fig. 3.   Electrophoretic analysis of cat tracheal mucin.

Concentrated samples of cat mucin were applied to cellulose acetate paper and subjected to electrophoresis under conditions of constant voltage (200v) for 45 min in 75 mM sodium barbitone buffer, pH 8.6.

The upper panel shows the distribution of radioactivity from ^{14}C-glucose-(x-x-x) or ^{35}S-sulphate-(o-o-o)labelled cat mucin. The lower panel shows an analysis of corresponding material stained by AB or PAS.   The major peaks of radioactivity co-migrate with the stained section.   In the lower panel:

a – unfractionated mucin
b and c – high and low molecular weight mucins respectively
          from Cl-4B chromatography
d and e – early and late (i.e. more anionic) eluting mucins
respectively from NaCl-gradient elution from DEAE-cellulose
chromatography.

the origin   (Fig.3).   The ^{35}S-sulphate label was slightly
displaced towards the more anionic region of the mucin zone.

These results seem to indicate considerable heterogeneity of
tracheal mucin. However, the influence of molecular weight on
electrophoretic mobility cannot be ignored: the smaller mucin
moved faster than the larger one and although a difference in
charge density of these two mucins is indicated from radio-
labelling data, the larger mucins may show some affinity for the
cellulose acetate paper and so be retarded by adsorption effects.
Interestingly however, when mucins were fractionated by salt
elution from DEAE-cellulose chromatography, they emerged as a
wide, fairly symmetrical peak at between 0.15M - 0.3M NaCl
(Gallagher et al, 1977). When the peak was split into two equal
halves and the concentrated fractions analysed by electrophoresis,
we found that the 'early' eluting material had a lower mobility
than the 'late' eluting components.   These properties roughly
corresponded to those of the high and low molecular weight
fractions off Cl-4B (Fig.3). The ion exchange data would seem
to be consistent with a higher charge density for the smaller
mucin fraction because, in principle, for macromolecules of
similar charge:mass ratios, the larger the molecule, the greater
the binding strength to an oppositely-charged resin.   Monosaccharide analysis showed a similar composition for the 'early'
and 'late' DEAE-fractions so presumably the enhanced charge
density is a consequence of a higher sulphate content.

Summary of biochemical findings.

The cat trachea secretes high and low molecular weight
mucins of similar carbohydrate composition but the smaller
component is enriched in sulphate. Pilocarpine selectively
promotes the secretion of gland cell mucin with a high ^{35}S:^{3}H
ratio.   This increased ^{35}S-sulphation is due principally to the
release of a high proportion of the smaller mucin and to the
presence of ^{35}S-sulphate rich glycopeptides in the high molecular
weight component. A homogeneous glycopeptide population characterised the ammonia stimulated mucin which probably originated
from the microvillus border of the surface epithelium.

Comparison with other tracheal mucins.

Comparative analysis is always difficult because of variation
in technique and possible involvement of disease. The major
difference between cat tracheal mucins and counterparts in other
mammals is their apparent resistance to thiol reagents (even
under denaturing conditions) and we are thus devoid of any
substantial information on the organisation of the high molecular
weight fraction. Heterogeneity of glycopeptide domains is
clearly evident in the cat mucin and distinctive glycopeptide

populations, characterised by significant differences in chemical
composition and polyanionic characteristics have been isolated
from human sputum (Havez et al, 1970: Roussel et al, 1978). Our
tentative conclusion that some of the highly sulphated mucins
which form part of the pilocarpine-sensitive pool, many not be
released unless the appropriate stimulus is provided, has an
interesting corollary in the results on mucins isolated from
bronchial washings of healthy human mucosa which was rich in
sialylated mucin but deficient in sulphated mucins (Lafitte et al,
1978).   The authors speculated that sulphated mucins may be
released only in the course of bronchial disease. Pilocarpine
may provide a stimulus to the healthy cat trachea comparable to
that in inflammatory states.   The question then arises of the
function of sulphated mucins. Since they are released in response
to stimulation we assume that they are of particular value in
mucosal protection against irritant dusts or pathogens.   They
may simply be more efficient at entrapment or perhaps the highly
charged sulphate residues stimulate ciliary activity, mucociliary coupling efficiency and hence airway clearance.

      We assume that the gel forming component in tracheal sec-
retions is the non-retarded mucin fraction isolated on Sepharose CL-4B. High molecular weight native mucins are also of clear
importance in sputum mucin gels (Creeth et al, 1977).   As
indicated previously, a disulphide-bond mediated aggregate of
mucin sub-units forms the active mucin but it should be borne in
mind that although in gastric mucin such bonds are located in the
'naked peptide' domains this feature is not established for
tracheal mucins. Creeth and co-workers (1977) also presented
powerful evidence for the role of cross-linking peptides in disulphide linking of sub-units and a similar conclusion was tentatively reached by Roberts (1976).   In this connection,low levels
of cysteine in purified mucin subunits might indicate that disulphide bonds do not mediate directly the assembly of mucin
subunits but that the aggregating or 'nucleating' properties of
cross-linking peptide requires a conformation stabilised by intra-
molecular disulphide links (Havez et al, 1968).   The crucial
experiments here will be to isolate native mucin from proteins
suspected of playing a cross-linking role: it should then be
possible, in reconstitution experiments, to examine the influence
of these proteins on the gel forming properties of the mucin.

      A further problem from our own studies is the physiological
function of the smaller, highly sulphated mucin (Fig.1). If the
large component is the gel forming species, the small one could
form part of a soluble phase which may be trapped in the interstices of the mucous gel.   When localised within the gel domain
its physical properties could have a considerable influence on
the rheological properties of the gel (see Bell et al, 1980, for
discussion of this point). Additionally, as a soluble component

the smaller mucin could form part of the 'sol' layer which bathes
the cilia and its polyanionic properties could, as indicated
earlier, affect ciliary mobility. Indeed, one of the outstanding
problems for future research concerns the role of charged
residues (sulphate and sialic acid)   in mucins. Highly sulphated
components have also been identified in human (Lafitte et al,
1977: Boat et al, 1976), dog (Ellis & Stahl, 1973) and rabbit
(Gallagher & Kent, 1975) respiratory mucin and it also appears
that sulphate and sialic acid are located predominantly on oligosaccharides of distinctive structural characteristics (Roberts,
1974: Roussel et al, 1975). We need to know how and why this
segregation occurs and the mechanisms employed by secretory
epithelia to modulate the polyanionic characteristics in response
to environmental variations. Our own studies (Gallagher et al,
1977) and those of others (Lamb & Reid, 1969) clearly indicate
heterogeneity in the types of mucin produced by different secretory cells and our findings with pilocarpine and nervous stimulation show that selectivity of the stimulus for particular target
cells is one way of altering the composition of extracellular
mucus.   However, the physiological requirement for structurally-
variable mucins is unclear and remains one of the outstanding
areas for future investigation.

REFERENCES

Allen, A. 1977. Structure and function in gastric mucus, in
    "Mucus in Health and Disease", Elstein, M. & Parke, D.V.
    Eds., Plenum Press, N.Y. p.283.
Bell, A.E., Allen, A., Morris, E. & Rees, D.A. 1980.
    The structure of native gastric mucus gel. Biochem.Soc.
    Trans. 8, 716.
Bensch, K.G., Gordon, G.B. & Miller, L.R. 1965.
    Studies on the bronchial counterpart of the Kulchitsky cell
    and the innervation of bronchial glands. J.Ultrastruct.Res.
    12, 668.
Boat, T.F. & Kleinerman, J.I. 1975.
    Human respiratory tract secretion: Effect of cholinergic and
    adrenegic agents on in vitro release of protein and mucous
    glycoprotein. Chest 67 (2 Suppl) 32S.
Boat, T.F., Cheng, P.W., Lyer, R.N., Carlson, D.M. & Polony,I.
    1976. Human respiratory tract secretions. Mucus glyco-
    protein of non-purulent tracheobronchial secretions and
    sputum of patients with bronchitis and cystic fibrosis.
    Arch.Biochem. 177, 95.
Camner, P., Mossberg, B. & Philipson, K. 1973a.
    Tracheobronchial clearance in chronic obstructive lung
    disease. Scand.J.Respir.Dis. 54, 272
Camner, P., Helström, P.A. & Philipson, K. 1973b.
    Carbon dust and mucociliary transport. Arch.Environ.
    Health, 26, 294.

Creeth, J.M., Bhaskar, K.R., Horton, J.R., Das, I., Lopez-
Vidriero, M.J. & Reid, L. 1977. The separation and characterisation of bronchial glycoproteins by density gradient
methods. Biochem. J., 167, 557.

Dahlamn, T. 1956. Mucous flow and ciliary activity in the
trachea of healthy rats and rats exposed to respiratory
irritant gases. Acta Physiol.Scand. 36, Suppl.123

Ellis, D.B. & Stahl, G.H. 1973. The biosynthesis of respiratory
tract mucins. Incorporation of radioactive precursors into
glycoproteins by canine tracheal explants in vitro. Biochem.
J. 136, 837.

Florey, H.W., Carleton, H.M. & Wells, A.Q. 1932. Mucus
secretion in the trachea. Br.J.Exp.Path. 13, 269.

Gallagher, J.T., Kent, P.W., Passatore, M., Phipps, R.J. &
Richardson, P.S. 1975. The composition of tracheal mucus
and the nervous control of its secretion in the cat. Proc.R.
Soc. B. 192, 49

Gallagher, J.T. & Kent, P.W. 1975. Structure and metabolism
of glycoproteins and glycosaminoglycans secreted by organ
cultures of rabbit trachea. Biochem.J. 148, 187

Gallagher, J.T., Kent, P.W., Phipps, R. & Richardson, P.1977
The influence of pilocarpine and ammonia vapour on the
secretion and structure of cat tracheal mucins. In "Mucus in
Health and Disease", Elstein, M. & Parke, D.V. Eds.
Plenum Press, N.Y. p.91.

Hall, R.L., Peatfield, A.C. & Richardson, P.S. 1978. The effect
of serum on mucus secretion in the trachea of the cat.
J.Physiol. 282, 47

Havez, R, Roussel, P., Degand, P. & Randoux, A. 1968.
Biochemical exploration of bronchial hypersecretions.
Prot.Biol.Fluids, 16, 343.

Havez, R. Degand, P., Roussel, P. & Randoux, A. 1970.
Definition biochemique du mucus bronchique. Le Poumon et le
Coeur, 26, 5.

Jones, R. 1978. The glycoproteins of secretory cells in airway
epithelium. In "Respiratory Tract Mucus", Ciba Foundation
Symposium 54 (New series)

Kaltreider, H.B. 1976 Expression of immune mechanisms in the
lung. Am.Rev. Resp. Dis. 113, 347

King, M., Gilboa, A.., Meyer, F.A. & Silberberg, A. 1974.
On the transport of mucus and its rheologic simulants in
ciliated systems. Am.Rev.Resp.Dis., 110, 740

King, M. & Vines, N. 1979. Effect of methacholine chloride on
rheology and transport of canine tracheal mucins. J. Appl.
Physiol. 47, 26.

Kokin, P. 1896. Veber die Secretorischen Nerven der Kehlkopf
und Luftröhrenschleimdrüsen. Arch.ges.Physiol.63, 622.

Lafitte,J.J., Lamblin, G., Lhermitte, M., Humbert, P., Degand,P.
& Roussel, P. 1977. Etude des mucines humaines obtenues pour
lavage de bronches macroscopiquement saines.Carbohyd.Res.56,383

Lamb, D. & Reid, L. 1969. Histochemical types of acidic glycoprotein produced by mucous cells of the tracheobronchial
glands in man. J. Path. 98, 213.

Lopez-Vidriero, M.T., Das, I., Smith, A.P., Picot, R. & Reid, L.
1977. Bronchial secretion from normal human airways after
inhalation of prostaglandin F2α , acetylcholine, histamine
and citric acid. Thorax, 32,734.

Lorenz, T.H., Korst, D.R., Simpson, J.F. & Musser, M.J. 1957.
A quantitative method of lyzozyme determination. I.An investigation of bronchial lyzozyme. J.Lab.Clin.Med. 49, 145.

Masson, P.L. & Heremans, J.F. 1966. Studies on lactoferrin, the
iron-binding protein of secretions. Prot.of Biol.Fluids.
14, 115.

Meyrick, J. & Reid, L. 1970. Ultrastructure of the cells in
the human bronchial submucosal glands. J.Anat. 107, 281

Murlas, C., Nadel, J.A. & Basbaum, C.R. 1980. A morphometric
analysis of the autonomic innervation of cat tracheal glands.
J.Autonomic.Nerv.Syst. 2, 23.

Phipps, R.J. & Richardson, P.S. 1976. The effects of irritation
at various levels of the airway upon tracheal mucus secretion
in the cat. J. Physiol. 261, 563.

Richardson, P.S., Phipps, R.J., Balfre, K. & Hall, R.C. 1978.
The roles of mediators, irritants and allergens in causing
mucin secretion from the trachea. In: "Respiratory Tract
Mucus" Ciba Foundation Symposium 54 (New series). p.111.

Roberts, G.P. 1974. Isolation and characterisation of glyo-
proteins in sputum. Eur.J. Biochem. 50, 265

Roberts, G.P. 1976. The role of disulphide bonds in maintaining
the gel structure of bronchial mucus. Archs.Biochem.Biophys.
173, 528.

Roussel, P., Lamblin, G., Degand, P., Walker-Nazir, E. &
Jeanloz, R.W. 1975. Heterogeneity of the carbohydrate chains
of sulphated bronchial glycoproteins isolated from a patient
suffering from cystic fibrosis. J.Biol.Chem. 250, 2114.

Roussel, P., Degand, P., Lamblin, G., Laine, A. & Lafitte, J.J.
1978. Biochemical definition of human tracheobronchial mucus.
Lung, 154, 241.

Sturgess, J. & Reid, L. 1972. An organ culture study of the
effects of drugs on the secretory activity of the human
bronchial submucosal gland. Clin.Sci. 43, 533

Williams, I.P., Phipps, R.J., Wright, N.L., Pack, R.J. &
Richardson, P. S. 1981. Sympathomimetic agonists stimulate
mucus secretion into human bronchi. Thorax, 36, 231.

CHARACTERISTICS OF HUMAN BRONCHIAL MUCUS GLYCOPROTEINS PREPARED IN THE ABSENCE OF REDUCING AGENTS

Geneviève Lamblin, Nicole Houdret, Pascale Humbert, Pierre Degand, Philippe Roussel and Henry Slayter*

Unité INSERM N°16, Place de Verdun, 59045 Lille France and *Sidney Farber Cancer, Boston, Mass, 02115

INTRODUCTION

Most of the procedures, already described to purify human bronchial mucus glycoproteins (mucins), comprise a reduction of bronchial mucus with disulphide bond breaking agents.

Purified mucins prepared after reduction of bronchial gel phase or soluble phase in phosphate buffer have a high carbohydrate and a low amino acid content. The molecular weight of such "reduced mucins" is generally less than 1×10^6 and electron microscopy has shown that these molecules consisted of rods with a distribution of lengths broader than would be expected for a homogeneous material[1].

Reduction under dissociative conditions (guanidine HCl) generally yields glycoproteins of higher molecular weight than if this reduction is carried under associative conditions in 0.075 M phosphate buffer. The larger decrease in size may be attributed to a mucolytic enzyme activated by reduction[2].

RESULTS

Bronchial mucus glycoproteins from a patient suffering from chronic bronchitis are extracted by dilution in water. Most of them are contained in a fraction which is excluded from a column of Sepharose CL2B eluted with a buffer containing 6 M guanidine HCl.

This fraction has a high carbohydrate content (galactose, fucose, N-acetylglucosamine, N-acetylgalactosamine and a little mannose). The proportion of hydroxyaminoacid is lower than in reduced mucins and the proportion of dicarboxylic acid is higher.

The molecular weight of this excluded mucin preparation is 1.3×10^6. EM shows that it has aggregating properties.

SDS polyacrylamide gel electrophoresis of this preparation shows glycoproteins which do not enter the gel and also components that have a much smaller molecular weight and which might play a role in the aggregating properties of this preparation.

EM of minor fraction of mucus glycoproteins which have a retarded elution from the Sepharose CL2B column, shows mainly rods.

CONCLUSIONS

Most of the bronchial mucus glycoproteins are extractible by water ; this confirms the works by Feldhoff et al.[3].

Chromatography under dissociating conditions leads to a glycoprotein preparation with a M.W. value more than 1×10^6.

Even in dissociating conditions, small components are still linked to the glycoproteins through non-covalent bonds.

REFERENCES

1. G. Lamblin, M. Lhermitte, P. Degand, P. Roussel and H. Slayter, Biochimie, (1980) 61 : 23.
2. N. Houdret, A. Le Treut, M. Lhermitte, G. Lamblin, P. Degand and P. Roussel, Biochim. Biophys. Acta, (1981) in press.
3. P.A. Feldhoff, V.P. Bhanavandan and E.A. Davidson, Biochemistry, (1979), 18 : 2430.

ISOLATION AND CHARACTERISATION OF NEUTRAL OLIGOSACCHARIDES

FROM HUMAN BRONCHIAL GLYCOPROTEINS

G. Lamblin, M. Lhermitte, A. Boersma, P. Roussel,
*H. Van Halbeek, *L. Dorland and *J.F.G. Vliegenthart

Unité INSERM N°16, 59045 LILLE, France and *University of Utrecht, The Netherlands

INTRODUCTION

Human acidic bronchial glycoproteins isolated from patients suffering from cystic fibrosis had been shown to be very heterogeneous with regard to acidity and molecular size of their carbohydrate chains. Neutral oligosaccharides from these glycoproteins have been isolated and purified ; their chemical structure have been determined.

RESULTS

Alkaline borohydride treatment of bronchial acidic glycoproteins leads to an heterogeneous population of reduced oligosaccharides and glycopeptides. The mixture, fractionated on ion exchange resin, gave a neutral oligosaccharide fraction (Ic) and three acidic oligosaccharide fractions.[1,2]
The neutral chains of fraction Ic were subsequently fractionated by DAX4 anion-exchange chromatography into eight oligosaccharides which were tested for purity by paper chromatography. Six oligosaccharides were pure and sugar analysis showed the components to be in the proper molecular proportions (Table I).

Table I : <u>Molecular composition of 6 major oligosaccharides eluted from paper chromatography.</u>

Oligosaccharides	GalNActitol	GlcNAc	Gal	Fuc
a	1	0.8		
b	1	0.8	1.2	
c	1		1.2	
d	1	1	1	
e	1	1	2	
f	1		1.2	1.2

Periodic oxydation, methylation analysis, gas liquid chromatography-mass spectrometry and 360-MHz ^{1}H-NMR of the 6 oligosaccharides lead to propose the following structures :

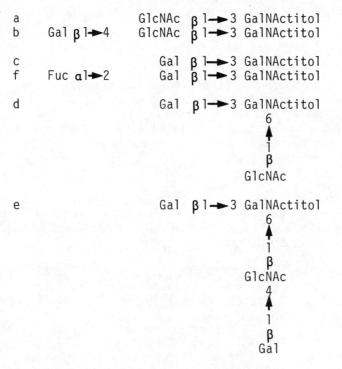

a   GlcNAc β 1→3 GalNActitol
b   Gal β 1→4   GlcNAc β 1→3 GalNActitol

c   Gal β 1→3 GalNActitol
f   Fuc α1→2   Gal β 1→3 GalNActitol

d   Gal β1→3 GalNActitol
                                    6
                                    ↑
                                    1
                                    β
                                  GlcNAc

e   Gal β1→3 GalNActitol
                                    6
                                    ↑
                                    1
                                    β
                                  GlcNAc
                                    4
                                    ↑
                                    1
                                    β
                                   Gal

In order to improve the purification of the oligosaccharides a convenient HPLC method has been worked out : it allows a rapid purification of the six oligosaccharides directly from fraction Ic.

CONCLUSION

     HPLC represents an interesting improvement among the methods used to purify neutral oligosaccharides.

     In the case of human bronchial mucus glycoproteins, at least three different cores exist in the carbohydrate chains.

     This HPLC procedure also enables a screening of neutral oligosaccharides from bronchial mucus glycoproteins secreted in different bronchial diseases.

REFERENCES

1 G. Lamblin, P. Humbert, P. Degand and P. Roussel, Clin. Chim. Acta, (1977) 79 : 425.

2 G. Lamblin, M. Lhermitte, A. Boersma, P. Roussel and V. Reinhold, J. Biol. Chem., (1980) 255 : 4595.

MACROMOLECULAR COMPOSITION OF SECRETIONS PRODUCED BY HUMAN BRONCHIAL EXPLANTS

Stephen J. Coles, K. R. Bhaskar, Donna Defeudis
O'Sullivan and Lynne M. Reid
Department of Pathology Research
The Children's Hospital Medical Center and
Harvard Medical School, Boston, MA 02115, USA

INTRODUCTION

In recent years, the successful maintenance of explants of human[1,2] and animal[3] airway mucosa in organ culture has contributed to our understanding of factors which regulate the secretion of tracheo-bronchial mucus. Such in vitro preparations, particularly of human airways, also provide the possibility for assessing the effects of pharmacologic stimuli on both the composition of airway mucus and the structural characteristics of its macromolecular components. This report describes the preliminary findings of an analysis of the macromolecular components of secretions produced by human bronchial explants.

METHODS AND RESULTS

Mucosal explants were prepared from specimens of one main and six lobar bronchi, obtained after lung resection, and maintained in organ culture for 48 h according to the method of Coles and Reid[1]. Each specimen provided 25-35 mucosal explants. In certain studies, methacholine chloride (5.10^{-6}M) was added to the culture medium to stimulate mucus secretion from submucosal glands[1]. Histologic assessment of explants after 48 h in culture showed that submucosal gland cells (mucous and serous) were well preserved with minimal necrosis. In the surface epithelium, goblet cells contained fewer secretory granules than before culture and, at the cut edges of the epithelium there was marked squamous cell proliferation. Autoradiographic analysis[1] showed that all mucous, serous and goblet cells were capable of incorporating, transporting and discharging ^{3}H-glucosamine.

357

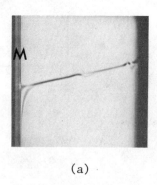
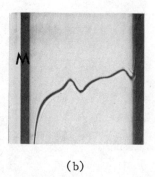

(a)   (b)

Fig. 1.   Analytical density gradient patterns of (a) unstimulated
          and (b) methacholine-stimulated human explant secretion,
          both in CsBr at an initial density of ∿1.5 g/ml.

     Media were harvested at 24 and 48 h and pooled and were seen
to contain strands of gelatinous material like mucus. Harvested
media were dialysed for 48 h and lyophilised yielding 2-6 mg dry
weight of non-dialysable material released by explants of one
specimen over a 48 h culture period. Lyophilised material was
dispersed in CsBr and examined by analytical density gradient
ultracentrifugation. The equilibrium patterns (Figure 1a and 1b)
showed the presence of the following components: (1) lipids, which
formed an insoluble layer at the top of the cell and partially
account for the dark bands at the left of Figure 1a and 1b, (2)
proteins, seen as a curvature in the meniscus (M) region, (3) in
the middle of the cell, a component of buoyant density ∿1.5 g/ml,
typical of bronchial mucus glycoprotein[4] - this component was only
barely detectable in secretions from unstimulated explants
(Figure 1a) but was increased substantially after methacholine
stimulation (Figure 1b), - and (4) at the base of the cell a
high density component of buoyant density ∿1.58 g/ml approaching
that of polysaccharides.

     The solutions were recovered from the cells and subjected to
preparative density gradient ultracentrifugation (DGU) in CsBr,
according to the method of Creeth et al.[4], to separate the four
components.

(1) Lipids   The lipid component was extracted into chloroform:
methanol (2:1, v/v) and fractionated on Unisil columns into non-
polar lipids (chloroform elution), glycolipids (acetone: methanol
9:1, v/v) and phospholipids (methanol elution). Each fraction was
subsequently examined by thin-layer chromatography[5]. The non-
polar lipid fraction separated into free glyceryl ethers, 1,3
diglycerides, free fatty acids, cholesterol and cholesteryl
palmitate. Two-dimensional TLC separated the phospholipid fraction

into three components: lysolecithin, phosphatidylcholine and phosphatidylethanolamine. While our preliminary studies have separated several glycolipids, we do not yet know which types are present.

(2) Proteins Radial immunodiffusion against appropriate antisera showed that albumin, IgA and IgG were present in the protein fraction. Lysozyme was also detected.

(3) Mucus glycoprotein Preliminary studies showed that this component contained only protein and carbohydrate. This, together with its characteristic buoyant density, suggests that it is a mucus glycoprotein. In addition, an increase in this component was observed after methacholine stimulation, indicating that it is derived predominantly from submucosal glands. In one experiment, we recovered a sufficient amount of this component to carry out gas chromatographic analysis of carbohydrates[6]. GalNAc was the predominant component; Fuc, Gal, Glc, GLcNAc, AcNeu, GalNAc were present in the molar ratios of 1:2:0.7:2.7:2.3:13.6.

(4) Dense component In the secretions from unstimulated explants, this component represented 50-60% of the total non-dialysable material. Gas chromatographic analysis indicated the presence of hexoses (12%), hexosamines (10%), xylose and glucuronic acid. Carbazole assay showed that uronic acids comprised 16% of this fraction. Sulfate ($\sim$5%) and protein (10%) were also present but no sialic acid or DNA. While only $\sim$60% of this component can be accounted for in terms of carbohydrate and protein, the presence of xylose and uronic acid suggest that it contains a proteoglycan.

In one experiment, the distribution of ^{14}C-glucosamine to the macromolecular components of human bronchial explant secretions was examined. Unstimulated explants were incubated for 48 h with 1 μCi/ml ^{14}C-glucosamine hydrochloride and the harvested secretions separated by DGU. Small amounts of label were found in the protein ($\sim$15%) and mucus glycoprotein ($\sim$10%) components but the highest concentration of bound label ($\sim$75%) was associated with the dense, proteoglycan-containing component.

CONCLUSION

Human bronchial mucosal explants maintained for 48 h in organ culture secrete lipids, proteins and serum-type glycoproteins, mucus glycoprotein and proteoglycan. In our studies, care was taken to prepare explants free from lung tissue and cartilage and so the components identified in the secretions are derived from the bronchial mucosa, though the presence of albumin indicates a small degree of contamination by blood serum. Contrary to our expectations, the major carbohydrate-protein complex synthesised and secreted by bronchial explants is not mucus glycoprotein but appears to be proteoglycan, though the mucus glycoprotein component is enriched by increasing submucosal gland secretion with metha-choline. While the presence of proteoglycans may be an artefact of the culture system (i.e. derived from connective tissue), this may not be the case since proteoglycans have been detected in canine bronchial aspirates in vivo[7].

REFERENCES

1.   S. J. Coles and L. Reid, Inhibition of glycoconjugate secretion by colchicine and cytochalasin B. An in vitro study of human airway, Cell Tissue Res., 214:107 (1981).

2.   T. F. Boat and J. L. Kleinerman, Human respiratory tract secretions. Effect of cholinergic and adrenergic agents on in vitro release of protein and mucous glycoproteins, Chest, 67(Suppl.):325 (1975).

3.   L. W. Chakrin, A. P. Baker, S. S. Spicer, J. R. Wardell, Jr., N. Desanctis and C. Dries, Synthesis of macromolecules by canine trachea, Am. Rev. Respir. Dis., 105:368 (1973).

4.   J. M. Creeth, K. R. Bhaskar, J. R. Horton, I. Das, M. T. Lopez-Vidriero and L. Reid, The separation and character-ization of bronchial glycoproteins by density gradient methods, Biochem. J., 167:557 (1977).

5.   P. O. Kwiterovich, H. R. Sloan and D. Fredrickson, Glyco-lipids and other lipid constituents of normal human liver, J. Lipid Res., 11:322 (1970).

6.   J. R. Clamp, T. Bhatti and R. E. Chambers, The examination of carbohydrates in glycoproteins by gas-liquid chromatography, in: "Glycoproteins, Their Composition, Structure and Function, 2nd edition", A. Gottschalk, ed., Elsevier, Amsterdam (1972).

7.   K. R. Bhaskar, unpublished observation.

CHARACTERISATION OF SOL AND GEL PHASES OF INFECTED AND MUCOID

SPUTUM SAMPLES FROM A CHRONIC BRONCHITIC PATIENT

K. R. Bhaskar, D. D. O'Sullivan, M. T. Lopez-Vidriero,
and L. M. Reid
Department of Pathology
The Children's Hospital Medical Center and
Harvard Medical School, Boston, MA 02115

Our recent application of density gradient methods to characterise the composition of sol and gel phases of sputum from asthmatic[1] and cystic fibrosis (CF)[2] patients showed differences between the two groups. Gel from the CF sputum dispersed readily in CsBr and soluble mucus glycoprotein separates from the insoluble lipids after the first preparative density gradient ultracentrifugation (DGU). Gels from the two asthmatic sputa, on the other hand, were very refractory to dispersion in CsBr and although the bulk of the serum proteins were removed in the first DGU, lipids remained bound to the mucus glycoprotein and were only removed by the third DGU. The mucus glycoprotein from asthma and CF had broadly similar chemical compositions yet at similar concentrations asthma glycoprotein gave far more viscous solutions than did the CF glycoprotein. Sol and gel of CF both contained mucus glycoprotein but in asthma this was confined to the gel phase. To investigate whether these differences can be attributed to the clinical variations or whether they arise because the asthmatic sputum samples were mucoid while the CF samples were infected, we have now examined infected and mucoid samples from the same chronic bronchitic patient.

The first sputum sample was collected when the patient was hospitalised with an infection, the second two weeks after the patient had been on antibiotic treatment for a week – by macroscopic examination the second sample was mucoid. Each sample was collected over a 1-2 hour period, and, immediately after collection, the samples were separated in sol and gel phases by ultracentrifugation at 160,000 g for 30 minutes. Sol and gel were in similar amounts in the mucoid sputum whereas only 33% of the infected sputum separated as gel. Viscosity of the sputum and of the

361

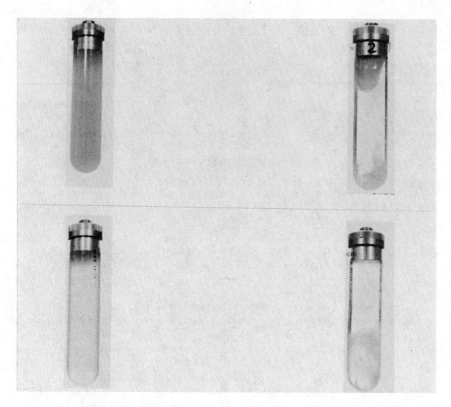

Fig. 1.   Centrifuge tubes containing gels from sputum samples of
          a chronic bronchitic patient dispersed in CsBr before
          (left) and after (right) DGU.
          Mucoid - upper
          Infected - lower

separated gel phases were recorded on the Ferranti Shirley rotation viscometer at $1350s^{-1}$: the values for the infected and mucoid
sputum were similar (0.58 and 0.51 respectively) but the gel from
the infected sputum had a higher viscosity (2.86) than the gel
from the mucoid sample (1.57).

     The sol and gel phases were then separately subjected to DGU
in CsBr. Gel from the mucoid sputum was very difficult to disperse
in CsBr and after the preparative experiment the bottom half of
the centrifuge tube, where mucus glycoprotein is expected to band
under the experimental conditions, contained an opaque gelatinous
precipitate (Figure 1). Analytical density gradient examination
of this fraction showed that it contained a soluble mucus glycoprotein and insoluble material of lower buoyant density (lipids).
Subsequent DGU enabled separation of the soluble glycoprotein

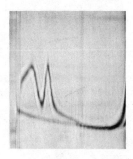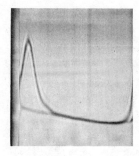

Fig. 2.   Sedimentation velocity patterns of protein fractions from
          the gel phase of mucoid (left) and infected (right) sputum
          samples of a chronic bronchitic patient. Patterns obtained
          16' after attaining a speed of 56,000 rpm.

from the insoluble lipids. The gel from the infected sample, on
the other hand, dispersed easily in CsBr and at the end of the
preparative run, the following zones were distinguishable in the
centrifuge tube: 1) a fleshy precipitate at the top containing
all the lipids, 2) a yellow liquid in the upper half with serum
proteins, 3) a clear liquid in the third quarter (glycoprotein)
and 4) a clear jelly in the bottom (DNA). Since there was considerable overlap between zones 3 and 4, they were recovered together
as a single fraction and subjected to further DGU in CsCl to
separate glycoprotein from DNA. While the centrifuge tubes containing the sol phases from the two samples did not show any visible
zones, apart from the lipids, all the components present in the
corresponding gel phases were present.

     Sedimentation velocity experiments of the protein fractions
from the two samples revealed a difference: both sol and gel of
the mucoid sample contained a substantial amount of 11S component
whereas this was practically absent from the infected sample
(Figure 2). Secretory IgA has a sedimentation coefficient of 11S
and since it is in the protein fraction, it is likely to represent
sIgA. Protein fractions from both mucoid and infected samples,
however, gave precipitation lines in immunodiffusion studies
against antiserum to secretory piece.

     Glycoprotein fractions from sol and gel of both samples have
been characterised as to buoyant densities, carbohydrate and amino
acid compositions. The density gradient patterns of sol and gel
glycoprotein of the infected sample included discrete components
of buoyant densities slightly higher and lower than that of the
main peak. Gas chromatographic analysis showed that Fuc, Gal,
GalNAc, GlcNAc and ACNeu were the only sugars present and together
accounted for 77-85% of the dry weight of the glycoprotein. In sol
and gel glycoprotein of the infected sample, molar ratios of sugars

with respect to GalNAc were similar (GalNAc:Fuc:Gal:GlcNAc:AcNeu,
1:1.3:3.6:2.5:2.0) as were those of the sol and gel glycoprotein
of the mucoid sample (1:2.0:4.5:3.3:2.0). The slightly higher
ratios in the latter suggest longer chains and we are currently
analysing the oligosaccharides released by alkaline borohydride.
Thin layer chromatography indicates that chains ranging in length
from 2-13 sugars and possibly higher are present in glycoprotein
from both gels.

Amino acids constituted 17-22% of the dry weights of all four
glycoprotein and their relative distributions showed small variations. The main difference was between the glycoprotein from the
gels of the mucoid (Ser + Thr + Pro, 38.7, Asp + Glu, 21.0) and
of the infected (Ser + Thr + Pro, 47.4, Asp + Glu, 13.2) samples.
Thiol reduction of human bronchial mucus glycoprotein is often
associated with enrichment of Ser + Thr + Pro at the expense of
Asp + Glu[1]. It is, therefore, possible that in the glycoprotein
from the mucoid gel more disulfide bonds are intact than in the
infected gel and that is responsible for the stronger association
between the mucus glycoprotein and lipids in the former gel.

In summary, this study of the sol and gel phases of infected
and mucoid sputum samples from the same chronic bronchitic patient
has shown differences similar to the differences we demonstrated
between CF and asthma sputum: lipids are readily removed from
the gel of the infected sample by DGU whereas, as in asthma gel,
lipids are strongly bound to the gel of mucoid chronic bronchitis.

Secretory IgA is found only in the mucoid sample but secretory
piece is found in both. Whether this represents a breakdown due
to infection or is a causal factor in increasing susceptibility
to infection is not known. Similarly it needs now to be established
whether the differences in the carbohydrate and amino acid compositions are due to enzymatic breakdown or reflect differences in
synthetic pattern.

REFERENCES
1.   K. R. Bhaskar and L. Reid, Application of density gradient
        methods for the study of mucus glycoprotein and other macro-
        molecular components of the sol and gel phases of asthmatic
        sputa, J. Biol. Chem., accepted for publication.
2.   K. R. Bhaskar and L. Reid, Application of density gradient
        methods to study the composition of sol and gel phases of
        CF sputa and the isolation and characterisation of epithe-
        lial glycoprotein from the two phases, in: "Perspectives
        in Cystic Fibrosis", J. Sturgess, ed., Imperial Press,
        Toronto (1980).

THE EFFECTS OF LOW MOLECULAR WEIGHT MUCUS MODIFYING SUBSTANCES ON

THE APPEARANCE OF MUCIN GELS IN THE SCANNING ELECTRON MICROSCOPE

Malcolm F. Beeson, Greg R. Parish*, David T. Brown[+]
and Christopher Marriott[‡]

Beecham Pharmaceuticals Research Division, Great Burgh
Epsom, KT18 5XQ and *Brockham Park, Betchworth, RH3 7AJ
[+]Department of Pharmacy, University of Nottingham
Nottingham, NG7 2RD, and [‡]Department of Pharmacy
Brighton Polytechnic, Brighton, BN2 4GJ; U.K.

The honeycomb structure which has been reported to exist in
cervical mucus (Chretien et al., 1973) and intestinal goblet cell
mucus (Forstner et al., 1977) has been investigated by Parish et al.
(1981), who showed the apparent structure to be an artefact
resulting from ice crystal formation during the freezing stage of
sample preparation for examination in the scanning electron
microscope (SEM). The artefact is modified by differences in
water content of the gel and freezing rate of the sample. We report
here the effects of alterations in the viscosity of the mucin gel,
produced by a variety of low molecular weight substances.

A 5% (w/v) purified mucin gel was incubated with the agents
listed in Table 1 for one hour at $25^{\circ}C$ before processing for SEM
assessment. Control mucin was treated with the vehicle only. The
samples were prepared for the SEM as described by Parish et al.
(1981).

Of the treatments studied, only calcium ions produced a change
in the SEM appearance of the gel (see figures 1 and 2). The
rheological properties of the gel were altered markedly by the
mucolytic (N-acetylcysteine and D-penicillamine) and mucus-
thickening (disodium tetraborate and oxytetracycline) agents added
(these effects have been fully documented with respect to human
respiratory mucin in the work of Marriott et al., 1981 and
Brown et al., 1981). Therefore, ice crystal nucleation and growth
appear to be independent of the rheology of the specimen. However,
with calcium ions changes were produced both on the surface and in

Table 1. Agents used for treatment of mucin gels

Substance	Concentration
N–Acetylcysteine	1.0% (w/v)
D–Penicillamine hydrochloride	1.0% (w/v)
Disodium tetraborate	1.0% (w/v)
Oxytetracycline hydrochloride	0.5% (w/v)
Calcium chloride	100 mM

the internal structure of the gel. Most noticeably, the septa between the pores increased to 1–2 μm in thickness.

Calcium, like tetraborate and the tetracyclines, is known to thicken mucus (Marriott et al., 1979), possibly by neutralisation of the negatively charged polar head groups of the glycoprotein sugar side-chains. The altered nature of the crystal voids as visualised in the electron micrographs may reflect this induced conformational change.

We conclude that the physical nature of the crystal voids produced by the freezing of mucin gels is controlled not only by the water content of the gel and its freezing rate, but also by the presence of calcium, and possibly other divalent ions. Alterations in the concentrations of such ions in vivo may help to explain the differences in pore size, reported to occur in human cervical mucus (Chretien et al., 1973).

REFERENCES

D. T. Brown, C. Marriott and M. F. Beeson, 1981, A Rheological study of mucus-antibiotic interactions, this publication.

F. C. Chretien, C. Gernigon, G. David and A. Psychoyos, 1973, The ultrastructure of human cervical mucus under scanning electron microscopy, Fertility and Sterility, 24:746.

G. Forstner, J. Sturgess and J. Forstner, 1977, Malfunction of intestinal mucus and mucus production, in: Mucus in Health and Disease, M. Elstein and D. V. Parke, eds., Plenum Publishing Corporation, N.Y.

C. Marriott, C. K. Shih and M. Litt, 1979, Changes in the gel properties of tracheal mucus induced by divalent cations, Biorheology, 16:331.

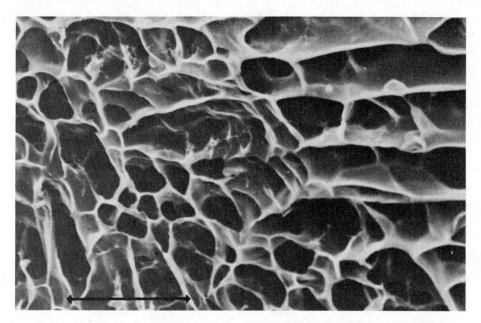

Fig. 1. SEM appearance of 5% mucin gel after treatment with
         penicillamine. This appearance is identical to that of
         untreated mucin. Size marker is 10 mµ.

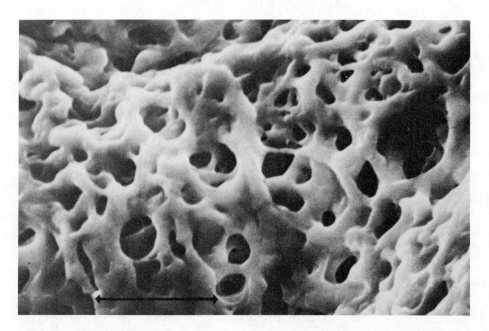

Fig. 2. Appearance of 5% mucin gel after treatment with calcium.
         Size marker is 10 mµ.

C. Marriott, D. T. Brown and M. F. Beeson, 1981, Evaluation of mucolytic activity using a purified mucus glycoprotein gel, this publication.

G. R. Parish, M. F. Beeson, D. T. Brown and C. Marriott, 1981, A freezing artefact associated with the preparation of mucin for examination using the scanning electron microscope, this publication.

CLINICAL ASPECTS OF RESPIRATORY MUCUS

L. Reid, K. Bhaskar, S. Coles

Department of Pathology
The Children's Hospital Medical Center
and Harvard Medical School, Boston, MA 02115

In the clinical setting it is impossible to separate respiratory mucus from the cilia that move it. Muco-ciliary clearance is
the silent watch-dog of the lung in health: in disease it is often
a noisy one. The 'bark' of a cough is usually associated with
mucus increased above normal in amount and often abnormal in composition.

The nature of normal bronchial mucus (taken here to be the
liquid in the airways - Reid and Clamp, 1978) is not known but
even under normal conditions its composition probably varies. In
the mucoid sputum of mild chronic bronchitis a circadian rhythm of
variation in concentration of glycoprotein (GP) and other macromolecules has been identified: it is likely that its pattern is
present in the normal subject also (Lopez-Vidriero et al.,
1977). It has recently been reported that muco-ciliary clearance
is significantly reduced during sleep (Pavia, personal communication).

Clinicians, physiologists and pharmacologists are currently
much occupied with the role of cough and the muco-ciliary system
in clearance of secretions from the lung - the factors that control
and modify these functions and their interaction in normal
subjects and in patients with various respiratory diseases. Chronic
bronchitis (CB) and cystic fibrosis (CF) are still the two major
problems but it seems that inherited abnormalities of ciliary
structure and function or of the bronchial mucus are more often
the reason for respiratory disease than previously suspected
(Afzelius, 1976; Sturgess et al., 1980).

The 'clinical' form of respiratory mucus is sputum. Even the

369

acute production of sputum is abnormal but if it is produced in
response to an acute challenge, such as infection, or a high concentration of an inhaled pollutant, then it can be considered the
'normal' response of a normal airway to an abnormal environment.
If the abnormal environment persists and an excessive secretion is
based on gland hypertrophy - as in the tobacco smoker - then it
can still be considered a 'physiological' response to an abnormal
environment: but the structure of the airway is now altered and
will often continue to be abnormal even if the 'acute' stimulus
is removed. The protective response is now disease. Such increase
in the amount of respiratory mucus produced is associated with
modulation of the epithelial cell population of the airway surface
and with hypertrophy of submucosal glands whose pattern of sensitivity to drugs is different from normal. The mixture of constituents in respiratory mucus changes and the composition of some
of its macromolecules. 'Mucus' describes the total bronchial
liquid recovered from the normal or diseased airway, in either
case representing a mixture of many macromolecules - glycoprotein
conjugates, lipid, proteins, DNA. Each category includes products
from several sources - eg. glycoproteins of serum or epithelial
type, and, in the case of the latter, derived from several cell
types in both surface epithelial cells and submucosal glands.

ROLE OF MUCUS IN AIRWAYS

     Airway mucus has many functions conveniently summarized in
the following seven. The main role of mucus is 1) to make its
contribution to the muco-ciliary system. No periciliary liquid,
no ciliary beat: no mucus, no ciliary clearance. The cilia and
bronchial mucus are essential to each other (Sleigh, 1977). The
mucus layer like the cilia is really part of the milieu interne.
Other functions include 2) sticking to particulate material, 3)
dissolving toxic gases, 4) humidification of inspired air, and
5) waterproofing the airway surface and so reducing fluid loss.
6) The airway wall synthesizes or transmits many mediators of
normal control and of inflammation into the airway lumen: these
are doubtless transported to more proximal airway levels by the
mucus carpet. Whether this is of functional importance is not
yet established. Many of the above can be considered aspects of
the protective function of mucus and 7) to increase in amount
under conditions of irritation is an additional function that
serves as the bridge between normal function and disease.

     The number and distribution of intra-epithelial nerve endings
in the airways of man is not known. In rat, it is known that they
are present in extrapulmonary, not intrapulmonary, airways (Jeffery
and Reid, 1973). Cells within the lung, such as those of the APUD
system (amine precursor uptake and decarboxylation system)
secrete biologically active humoral agents, which transported in

the mucus, provide a feedback between peripheral and central airways.

In the present state of our understanding it seems best to distinguish 3 layers of airway mucus:

a) the <u>periciliary</u> layer
b) the <u>surface mucus</u> '<u>film</u>' that under conditions of stimulation may be thick enough to justify the term '<u>blanket</u>'
c) rafts or gobs of secretion that float on the surface of <u>b</u> and are transported by it.

Layer <u>c</u> is included in sputum, probably <u>b</u> is also. Bronchial aspirate will include <u>b</u> and <u>c</u> if it is present: it may be that some <u>a</u> is also aspirated but the contribution of the periciliary layer either to sputum or bronchial aspirate is by no means clear. Probably both layers <u>b</u> and <u>c</u> include sol and gel. The sol of bronchial mucus may be the closest we come to recovering the periciliary layer. Cough clears layer <u>c</u>, to what extent it moves <u>b</u> also is not known.

Normally the muco-ciliary system transports airway content and alveolar constituents, such as the lipid and apoprotein of surfactant and macrophages from alveolar space. There are differences in components of the muco-ciliary system at different airway levels. At the periphery the ciliated cells are more numerous but the cilia are shorter than centrally. Muco-ciliary transport in small airways is slower than in the large. The ciliary system - escalator or conveyor belt - starts on the airway epithelium of respiratory bronchioli. It is in the terminal bronchiolus that a complete or continuous ciliary carpet around the airway circumference starts: the real foot of the stairway is represented by epithelial strips that spread into the acinus and make up part of the wall of the respiratory bronchioli. Thus strips of the ciliary carpet stretch into the alveolar region.

The <u>periciliary liquid</u> layer is present at all levels of the airway whether or not there are submucosal glands, although the contribution to it of the special secretory cells is not known. We suggest that the ciliated cell itself is probably responsible for this layer - or at least makes a major contribution to it. Studies with lectins show diffuse staining of the ciliated cell for glycoconjugates (GC) (Conner, personal communication) and autoradiographic studies show active uptake and transport of GC precursors by these cells (Coles, personal observation). It can be predicted that this layer has low viscosity. It probably 'shifts' very little since the liquid mantle of a cilium moves with the cilium, but is not moved by it (Sleigh, 1977). Close control of the amount of this liquid is necessary to normal function. Factors controlling the shift of water to the lumen which, at least in the dog, is coupled to Cl^- shift are important (Olver et al., 1975).

Judged from its position it is likely that the brush cell influences
this layer. The processes of the brush border of this airway cell
have dimensions like those of the brush cell of gut. They are much
shorter than and project much less than the cilia: they are bathed
by the deeper layers of the periciliary liquid.

The mucus 'film' or surface layer varies at different levels:
like the periciliary layer it does not seem to need the presence of
submucosal glands. In the rabbit Sturgess (1977) has shown, by
freeze fracture studies, that at the periphery the film is streaky
and incomplete and, when present, is thinner than centrally. At
the periphery mucus glycoprotein could derive from the Clara cells.
More centrally where glands are present the layer usually appears
complete, thicker but still with a mesh-like structure. Since in
many species glands are present in these airways in considerable
quantity and have a basal level of secretion not inhibited by the
usual antagonists, it is likely that they have a role in producing
the normal surface film. The surface film probably includes sol
and gel. Nadel (personal communication) reports that in the ferret
and cat stimulation of gland secretion does not cause hillocks as
happens in the dog, suggesting that in the former species the gland
secretion quickly spreads over the airway surface, in the way the
film would behave, and does not form a raft or gob. If it is
accepted – and it is not an unreasonable hypothesis – that serous
cells discharge a thinner liquid than mucous, then the secretion
represented by the surface film may be largely from the serous
cells. A simple hypothesis would be that the serous cells secrete
the GP of the sol (as well as lysozyme) and the mucus GP of the
gel. In the gland the serous cell product always passes over the
mucous cell, an arrangement in which gel mixes with sol to give the
single product that emerges from the gland.

Freeze fracture studies show how, during its effective stroke,
the cilium is extended and contacts the mucus film, while the bend-
ing lateral movement of the recovery stroke draws it well below
the surface of the periciliary liquid: under conditions of disease
the film increases in thickness to form a veritable blanket. Freeze
fracture studies of both the normal and hypersecretory airway show
the 'periciliary layer' as empty space between the cilia and the
surface 'film' as a mesh-like layer of fibers, gel – probably – with
spaces or interstices in which presumably the sol would be found.
The role of lipid or lysozyme in the spreading of the film is not
known but, a priori, the presence of lipid – it separates with the
gel – may well contribute to the formation of the film, and to the
sol/gel interface. That the film thickens to a blanket in diseases
associated with gland hypertrophy is also in favour of the glands
being the source of this layer. The gland hypertrophy in human
disease is associated with hypertrophy of mucous cells but serous
cells seem relatively unchanged (Reid and deHaller, 1967). Increase
in the mucus or gel part of the blanket is the essential feature of

disease. How the balance between the various constituents of the
mucus blanket is maintained is not known.

The gobs or rafts that move as localized collections of thick
secretion probably come from hypertrophied glands and also goblet
cells of the surface. The discharging goblet cell is often seen
with its concentrated mucus discharge forming a raft above the
tips of the cilia rather like a mushroom.

The speed of muco-ciliary transport is often unchanged in
disease but sometimes impaired - eg. as demonstrated in CB by transport of teflon discs along the trachea (Pavia et al., 1980). Lung
clearance is a much more difficult function to test since penetration is often so different in disease from the normal. In the
normal the small peripheral airways are free of secretion so that
penetration of the test substance occurs far to the periphery.
Clearance must then take place over the maximum distance. In hypersecretion the first effect of the disease is blockage of peripheral
airways so that penetration is less. 'Clearance' involves less
airway distance, so delayed rates of transport are compatible with
normal clearance, and normal rates may produce the surprising
result of faster clearance. Using a patient as his own control,
the effect of drugs and environment on the rate of clearance can
be examined (Pavia et al., 1980; Clarke, 1980).

ANALYSIS OF SPUTUM OR BRONCHIAL ASPIRATE

While use of bronchial aspirate for analysis avoids contamination by saliva, it is usually drawn through a tube followed by
saline, or if the volume of airway secretion is small then lavage
with saline is often the only way enough secretion is recovered for
study.

In the past the serious and undesirable 'artefact' associated
with chemical studies of mucus, in particular the glycoprotein, has
been the methods used to solubilize the sputum, a necessary step
to biochemical analysis. Proteolysis (Lamblin et al., 1977) or
thiol reduction (Roussel et al., 1975; Boat et al., 1976), procedures
commonly used, lead to degradation. The use of urea to solubilize
gels from sputum (Roberts, 1974) was an improvement over other
earlier methods and had the added advantage that enzymes originating from disintegrating cells and saliva are inactivated by 6M urea,
thus preventing enzymatic degradation of the bronchial glycoprotein
during isolation. It has the drawback, however, of exposing the
glycoprotein to denatured protein trapped in the gel, facilitating
thiol-disulfide exchange. Recently we have used equilibrium
density gradient ultracentrifugation (DGU) in caesium bromide (CsBr)
to isolate the epithelial glycoprotein (Creeth et al., 1977). The
glycoproteins isolated by this technique are much larger than those

isolated by earlier techniques. Our studies also established the presence and importance of disulfide bond aggregation in bronchial glycoprotein. Besides yielding undegraded glycoprotein, it has the major advantage that it permits quantitative recovery, as separate fractions, of the major macromolecular classes found in bronchial mucus - lipids, proteins, other glycoproteins and glycoconjugates as well as DNA.

Before Bhaskar and Creeth introduced DGU for the study of bronchial mucus, we concentrated on the use of 'marker' substances to identify the various components of sputum - eg. albumin for tissue fluid, fucose or sulfate for epithelial glycoprotein, mannose for serum-type glycoprotein. Since neuraminic acid (NANA) is found in serum and epithelial type GPs, the NANA/fucose ratio gives an indication of the proportions of epithelial to serum GP (Lopez-Vidriero, et al., 1977; Lopez-Vidriero and Reid, 1978). These provided a basis for comparison between disease and for relating visco-elastic properties to chemical components.

In mucoid sputum, viscosity increases with the concentration of epithelial glycoprotein. Certain characteristics of disease emerge such that sputum from patients with asthma is sometimes very viscous and that this is independent of the macromolecular or GP content: it seems that the nature of cross-linkage or the physical characteristics of the sputum are significant rather than the con-centration of constituents. Purulence also increases viscosity. In cystic fibrosis (CF) the sputum is usually very viscous but not more so than would be expected from its macromolecular content. Purulent CF sputum tends to have higher concentrations of DNA than CB sputum. Studying whole sputum we found that the major differ-ences are between mucoid and purulent sputum, regardless of the disease that has caused sputum production.

Recently Dr. Lopez-Vidriero by separating sol from gel has been able to demonstrate differences between diseases, comparing either mucoid or purulent samples (Lopez-Vidriero et al., 1979). She has used 160,000 g for separation, a higher speed than pre-viously applied in any systematic study of sputum. It gives com-plete separation of gel in all disease groups studied, although it is not necessary in each case.

Bronchorrhea (synonyms: pituitous catarrh or bronchitis serosa) describes a condition characterized by expectoration of a large amount of a thin sputum like egg-white. By convention we define it as 'more than 100 ml per day' but it is often much more: its physicochemical characteristics are also seen with smaller daily production (Keal, 1971). Centrifugation of this sputum yields a very small volume of gel - we call it a 'bronchorrhea' pattern of separation if the sol is more than 85% by volume (Lopez-Vidriero, et al., 1979). By contrast, mucoid sputum from chronic bronchitis

gives more than 40% gel by volume - a bronchitic pattern. In
chronic bronchitis there is gland hypertrophy, in bronchorrhea not
necessarily so. Bronchorrhea secretions have a low macromolecular
content and viscosity and perhaps represent, or at least include,
periciliary liquid in a pathological amount. The sol-gel pattern
of separation from purulent sputum of CF can follow either of these
patterns.

Recently Dr. Bhaskar has applied the density gradient method
to analyze the sol and gel phases of sputum samples from CF (Bhaskar
and Reid, 1980) and asthmatic (Bhaskar and Reid, 1981) patients.
Of the two asthmatic samples, one was from a patient with extrinsic
asthma (AsEXT) and the other from a patient with intrinsic asthma
(AsINT). Sol and gel were equal in volume in AsINT while 73% of
AsEXT sputum separated as gel. On the other hand, viscosity of the
whole sputum was similar as is that of the gels - in each case the
gel was almost three times as viscous as the whole sputum.

In both AsEXT and AsINT, sol consisted of only serum components
- no GP was detected. The gels also contained substantial amounts
of serum proteins but had, in addition, all the GP and also all the
lipids, which make the gel opaque. While the bulk of the serum
components were released from the gels in the first DGU, the lipids
were strongly bound to the GP, the two occurring together as insoluble
fractions after the first DGU. Repeated DGU of this insoluble fraction results in the separation of soluble GP ('native') from the
lipids. In both AsEXT and AsINT a portion of the insoluble fraction
was reduced with dithiothreitol (DTT) and treated with iodoacetamide
before DGU and the 'reduced' GP thus isolated was also characterized.

In both AsEXT and AsINT, the 'reduced' GP has lower sedimentation coefficients and solution viscosities than the 'native' GP.
Buoyant densities of the 'reduced' GP are higher than those of the
corresponding 'native' GP, suggesting that treatment with DTT
removes a peptide portion. This was confirmed by amino acid
analysis. Molar ratios of carbohydrates show no significant difference between 'native' and 'reduced' GP, but the relative distribution of amino acids does: in both cases, Ser, Thr, and Pro,
especially Thr, increase at the expense of Asp and Glu, indicating
that, in the 'native' GP, a crosslinking peptide/protein, relatively
richer in Asp and Glu, takes part in the formation of aggregates
through disulfide bonds.

Of four CF sputum samples studied, two showed a bronchorrhea
pattern, of gel 30% or less, while in the other two gel was 50%
or more. The difference between the viscosity of the sputum and the
gel is greater in the bronchorrhea pattern - eg. in one of the
'bronchorrhea' separations, the gel is about 13 times as viscous
as whole sputum, whereas in a bronchitic type the gel is only about
4 times that of sputum: the absolute value for the gels is similar.

It is not clear whether this represents a concentration difference in total sputum or interaction or lack of it between the macromolecules of the gel.

The GP from sol and gel of two sputum samples was analyzed in some detail. There was little difference in the chemical composition of the GPs from sol and gel, including their sugar and amino acid composition. In one case sedimentation velocity experiments at similar concentrations of the GP showed a difference. The sol GP migrates as a single peak, the gel gives two of about equal area, the slower being similar to the sol (7S), the faster having an S value of 13S. On addition of dithiothreitol (DTT), the fast peak is lost and the gel GP now migrates as a 7S peak, indicating that the GP includes aggregates formed through disulfide bonds. Study of the 'reduced' GP shows that its carbohydrate composition is scarcely altered but the amino acid distribution again shows increase in threonine and serine at the expense of aspartic and glutamic acids. This further lends support to the idea that a crosslinking protein/peptide relatively rich in aspartic and glutamic acids is responsible for disulfide aggregation.

In the CF sputum as in the asthma, lipids are confined to the gel and have the highest concentration of any macromolecule. In the preparative tube they are seen as a fleshy precipitate at the top of the tube. In contrast to the gels from asthmatic sputum the lipid in CF gels is readily separated in the first preparative DGU (Figure 1).

DNA occurs in both sol and gel but mainly in the gel. Its concentration can be as high or higher than that of the GP. It forms a clear and viscous jelly. Colorimetric assay indicates that DNA constitutes less than 50% of this jelly and analytical density gradient shows that the other half is even denser than DNA. After removal of the DNA by DNAse, the remaining material, when examined by GC, is found to contain up to 40% carbohydrate. In addition to the sugars expected in glycoprotein, this fraction contains significant amounts of xylose and glucuronic acid, both components of proteoglycans.

We have also identified components of proteoglycans in canine airway secretion both pre and post SO_2 exposure. This, if confirmed, is the first identification of such a component in airway secretion. It is also present in secretion obtained from explants of human airways in OC.

In the model of SO_2 induced canine bronchitis, the effect of drugs on baseline secretion is being followed in normal and bronchitic animals: interesting differences are being found (Figure 2). With exposure to SO_2 an increased amount of a dense component is found. As in CF DNA does not entirely account for

this fraction. Preliminary studies indicate the presence of uronic
acid and also sugars usually associated with glycoproteins.

Comparing the sol/gel constituents in purulent CF sputum with
purulent CB - in CF the IgA is found mainly in the gel, whereas
in CB it divides more evenly: in CF the IgA seems more closely
bound to the GP.

The use of DGU to analyze sputum reveals that macromolecules
other than GP are present in significant amount, and are sometimes
tightly bound to the GP. A proteoglycans previously unsuspected
is likely to be important to the internal structure of sputum and
to its visco-elastic behaviour since proteoglycans are very viscous.
The lipid more likely will allow spread of the secretion and its
formation to a film. The obvious 'stickiness' of the proteoglycans and the DNA suggests a further look a this physical
property of airway secretion. The monomer of the airway epithelial glycoprotein forms a polymer through disulfide bonds: it
seems that this involves also a small peptide. Does this occur
within or without the cell? If a separate small peptide takes
part then an extracellular site is possible.

The interaction between molecules is likely to be critical

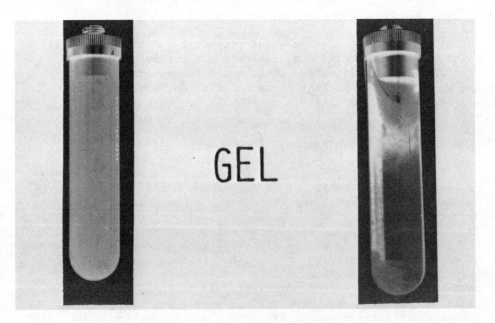

Fig. 1. Preparative centrifuge tubes containing gel from a CF
         sputum dispersed in CsBr (ρe = 1.41 g/ml) before (left)
         and after (right) ultracentrifugation at 40,000 rpm for
         48 h.

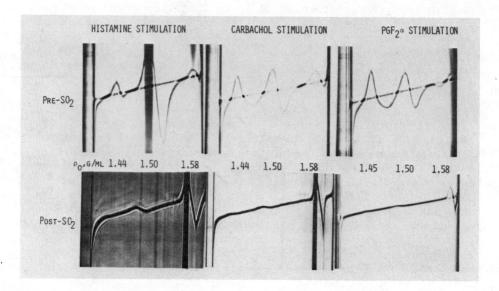

Fig. 2.   Equilibrium density gradient patterns in CsBr ($\rho e \simeq 1.5$
g/ml) of glycoprotein and 'denser' fractions of canine
bronchial mucus - aspirates from the same dog.

in establishing the visco-elastic properties of the secretion.
Which molecules are responsible for the fluctuating viscosity
seen at low shear rates and which molecules or linkages determine
the absolute level at high shear rates is not clear (Lopez-
Vidriero et al., 1977).

ORGAN CULTURE OF HUMAN AIRWAYS

     Because of major species differences in the number and type
of mucus-secreting cells, organ culture of human airways is a key
technique for understanding the control mechanisms of mucus
secretion in human airways, particularly its glycoconjugate
components. The surface epithelium can be separately studied from
the submucosal glands as can the variety of secretory cell types
at each site: and studies can be made of the normal and hypertrophied bronchitic gland.   Using labeled precursors of the glycoconjugate combined with autoradiographic analysis of secretory
cells, we can trace cell events, and by varying the labeling
protocol can analyze precursor uptake, glycoconjugate synthesis,
secretory granule transport, storage and discharge.   Scintillation
counting of labeled secretions released into the medium gives
additional information and the secretory product can be collected
for biochemical analysis.

In earlier studies a wide range of drugs had been applied to the glands (Sturgess and Reid, 1972). Serous and mucous cell discharge is augmented by cholinergic agonists and reduced but never completely inhibited by their antagonists. The adrenergic drugs then available have no effect on the cycle of synthesis and secretion which in the human gland takes about 4 hours.

None of the pharmacological agents, that have been used, stimulate secretion from the goblet or other secretory cells of the surface epithelium, although in animals, in vivo, particulate irritants cause discharge of surface cells. This limits the results obtained for surface cells since a model of stimulated secretion is not available for study. The serous and mucous cells of the submucosal glands usually resemble each other in their response and in the pattern of their control; the surface cells are different. In the glands the pattern of control of the hypertrophied gland is similar to that for the normal although sensitivity to drugs is different (Sturgess and Reid, 1972; Coles and Reid, 1978).

Recently Dr. Coles has developed a radiochemical assay for labeled glycoconjugates secreted by bronchial explants and incubated with ^{14}C-glucosamine. The tissue, especially the glands, remains in a good state of preservation up to 48 hours and quantitative recovery of labeled mucus glycoconjugates is possible. After 12 hours incubation with ^{14}C-glucosamine, labeled glycoconjugate secretion becomes linear with time and this is maintained for a further 12 to 24 hours. Comparison of drug effects is made during this linear phase. Medium is harvested over two consecutive 60-minute periods and the amount of labeled glycoconjugate (GC) secreted in the second period is expressed as a percentage of the first. Normalization of data in this way is necessary because of the differences in the amount of glandular tissue included in explants.

Cytochalasin B and Colchicine

In explants of main or lobar bronchi from 14 'normal' patients, the effects of cytochalasin B (a microfilament disruptor) and colchicine (which inhibits the assembly of microtubles) have been studied (Coles and Reid, 1981).

Cytochalasin B has no effect on baseline release of ^{14}C-labeled GC but inhibits bethanechol-stimulated discharge (Figure 3). Quantitative autoradiographic analysis of explants shows that in goblet cells of surface epithelium, or mucous or serous cells of gland, cytochalasin B has no effect on baseline incorporation of labeled precursors or intracellular transport and discharge of labeled macromolecules. It does block the increased discharge produced by bethanechol. This suggests that intact microfilaments are not necessary for baseline secretion but may be essential to agonist-induced discharge.

     Colchicine, like cytochalasin B, inhibits bethanechol-stimula-
ted release of ^{14}C-labeled GCs but has no effect on baseline
secretion. Cell studies show, however, that while baseline precursor incorporation and discharge are unaffected, colchicine
reduces intracellular transport in each cell type. In addition,
colchicine inhibits bethanechol-induced discharge from mucous and
serous cells. These studies suggest that, in each cell type,
microtubules are important in intracellular transport of secretory
granules, while in the submucosal gland, they are essential for
agonist-induced discharge.

Vasoactive Intestinal Polypeptide (VIP)

     Some of the nerve endings adjacent to the human bronchial
submucosal glands have been shown to contain the octosapeptide
known as vasoactive intestinal polypeptide (VIP) (Dey and Said,
1980). Dr. Said, who isolated this substance from hog duodenum

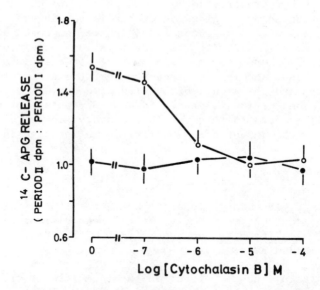

Fig. 3. Effect of cytochalasin B on baseline and methacholine-
         stimulated release of ^{14}C-labeled glycoconjugates (^{14}C-APG)
         by human airway in vitro. Cytochalasin B was added alone
         to incubation medium or together with 5.10^{-5}M methacholine
         chloride. Each point is the mean ± SE of 6-8 determina-
         tions. Baseline release •-•; methacholine-stimulated
         release o-o.

(Said and Mutt, 1970), invited us to investigate its effect on airway secretion. We have studied it (Coles et al., 1981) in the system described above using also an assessment of the lysozyme in
the secretion recovered from the organ culture medium (explants
were obtained from seven subjects with glands of normal size and
five with the hypertrophied glands of chronic bronchitis).

In the explants of normal airways, VIP (10 ng to 1 μg/ml)
causes a dose-dependent reduction of baseline release of GCs and
lysozyme (Figure 4). VIP (1 μg/ml) also causes a partial inhibition of methacholine-stimulated release of GCs and lysozyme. Cell
studies showed that the concentration of VIP (1 μg/ml) which causes
maximal inhibition ($\sim$ 50% of baseline) of GC release partially
inhibits baseline and methacholine-stimulated discharge from mucous
and serous cells but not from goblet cells. In explants from CB
patients, doses as high as 10 μg/ml do not inhibit baseline (Figure
4) or methacholine-stimulated GC release, or mucous cell discharge.
VIP does reduce baseline and methacholine-stimulated lysozyme
release and reduces serous cell discharge: but this is less marked
than in normal airways. This is the first such dissociation we
have found between serous and mucous cells.

Thus the effect of VIP is inhibitory. Since VIP is a naturally
occurring substance it may well have a role in the neurohumoral
regulation of mucus secretion by the human airway. These studies
also indicate a fundamentally different regulation of secretory
cell types in CB. The absence of any inhibitory effect of VIP on
the bronchitic airway may well represent part of the mechanism
by which the hypersecretory state is induced. In CB, whereas the
mucous acini hypertrophy, the serous do not, perhaps reflecting
the maintained inhibitory activity of VIP nerves.

Prostaglandins

Preliminary results of the effects of $PGF_2\alpha$ and PGE_2 show that
both drugs have a stimulatory effect, but apparent only in some
explants, not in all. It is not due to patient sensitivity. These
agents affect mucous and serous cells, not goblet cells. When a
PG is seen to stimulate, only some cells and acini are affected.
The effect of any stimulatory drug is focal. What is peculiar to
the PGs is the 'size' of the territory that, at a given time, is
sensitive or insensitive. A much larger group of cells or acini
seems to be in the same 'phase of sensitivity' to PGs than is the
case for other drugs.

Calcium Ions (Ca^{2+})

Although accepted that Ca^{2+} usually have a role in exocrine
secretion (Douglas, 1968), this has not yet been established for

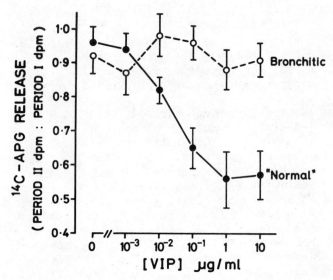

Fig. 4. Effect of VIP on baseline release of ^{14}C-labeled, acid-
         precipitable glycoconjugates (^{14}C-APG) by 'normal' and
         bronchitic human airways in vitro. Each point is the
         mean ± SE of 6-8 separate determinations.

airway secretory cells. Our preliminary studies using dog tracheal
explants have revealed somewhat surprising results. Incubation of
explants in Ca^{2+}-free medium (+ 1mM EGTA) causes increased release
of ^{14}C-labeled glycoconjugates which is linear with time for at
least 60 minutes (Figure 5). Progressive surface epithelial
exfoliation develops later and the glandular mucous cells change
shape and then separate from their basement membranes. The increase
in GC release occurs before exfoliation and so cannot be blamed on
shed cells. The effect of Ca^{2+} can be reversed by addition of Ca^{2+}
or Sr^{2+} but not by Mg^{2+}. Since from the preliminary autoradiographic
cell analysis it seems that the Ca^{2+}-free medium does not cause
discharge from submucosal gland or goblet cells, it suggests
another source of GC. This may result from discharge of material
already in the duct perhaps by a conformational change of the mucus
leading to increased solubility (Forstner et al., 1977).

It seems that Ca^{2+} are necessary for methacholine-stimulated
discharge. These studies are continuing but suggest that Ca^{2+}
have a dual role in mucus secretion, one concerned with its release,
the other with its solubility.

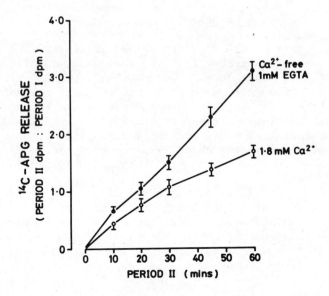

Fig. 5.   Effect of removal of calcium ions on time course of [14]C-
labeled, glycoconjugate (^{14}C-APG) release by canine
tracheal explants. Explants were incubated in Earle's
salts solution containing either 1.8mM $CaCi_2$ or in Ca^{2+}-
free Earle's containing 1mM EGTA. Each point is the
mean ± SE of 10-12 separate determinations.

Proteoglycan Secretion
_____

     DGU analysis of secretion harvested from organ culture of
explants from five human specimens gave the unexpected result of a
relatively small amount of the mucus GP fraction (Fraction I - DGU)
and an unexpectedly large one of higher bouyant density where DNA,
proteoglycans or polysaccharides are expected. Fraction I, the
lightest fraction, contained protein and lipid. Whereas 10% of the
total material was in GP fraction, 65% was in III: 75% of bound
[14]C-glucosamine was associated with Fraction III. Fraction III
contains negligible DNA, its main component seems to be a proteoglycan containing, as a percent of its dry weight, 15% uronic
acid, 9-10% protein, 10-12% hexose, 8% hexosamine, 2-3% sialic
acid and 3-5% sulfate.

     Since the organ culture obviously includes connective tissue
as well, the importance of this substance would be less were it
not that it is similar to what we recover from sputum and canine
bronchial aspirate.

Canine Model of Chronic Bronchitis

In collaboration with Drs. Ingram and Drazen, we are currently
studying secretions collected from the canine airway of normal dogs
and dogs with sulfur-dioxide-induced CB (SO_2CB), in each case before
and after stimulation with drugs. The material is collected after
20 ml of saline are introduced into the right main bronchus below
the first branch. The technique is standardized and 15-18 ml are
usually recovered. The total macromolecular yield is calculated
and results of analysis expressed as percent of macromolecular
total. Certain differences between drugs can be detected - basal
secretion includes 2 GP peaks, both of which are increased by
carbachol and $PGF_2\alpha$; histamine increased mainly the denser peak.
In all a dense component resembling the proteoglycan (PG) described
above is also present. On exposure to SO_2 all three drugs produced
secretions in which the PG predominates. This component is a clear
and viscous jelly and contains uronic acid. The amount of secretion
increases with exposure to SO_2 as does its macromolecular concentration. Physiological studies of airway function and sensitivity are
also being made.

ANIMAL MODELS OF CHRONIC BRONCHITIS

The changes of chronic disease ultimately bring us to the
changes in the epithelial cell population that are responsible for
the excessive and abnormal secretions of disease. Hypertrophy of
submucosal gland and increase in secretory cells of the surface
epithelium are the two chief hallmarks of chronic bronchitis and
cystic fibrosis.

A variety of stimuli and irritants can produce this in a number
of species (Reid, 1978; Jones and Reid, 1978; Reid et al., 1981).
The three models in which we have been interested, largely because
of their relevance to human disease, are produced by irritants
such as tobacco (Lamb and Reid, 1969; Jones et al., 1972, 1973;
Jeffery and Reid, 1977; Jones and Reid, 1978) or sulphur dioxide
(Reid, 1963; Lamb and Reid, 1968; Jeffery and Reid, 1977), drugs
(Sturgess and Reid, 1973; Bolduc and Reid, 1978; Jones and Reid,
1979a and b) and infection (Jones et al., 1975). Whereas earlier
studies were concerned with chronic exposure, our recent studies
show that the effect on the epithelium is surprisingly rapid. Two
studies, one with the β adrenergic agents isoproterenol and
salbutamol, and the other with tobacco smoke illustrate the new
insights into the intracellular biological events.

β Adrenergic Agents

Whereas we have not shown any effect of β adrenergic agents

on the airways in culture (Sturgess and Reid, 1972), in vivo studies
in the rat show these drugs rapidly change both the cell population,
causing increase in secretory cells and the secretion found within
the cells,both the pre-existing ones and the newly induced ones
(Jones, 1979). Furthermore differences in the effect of these two
drugs indicate that the β_1 and β_2 receptors have a different and
characteristic regional distribution (Figure 6).

Isoproterenol is a non-selective β agonist, salbutamol more
selective for β_2 receptors. Quantitative histochemical methods
make it possible to follow changes in the type of intracellular
glycoprotein at all airway levels. The distribution of secretory
cells and the proportion of the various types of GP normally varies
with the level pointing to a regional difference in modulation
and receptors (Jones and Reid, 1979a). This study indicates that
β_1 and β_2 receptors (others, of course, will doubtless be identified) are present in airways and show wide regional variation. The
maximum effect is present after 1 or 2 doses. Recovery from the
isoproterenol effect occurs, but is slow and follows a fluctuating
course (Jones and Reid, 1979b). Even after one month it is not
complete except at one level, the proximal trachea. The regional
differences in the normal indicate that there is a homeostatic
mechanism. This can be reset in a variety of ways. The effect
of the β adrenergic system has implications for choice of drugs
and their surveillance, although the effect of drugs on the mucus
secretory cells of airways is not one that is usually considered
in toxicological testing.

Tobacco Smoke

Studies of the early effects of tobacco smoke have shown that
here also maximum effect is produced very early. The first exposure causes fall in surface secretory cell count because cells
have discharged (Jones and Reid, 1978). By the second day the
number of secretory cells has increased and by the third day it
reaches its greatest concentration. The epithelium has adapted
to the stimulus in that (i) after Day 1 the 'massive' emptying is
not apparent, (ii) cells have been switched on from non-secretory
to secretory cells, (iii) secretory cells are changing the nature
of the GP they package into granules and (iv) the nature of the
granules. The last refers to the 'metaplasia' or conversion of
Clara or serous cells of the surface epithelium to a mucous or
goblet cell type. This is apparent from counts of the various
cell types and identification of transitional cell forms (Jeffery,
1973). The combination of the tobacco with the non-steroidal
anti-inflammatory (NSAI) agent phenylmethyloxadiazole (PMO) protects,
at least at first, from the increase in secretory cell number, it
does not prevent the switch in GP type (Jones et al.,1972 and 1973).
It seems that the most sensitive marker of cell injury is the shift

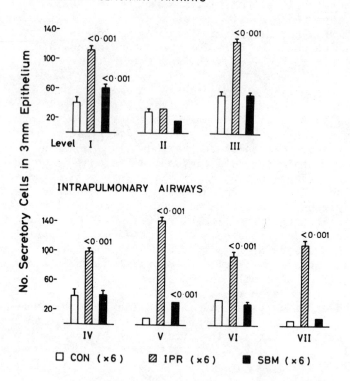

Fig. 6.   Effect on secretory cell number of isoproterenol (IPN) and
          salbutamol (SMB) (10 mgs/100 gm body weight given sub-
          cutaneously daily for 6 days). Airway levels I to VII
          are proximal to distal. (For detail see Text)

in the type of GP synthesized. The protection offered by the PMO
is also given by another NASI agent — phenylbutazone (Coles,
personal communication).

     The 'tolerance' to the tobacco is quickly lost (Figure 7).
After one day's rest from exposure to smoke, the next day's dose
causes a similar overwhelming discharge to that seen on Day 1.

     It has been known that exposure to tobacco smoke causes a
striking increase in cell mitosis within the first 24 hours
(Wells and Lamerton, 1975). The recent studies of Drs. Bolduc
and Jones show that if there is a forty-eight hour break between
exposures a second wave occurs. PMO delays but does not prevent
these mitotic bursts (Bolduc, 1976; Bolduc et al., 1981).

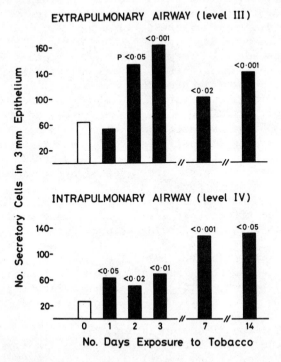

Fig. 7. Development of new secretory cells over 2 weeks of daily
         exposure to cigarette tobacco smoke. In extrapulmonary
         airways maximum response occurs by Day 3, in the intra-
         pulmonary by Day 7.

     We have lived through several decades when millions of
dollars/pounds were spent on mucolytic agents with, I would
submit, little evidence that they were what the patient needed
or wanted. Now we seem to be entering an era when speed-up of
muco-ciliary clearance is the earnest hope of the pharmaceutical
industry. The possibility of 'protecting' the airway epithelium
from the transformation in disease is offered by the effect of
the NSAI agent. It is surprising that so little interest has been
shown in them both by the medical profession and the pharmaceutical
industry.

REFERENCES

     Afzelius, B.A., 1976, A human syndrome caused by immotile

cilia, Science, 193:317.

Bhaskar, K. and Reid, L., 1980, Application of density gradient methods to study the composition of sol and gel phases of CF sputa and the isolation and characterisation of epithelial glycoprotein from the two phases, in: "Perspectives in Cystic Fibrosis", Sturgess, J., ed., Imperial Press, Toronto.

Bhaskar, K. and Reid, L., 1981, Application of density gradient methods for the study of mucus glycoprotein and other macromolecular components of the sol and gel phases of asthmatic sputa, J. Biol. Chem., accepted for publication.

Boat, T., Cheng, P., Iyer, R., Carlson, D. and Polony, I., 1976, Respiratory tract secretions. Mucous glycoproteins of non-purulent tracheobronchial secretions and sputum of patients with bronchitis and cystic fibrosis, Arch. Biochem. Biophys., 177:95.

Bolduc, P., 1976, Cell turnover of the bronchial and alveolar lining in the rat lung in various types of hypertrophy, Ph.D. Thesis, University of London.

Bolduc, P., Jones, R. and Reid, L., 1981, Mitotic activity of airway epithelium after short exposures to tobacco smoke and the effect of the anti-inflammatory agent phenylmethyloxadiazole (in preparation).

Bolduc, P. and Reid, L., 1978, The effect of isoprenaline and pilocarpine on mitotic index and goblet cell number in rat respiratory epithelium, Brit. J. Experiment. Pathol., 59:311.

Clarke, S., 1980, (Ed.) Muco-ciliary Clearance - A Symposium, Chest, Suppl.

Coles, S., personal communication.

Coles, S. and Reid, L., 1978, Glycoprotein secretion in vitro by human airway: normal and chronic bronchitis, Experiment. Mol. Pathol., 29:326.

Coles, S. and Reid, L., 1981, Inhibition of glycoconjugate secretion by colchicine and cytochalasin B. An in vitro study of human airway, Cell Tissue Res., 214:107.

Coles, S., Reid, L. and Said, S., 1980, Effect of vasoactive intestinal peptide on glycoprotein secretion by submucosal glands and surface epithelium of human airway in vitro, in: "Perspectives in Cystic Fibrosis", Sturgess, J., ed., Imperial Press, Toronto.

Conner, M., personal communication.

Creeth, J., Bhaskar, K., Horton, J., Das, I., Lopez-Vidriero, M. and Reid, L., 1977, The separation and characterization of bronchial glycoproteins by density-gradient methods, Biochem. J., 167:557.

Czegledy-Nagy, E., and Sturgess, J., 1976, Cystic fibrosis. Effects of serum factors on mucus secretion. Lab. Invest., 35:588.

Dey, R. and Said, S., 1980, Immunocytochemical localization of VIP-immunoreactive nerves in bronchial walls and pulmonary vessels, Fed. Proc.,39:1062.

Douglas, W., 1968, Stimulus-secretion coupling: the concept and clues from chromaffin and other cells, Brit. J. Pharmacol.,

34:451.

Forstner, G., Sturgess, J. and Forstner, J., 1977, Malfunction
of intestinal mucus and mucus production, in: "Mucus in Health and
Disease, Advances in Experimental Medicine and Biology," Elstein,
M. and Parke, D., eds., Plenum Press, New York and London.

Jeffery, P., 1973, Goblet cell increase in rat bronchial
epithelium arising from irritation on drug administration - An
experimental and electron microscopic study, Ph.D. Thesis,
University of London.

Jeffery, P. and Reid, L., 1973, Intra-epithelial nerves
in normal rat airways: a quantitative electron microscopic study,
J. Anat., 114:35.

Jeffery, P. and Reid, L., 1977, The respiratory mucous membrane,
in: "Respiratory Defense Mechanisms," in Lung Biology in Health
and Disease, Lenfant, C., exec. ed., Marcel Dekker, Inc., New York
and Basel.

Jones, R., 1979, Secretory cell hyperplasia and modification
of intracellular glycoprotein induced in rat airway epithelium by
β-agonists: A quantitative histochemical study, Ph.D. Thesis,
University of London.

Jones, R., Baskerville, A. and Reid, L., 1975, Histochemical
identification of glycoprotein in pig bronchial epithelium: (a)
normal and (b) hypertrophied from enzootic pneumonia, J. of Pathol.,
116:1.

Jones, R., Bolduc, P. and Reid, L., 1972, Protection of rat
bronchial epithelium against tobacco smoke, Brit. Med. J., 2:142.

Jones, R., Bolduc, P. and Reid, L., 1973, Goblet cell glycoprotein and tracheal gland hypertrophy in rat airways: the effect
of tobacco smoke with or without the anti-inflammatory agent phenylmethyloxiazole, Brit. J. Experimental. Pathol., 54:229.

Jones, R. and Reid, L., 1978, Secretory cells hyperplasia and
modification of intracellular glycoprotein in rat airways induced
by short periods of exposure to tobacco smoke, and the anti-inflammatory agent phenylmethyloxiazole, Lab. Invest., 39:41.

Jones, R. and Reid, L., 1978, Secretory cells and their glycoproteins in health and disease, Brit. Med. Bull., 34:9.

Jones, R. and Reid, L., 1979a, β-agonists and secretory cell
number and intracellular glycoprotein in airway epithelium, Am. J.
Pathol., 95:407.

Jones, R. and Reid, L., 1979b, β-agonists and rat airway
secretory cell hyperplasia, Fed. Proc., 38:1155.

Keal, E., 1971, Biochemistry and rheology of sputum in asthma,
Postgrad. Med. J., 47:171.

Lamb, D. and Reid, L., 1969, Mitotic rates, goblet cell increase
and histochemical changes in mucus in rat bronchial epithelium during exposure to sulphur dioxide, J. Pathol. Bacteriol., 96:97.

Lamb, D. and Reid, L., 1969, Goblet cell increase in rat
bronchial epithelium after exposure to cigarette and cigar smoke,
Brit. Med. J., 1:33.

Lamblin, G., Lafitte, J., Lhermitte, M., Degand, P. and Roussel, P., 1977, Mucins from cystic fibrosis sputum, Mod. Probl. Paediat., 19:153.

Lopez-Vidriero, M. and Reid, L., 1978, Chemical markers of mucus and serum glycoproteins and their relation to viscosity in mucoid and purulent sputum from various hypersecretory diseases, Am. Rev. Resp. Disease, 117:465.

Lopez-Vidriero, M., Das, I. and Reid, L., 1977, Airway secretion: source, biochemical and rheological properties, in: "Respiratory Defense Mechanisms", in Lung Biology in Health and Disease, Lenfant, C., exec. ed., Marcel Dekker, Inc., New York and Basel.

Lopez-Vidriero, M., Das, I. and Reid, L., 1979, Bronchorrhoea-separation of mucus and serum components in sol and gel phases, Thorax, 34:447.

Nadel, J., personal communication.

Olver, R., Davis, B., Marin, M. and Nadel, J., 1975, Active transport of Na$^+$ and Cl$^-$ across the canine tracheal epithelium in vitro, Am. Rev. Resp. Disease, 112:811.

Pavia, D., personal communication.

Pavia, D., Bateman, J. and Clarke, S., 1980, Deposition and clearance of inhaled particles, Bull. Europ. Physiopath. Resp., 16:335.

Reid, L., 1963, An experimental study of hypersecretion of mucus in the bronchial tree, Brit. J. Experiment. Pathol., 44:437.

Reid, L., 1978, Animal models in clinical disease, Fed. Proc., 36:2703.

Reid, L. and Clamp, J., 1978, Biochemical and histochemical nomenclature of mucus, Brit. Med. Bull., 34:5.

Reid, L. and deHaller, R., 1967, The bronchial mucous glands - their hypertrophy and change in intracellular mucus, Mod. Probl. Paediat., 10:195.

Reid, L., Jones, R. and Lopez-Vidriero, M., 1981, Experimental chronic bronchitis, in: "International Review of Experimental Pathology," Academic Press, in preparation.

Roberts, G., 1974, Isolation and characterisation of glycoproteins from sputum, Europ. J. Biochem., 50:265.

Roussel, P., Lamblin, G., Degand, P., Walker-Nasir, E. and Jeanloz, R., 1975, Heterogeneity of the carbohydrate chains of sulfated glycoproteins isolated from a patient suffering from cystic fibrosis, J. Biol. Chem., 250:2114.

Said, S. and Mutt, V., 1970, Polypeptide with broad biological activity: isolation from small intestine, Science, 169:1217.

Sleigh, M., 1977, The nature and action of respiratory tract cilia, in:"Respiratory Defense Mechanisms", in Lung Biology in Health and Disease, Lenfant, C., exec. ed., Marcel Dekker, Inc., New York and Basel.

Sturgess, J., 1977, The mucous lining of major bronchi in the rabbit lung, Am. Rev. Resp. Disease, 115:819.

Sturgess, J., Chao, J., Peter Turner, J., 1980, Transposition of ciliary microtubules: another cause of impaired ciliary motility,

N. Engl. J. Med., 303:318.

  Sturgess, J. and Reid, L., 1972, An organ culture study of
the effect of drugs on the secretory activity of the human bronchial
submucosal gland, Clin. Science, 43:533.

  Sturgess, J. and Reid, L., 1973, The effect of isoprenaline
and pilocarpine on (a) bronchial mucus-secreting tissue and (b)
pancreas, salivary glands, heart, thymus, liver and spleen,
Brit. J. Experiment. Pathol., 54:388.

  Wells, A. and Lamerton, L., 1975, Regenerative response of
the rat tracheal epithelium after acute exposure to tobacco smoke:
a quantitative study, J. Nat. Cancer Institute,55:887.

CHEMICAL COMPOSITION OF PATHOLOGICAL HUMAN TRACHEOBRONCHIAL

MUCUS COLLECTED BY FIBEROPTIC BRONCHOSCOPY

Raoul Carubelli, Goverdhan P. Sachdev, Owen F. Fox,
Gary Wen, Frank O. Horton and Robert M. Rogers

Biomembrane Research Laboratory, Oklahoma Medical
Research Foundation; and Pulmonary Disease Section
Department of Medicine, University of Oklahoma Health
Sciences Center, Oklahoma City, OK 73104, U.S.A.

INTRODUCTION

In our previous work with normal tracheobronchial secretions, we have shown that fiberoptic bronchoscopy allows the collection of uncontaminated specimens in amounts suitable for chemical analysis and initial fractionation (Sachdev, et al., 1980). In this communication, we describe the chemical composition and electrophoretic profiles of respiratory mucus collected by fiberoptic bronchoscopy from patients with various chronic pulmonary diseases.

RESULTS

Respiratory mucus was collected by gentle suction from the trachea and bronchi using a polyethylene catheter inserted through the biopsy channel of an Olympus Bf, type B2, fiberoptic broncho-scope. The dry weight of dialyzed and lyophilized specimens from male patients, 20-40 years of age, with chronic bronchitis, emphysema, chronic obstructive pulmonary disease (combined symptoms of emphysema and chronic bronchitis), and asthma ranged from 20 to 100 mg; this is 5 fold higher than the yield from healthy, non-smoking volunteer male subjects of the same age group, which ranged from 5 to 20 mg. Only specimens from donors with blood group O were used for these studies in order to minimize variability of the carbohydrate moiety of the glycoprotein components. Mucus solutions in distilled water (for chemical analyses) or in 0.1 M phosphate buffer, pH 7.0 (for column chromatography), were prepared by mild sonication, 20 sec, in an ice bath. The results of the analyses for DNA, protein, neutral hexoses, hexosamines, sialic acid and sulfate

393

monoester, conducted by the methods previously described (Sachdev, et al., 1978), are shown in Table 1. The data indicate that these specimens of pathological tracheobronchial mucus contain higher levels of DNA, neutral hexoses, hexosamines and sialic acid than the normal mucus from healthy donors. The contents of protein and sulfate monoester, on the other hand, do not show any definite pattern of changes. The glycoprotein profiles, investigated by sodium dodecyl sulfate (SDS)-gel electrophoresis on composite gels (polyacrylamide 2%/agarose 0.5%), showed that the main mucin band of normal mucus migrated slightly ahead of the corresponding band of the pathological secretions (Fig. 1). In addition, variability in the intensity of the periodic acid-Schiff (PAS)-stain of some of the small molecular weight glycoprotein bands indicated differences in the relative proportion of these components. The main difference observed in gels stained with Coomassie Blue (data not shown), was a marked increase, in pathological mucus secretions, of two distinct low molecular weight components migrating at the level of the tracking dye (bromophenol blue).

Fractionation of mucus specimens from patients with chronic bronchitis and chronic obstructive pulmonary disease on columns of Sephadex G-200 showed that 30-45% of the total material was eluted as an included peak; versus 10-15% for normal mucus. Using sputum solubilized with mercaptoethanol and columns of Sepharose 4B, Barton et al. (1977) also found a marked increase of the low molecular weight components in the secretions from patients with chronic bronchitis.

CONCLUSIONS

The higher carbohydrate content observed in respiratory mucus from patients with chronic bronchitis, emphysema, chronic obstructive pulmonary disease and asthma suggests either more glycoprotein

Table 1. Chemical Composition of Normal and Pathological Human Tracheobronchial Mucus Collected by Fiber-optic Bronchoscopy from Donors with Blood Type O

Mucus Component[a]	Normal	Chronic Bronchitis	Chronic Obstructive Pulmonary Disease	Emphysema	Asthma
DNA	0.59 ± 0.19 (4)	1.34 ± 0.11 (3)	0.71 ± 0.18 (3)	1.03 ± 0.22 (4)	2.22 (2)
Protein	48.03 ± 6.92 (4)	48.00 ± 8.62 (3)	43.13 ± 2.22 (9)	35.55 ± 3.27 (4)	51.85 (2)
Hexoses	14.73 ± 2.06 (7)	18.33 ± 1.83 (3)	18.46 ± 2.07 (9)	22.93 ± 1.51 (4)	18.45 (2)
Hexosamines	5.24 ± 0.47 (4)	7.53 ± 1.29 (3)	9.76 ± 2.05 (3)	7.14 ± 1.39 (4)	6.58 (2)
Sialic Acid	2.94 ± 0.24 (4)	4.88 ± 0.95 (3)	5.63 ± 0.29 (9)	5.62 ± 0.26 (4)	4.16 (2)
Sulfate	1.39 ± 0.95 (4)	1.33 ± 0.16 (3)	1.42 ± 0.23 (3)	1.38 ± 0.12 (3)	1.60 (2)

[a]The values represent Means ± S.E.M.; number of specimens analyzed shown in parenthesis.

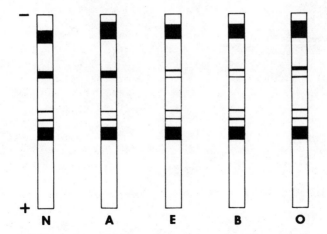

Fig. 1.   SDS-gel electrophoretic patterns of normal (N) and patho-
          logical human respiratory mucus from patients with asthma
          (A), emphysema (E), chronic bronchitis (B) and chronic
          obstructive pulmonary disease (O). Gels stained with PAS.

or abnormal glycosylation. The slight decrease in the electrophoretic mobility of the major mucin of the pathological secretions
suggests the possiblity of a larger molecular weight, or the formation of aggregates. As for the low molecular weight components,
the combined electrophoretic and chromatographic data of the pathological secretions indicates increased amounts and changes in the
relative proportions of the various components. Abnormal interactions among glycoproteins and/or between glycoproteins and other
components of respiratory secretions could account for the altered
rheological properties of pathological tracheobronchial mucus.
(This research was supported in part by NIH Grant HL 23110.)

REFERENCES

Barton, A. D., Weiss, S. G., Lourenço, R. V., Dralle, W. M., and
        Shamsuddin, M., 1977, Mucus Glycoprotein Content of Chronic
        Bronchitis Sputum, Proc. Soc. Exp. Biol. Med. 156:8.
Sachdev, G. P., Fox, O. F., Wen, G., Schroeder, T., Elkins, R. C.,
        and Carubelli, R., 1978, Isolation and Characterization of
        Glycoproteins from Canine Tracheal Mucus, Biochim. Biophys.
        Acta 536:184.
Sachdev, G. P., Myers, F. J., Horton, F. O., Fox, O. F., Wen, G.,
        Rogers, R. M., and Carubelli, R., 1980, Isolation, Chemical
        Composition, and Properties of the Major Component of Normal
        Human Tracheobronchial Secretions, Biochem. Med. 24:82.

Spinability and transport properties of

sputum in chronic bronchitis

PUCHELLE E.[*], ZAHM J.M.[*], and A. PETIT[**]

[*] Unité de Physiopathologie
Respiratoire 54500 Vandoeuvre-les-Nancy INSERM
[**] Laboratoire d'Histologie
Faculté de Médecine, 51100 Reims, France

Besides the viscoelastic properties of bronchial mucus, other
rheological characteristics such as spinability may interfere in
the mucociliary transport process. Although this rheological
property has been widely investigated on the cervical mucus, no
study has been devoted to the spinability of bronchial mucus and
to its role in mucociliary transport.
The spinability of sputum has been measured with an automatic
apparatus derived from that developed by BURNETT et al. (1967)
and CHRETIEN et al. (1977) for cervical mucus. Calibrated samples
(20 μl) of sputum were placed on a plastic slide, gripped with a
clip fixed to a rack-pinion system and raised at a constant speed
of 1 cm.sec^{-1}. The spinability of the sample, corresponding to the
maximal length of the thread that could be drawn before rupture,
was directly obtained on a digital display. The spinability was
correlated to the elasticity and apparent viscosity measured with
a concentric cylinder rheometer (PUCHELLE et al. 1972) and to the
ciliary transport rate measured on the depleted frog palate
(KING et al. 1974).
The mean spinability of sputa collected in 20 patients with
chronic bronchitis was 50.9 ± 25.4 mm. In mucoid sputa the spinability was significantly ($p < 0.05$) higher (77.8 ± 26.7 mm) than
in purulent sputa (44.2 ± 20.8 mm). Spinability was found to
increase significantly as elasticity increased ($r = 0.54$, $p <
0.02$) whereas a negative relationship was obtained with the
viscosity ($r = -0.50$, $p < 0.05$) and the elastic modulus ($r =
-0.72$, $p < 0.01$). A significant correlation ($r = 0.61$, $p < 0.01$)
was demonstrable between the spinability and the ciliary trans-
portrate of the sputum samples. The frog mucus used as reference

for defining normal transport rate had an average spinability of
38.6 ± 3.6 mm. In 5 of the 7 patients were the spinability of
sputum was lower than that of the reference frog mucus, the
relative transport rate was markedly decreased. Furthermore, in
patients with abnormally low values of elasticity, the mucociliary
transport remained effective provided the spinability was higher
than 40 mm.
These results suggest that the spinability of bronchial mucus
might be an important rheological property controlling the
efficiency of mucociliary clearance.

References

BURNETT J., GLOVER A. and SCOTT-BLAIR G.W. (1967) Biorheol.
4, 41.

CHRETIEN F.C., OZENDA B., VOLOCHINE B. (1977) Med. & Biol. Eng.
& Comput. 15, 673.

PUCHELLE E., ZAHM J.M., SADOUL P. (1972) Biorheol. 10, 481.

KING M., GILBOA A., MEYER F.A. and SILBERBERG A. (1974)
Am. Rev. resp. Dis. 110, 740.

THE MECHANISMS AND CONTROL OF BRONCHIAL MUCOUS CELL HYPERPLASIA

Peter K. Jeffery, Margaret Ayers and Duncan Rogers

Department of Lung Pathology,
Cardiothoracic Institute, Brompton Hospital,
London, SW3 6HP, England.

Epithelial mucous (i.e. goblet) cell hyperplasia is a consistent observation in the airways of the lung in both human hypersecretory diseases (Reid 1954) and in animal models of airway's hypersecretion (Reid 1963; Lamb & Reid 1968; 1969). The specific pathogen-free (SPF) rat when exposed to an atmosphere of tobacco smoke is a well characterized and suitable animal model to study the pathogenesis of hypersecretion (Reid 1970).

The epithelium lining the airways of many species is more complex than formerly realized: the normally sparse number of

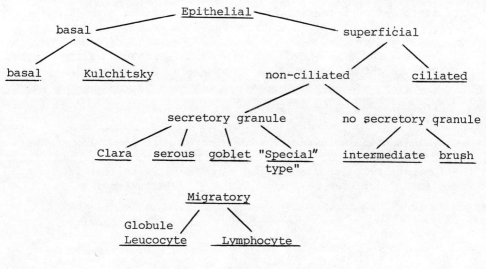

mucous cells are surrounded by a cellular environment consisting of
at least eleven cell types (fig.1) (Jeffery & Reid 1975, Jeffery
1978).

The results of early studies indicated that the basal cell was
a progenitor (stem) cell from which the more specialized (in terms of
function) superficial cells arose firstly by division and subsequently by differentiation. In the normal SPF rat the cells specialized
for secretion are comprised of the serous (predominent) and mucous
cells of the proximal (rostral) airways and the Clara cells of the
distal (peripheral) bronchioli (Jeffery & Reid 1975). When these
animals (free from respiratory disease) are passively exposed to
tobacco smoke for 4 hours daily for up to 14 days there is a highly
significant increase in the number of epithelial mucous cells at all
airway levels studied and a change in the histochemical types of
mucus produced by them (Jones, Bolduc & Reid 1972; 1973). By
electron microscopy there is a sharp decrease in the mumber of
serous cells and a gradual increase (from 2 - 18% of the total cell
population) in the number of mucous cells: there is morphological
evidence that, at least in part, this is brought about by transformation of existing serous to newly formed mucous cells (Jeffery &
Reid 1981).

The aims of the studies presented here were to investigate:
(i) the relative importance of cell division, differentiation and
transformation in the observed mucous cell increase and (ii) the
extent to which different anti-inflammatory agents might inhibit or
modify these tobacco smoke-induced changes.

MUCOUS CELL INCREASE

The contributions of cell division and differentiation to the
observed mucous cell increase were determined by pulse labelling the
cell nuclei of experimental animals with tritiated thymidine
(^{3}H-T) (i.p. injection: 1μCi/gm body weight) and quantifying the
number of each cell type so labelled or not following autoradiography
of 1μm thick resin-embedded sections: the left main extrapulmonary
bronchi were chosen for study and animals killed sequentially during
the 14 days of exposure to smoke.

a) Cell Division

To determine which cells were entering division, animals were
given a pulse label of ^{3}H-T 1 hour prior to their death: tissue was
then prepared for autoradiography (Rogers 1979) and the labelling
index (L.I.) for each cell type (L.I.$_{cell}$) calculated:

$$\text{L.I.}_{cell} = \frac{\text{number of each cell type labelled}}{\text{number of each cell type counted}}$$

The overall labelling index of the epithelium (i.e. $L.I._{epi}$ = number of cells labelled/total number cells present) expressed per 1000 cells is normally 3.3 and extremely low in comparison with many other renewing tissues such as epidermis and duodenal epithelium (L.I. X 1000 = 19 & 154 respectively (Messier & LeBlond 1960). This makes the counting of large numbers of cells imperative for adequate sampling of airway epithelium.

The normal proportions of each cell type which could be identified in 1μm sections and their respective labelling indices are shown in table I. The L.I. gives a "feel" for the frequency with which each cell type is found to divide in relation to the total number of that particular cell type. The "% labelled cells" shows what percentage of all dividing cells were ciliated, serous, mucous, etc. and thus gives information only on the relative contribution of each type of dividing cell to the total fraction of dividing cells.

Table I   Cell Types in Normal Epithelium: their proportions, labelling indices, and % labelled cells.

Cell type	% cell type in epithelium	$L.I._{cell}$ (X 10^3)	% labelled cells
ciliated	41.2	--	--
serous	29.8	$3.0 (\pm 0.6)$	27
transitional	--	--	--
mucous (goblet)	2.9	--	--
intermediate	6.8	$4.4 (\pm 1.9)$	7
basal	22.1	$10.1 (\pm 0.3)$	66

In agreement with early findings most of the dividing cells are found to be basal but interestingly a substantial proportion (27%) are secretory and of the serous type. The implication is that cells other than the basal may have a stem cell role and contribute to normal maintenance and repair of airway epithelium.

Following 14 days exposure to tobacco smoke the normal pattern of cell proportions and their labelling indices significantly changes (Table II).

Table II   Cell types in tobacco smoke - exposed epithelium:(14 days):
           proportions, labelling indices, and % labelled cells

Cell type	% cell type	L.I. X 10^3	% labelled cells
ciliated	51	0.2	2.6
serous	8	9.8	10.3
transitional	2	-	0
mucous(goblet)	12	8.1	15.4
intermediate	5	-	0
basal	22	9.6	64.1

As previously shown (Jeffery & Reid 1981) the proportions of mucous
and ciliated cells increase mainly at the expense of serous.   The
results for L.I.$_{epi}$ show that there is a significant increase above
the normal to 6 dividing cells per 1000 counted at between 1 and 3
days.   While they may be fewer in number at this time, dividing
serous cells (L.I.$_{cell}$) are found with greater frequency than in the
normal.   Interestingly the same is now true for the mucous cell
which by 14 days has increased its numbers to account for 12% of the
total cell population, (Ayers and Jeffery 1981). Furthermore 15% of
the labelled (i.e.dividing) cells are identifiable as mucous indicating that this cell type, once thought of as an "end" cell (subserv-
ing only a secretory function) is itself capable of self-replication
in response to irritation by tobacco smoke. Fig.2(a & b) shows the
L.I.$_{cell}$ for each cell type during the 14 days of exposure: while
there were proliferative responses seen in intermediate and transitional cell types (a) the most marked proliferative response was
in the mucous (b) and this was seen at 7 days of exposure.

Two main points emerge immediately from this study:

1) the basal cell can no longer be thought of as the only cell
   type forming the proliferative 'compartment' of mature airway
   epithelium but must be considered along with the significant
   contribution made by the serous cell:
   and
2) the proliferative response to irritation shifts not only the
   proportional balance of airway cell types but also the balance
   of cell types dividing: both these shifts favour the appearance
   of the epithelial mucous cell which is the hallmark of hypersecretion.

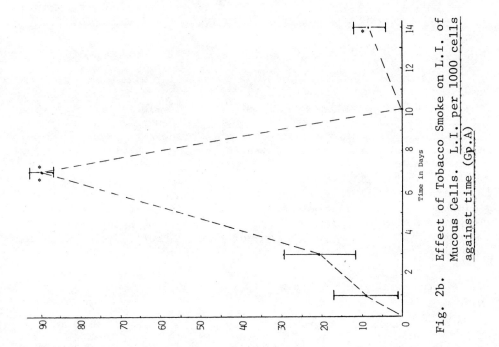

Fig. 2b. Effect of Tobacco Smoke on L.I. of Mucous Cells. L.I. per 1000 cells against time (Gp.A)

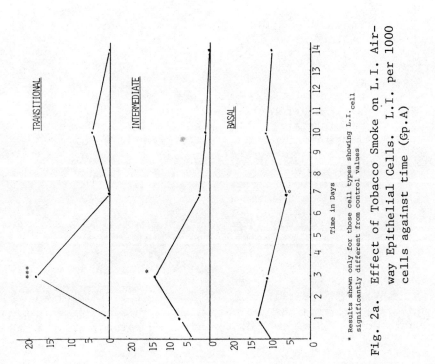

* Results shown only for those cell types showing L.I.$_{cell}$ significantly different from control values

Fig. 2a. Effect of Tobacco Smoke on L.I. Airway Epithelial Cells. L.I. per 1000 cells against time (Gp.A)

If self-replication of mucous cells is a marked feature, what role then does differentiation play in mucous cell hyperplasia?

b) Cell Differentiation

To determine the normal pathways of cellular differentiation and their modification by tobacco smoke all animals were given a single pulse of ^{3}H-T (i.p. injection of 1μCi/gm body weight) but only at the beginning of the experiment. Subsequent and sequential killing of control and tobacco smoke-exposed animals allows the passage of ^{3}H-T from the nuclei of the originally labelled cohort of 'stem' cells to be followed through their progeny and into the fully differentiated "end" cell compartment before its loss along with cells from the epithelium . The results are at a preliminary stage and require verification with tritiated-uridine which suffers less from the problems of reutilization. Fig.3. shows a working hypothesis suggested from the results of the shift in % labelled cells with time in normal animals.

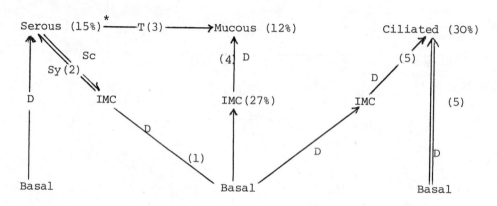

IMC = "Intermediate" cell
D   = differentiation
T   = transformation via transitional cell type
Sc   = discharge of secretory granules
Sy   = synthesis
*   figures are the final % of labelled cells at 14 days after
        a single pulse of ^{3}H-T.

Fig.3. Preliminary Interpretation of Cell Differentiation
        Results for Normal Bronchial Epithelium

Several pathways of differentiation may operate, the inter-mediate having a central role. The rise in % of labelled cells which are intermediate (corresponding with a decrease in basal) is partly due to basal cell differentiation pathway (1). We suggest

also that a contribution is made by the discharge of secretory
granules from serous cells (2) which then are counted as "intermediate".
Labelled serous cells are then replaced by subsequent synthesis of
granules. Serous cells may also contribute to the increase in % of
labelled cells which are mucous by synthesis of altered secretory
granules (i.e. transformation - (3)). There also appears to be a
contribution to the mucous cell by differentiation via the intermediate (pathway (4)). Differentiation of basal mainly favours the
production of ciliated cells either directly or via the intermediate
cell ((5)).

     Tobacco smoke alters the normal pattern of differentiation.There
is now a more early (by 3 days) and rapid movement of label out of
the basal cell compartment (i.e. faster differentiation of stem cells).
Intermediate cells are 22% of labelled cells at 3 days and this gradually decreased to 11% labelled cells at 14 days. While ciliated
cells show some early labelling there is the normal sharp rise at
between 3 and 7 days and this maintained to 14 days. Also while
mucous cells are labelled earlier than normal (i.e. at 3 days) they
only represent 7% of labelled cells (normally 12%) at 14 days. By far
the most interesting alteration is between 10 and 14 days when the
transitional cell (by definition containing a mixture of serous and
mucous secretory granules) is labelled and increases to 27% of
labelled cells. At the same time there is a marked fall in both
labelled basal and serous cells. A hypothetical scheme is shown in
fig.4 which shows that whereas pathways (1), (2), and (5) appear to
operate normally, irritation emphasizes pathway (3) in favour of the
transitional cell.

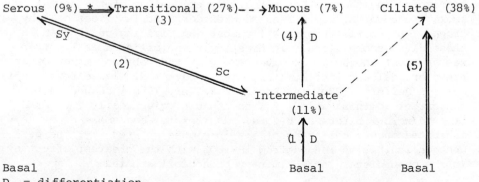

D = differentiation
T = transformation via transitional cell type
Sc = discharge of secretory granules
Sy = synthesis
* figures are the final % of labelled cells at 14 days after a single
   pulse of ³H-T and the beginning of exposure.

Fig.4 Preliminary Interpretation of Cell Differentiation - Results
                for Tobacco Smoke - Exposure Bronchus

These observations whilst preliminary, indicate that:-

(i) differentiation of basal to ciliated cells (via the intermediate)
is the normal method of maintaining ciliated cell number and this is
slightly accelerated with irritation by tobacco smoke.
(ii)direct differentiation of basal cells does not appear to play a
major role in the increase of mucous cell number which is more likely
to be due to self replication.
(iii)Furthermore the concurrent reduction in the % of labelled cells
which were serous (at 14 days) is supportive evidence for a transformation of serous to mucous cells via the transitional cell type.

In conclusion, the observed increase in mucous cell number after
tobacco smoke would appear to take place by a combination of mucous
cell self-replication and transformation of existing serous cells.

Anti-inflammatory Drugs

Can the tobacco smoke-induced changes to bronchial epithelium be
modified by anti-inflammatory drugs? Dalhamn (1966; 1969) and
Dalhamn and Rylander (1971) showed that the anti-tussive agent oxala-
mine citrate, whose active but stable component is phenylmethyloxadiazole (PMO), increased the time taken for tobacco smoke to arrest
ciliary motion. Jones et al (1972) showed that inhalation of smoke
from cigarettes, whose tobacco included PMO as 2% by weight, inhibited the mucous cell hyperplasia seen after tobacco smoke alone.
Jeffery & Reid (1981) examined the tobacco smoked-induced changes
by light microscopy of 1μm resin-embedded sections and electron
microscopy and showed that the airway epithelium of the animals whose
smoke included PMO had a less thickened epithelium, less cell hypertrophy, epithelial mucous cell number and dividing cells than in
those given smoke alone. Interestingly the ciliated cell number
was greatest in the group given PMO with the tobacco smoke. Other
non-steroidal anti-inflammatory agents have a similar effect. Greig,
Ayers & Jeffery (1980) and Rogers & Jeffery (1982) showed that
concurrent administration of Indomethacin, given(daily I.P. at 2
or 4 mg/kg/bw) before or before and after tobacco smoke exposure,
inhibited mucous cell hyperplasia in a dose related way: the percentage inhibition ranged from 18-100% with the greatest effect seen
in the most distal intrapulmonary bronchioli studied.

The effect of the steroidal anti-inflammatory agents, dexamethasone, prednisolone and hydrocortisone have been investigated in a
similar way and their effects compared with the effects of Indomethacin. Each drug was given at a daily (I.P.) dose of 4 mg/kg/bw as
2 mg before and 2 mg after tobacco smoke exposure. After 5 days of
treatment the daily dose of dexamethasone was reduced by half due to
its apparent toxicity as measured by a loss in animals' weights.
Table 3   shows the percentage increase in epithelial mucous (goblet)

cell number induced by exposure to tobacco smoke and the amount
by which each drug inhibited this response (expressed as a percentage).

Table 3. The effect of Tobacco Smoke (TS) with or without anti-
          inflammatory Drug on mean mucous cell number

	% increase by T.S.	% inhibition of increase			
Airway level		Dexameth-asone	Prednis-olone	Hydrocort-isone	Indometh-acin
TRACHEA	137	2	50	0	121
AXIAL					
Upper	101	68	70	86	68
Lower	153	62	78	74	100
PROXIMAL LATERALS					
Upper	89	69	38	62	87
Lower	92	53	50	80	98
DISTAL LATERALS					
Upper 2	142	170	125	0	100
Lower 2	240	91	35	0	5

Tobacco smoke produced a significant increase in the numbers of
mucous cells at all airway levels examined ($P<0.001$) Apart from
hydrocortisone which showed no inhibition in the trachea and most
distal bronchioli, all the anti-inflammatory drugs used showed
degrees of inhibition varying from 2 - 170%. Comparing the drugs,
Indomethacin had a significantly greater effect than the others in
the trachea ($P<0.001$), and proximal halves of the upper and lower
lateral bronchioli ($P<0.05$): dexamethasone had a greater effect than
the other drugs in the distal halves of the lateral bronchioli
($P<0.05$). While all drugs inhibited the mucous cell increase, all
had a similar degree of inhibition in both halves of the intrapulmonary axial airway.

     Considering the effects of each drug on the animals overall
weight gain, animals given injections of sterile water and those
given Indomethacin alone gained weight similarly. Hydrocortizone
and prednisolone both alone significantly decreased the normal weight

gain (P<0.001) whereas dexamethasone caused overall loss in weight. No drug inhibited the decrease in weight gain of rats exposed to tobacco smoke alone and once again dexamethasone caused a loss in weight in the exposed animals. In summary the steroidal anti-inflammatory drugs studied gave no particular advantage in inhibiting the tobacco smoke-induced mucous cell hyperplasia over and above the inhibition afforded by the nonsteroidal anti-inflammatory agent, Indomethacin.

Summary

The studies presented here describe the involvement of cell proliferation, differentiation and transformation in the epithelial mucous cell increase observed in the airways of experimental animals exposed to tobacco smoke (TS) for 14 days: the inhibitory effect of concurrent administration by injection of non-steroidal and steroidal anti-inflammatory drugs is also presented. In response to T.S. mucous cells arise mainly by self-replication and transformation of existing serous (secretory) cells while the production of ciliated cells appears to be mainly by differentiation of basal cells. Indomethacin inhibits the T.S.-induced mucous cell increase in a dose related way, the extent of which is dependent on the initial degree of response to T.S. and the airway level studied. The steroids dexamethasone, prednisolone and hydrocortisone, whilst effective in their inhibition gave no particular advantage over and above that seen with Indomethacin.

Acknowledgements

We acknowledge the support given to Margaret Ayers by The Cystic Fibrosis Research Trust and to Duncan Rogers by the Medical Research Council and thank Sonia Neil for her help with the preparation of the figures and typing of the manuscript.

References

Ayers, M., and Jeffery, P.K., 1981, Cell division and differentiation in bronchial epithelium, Proc. 5th School Thoracic Medicine Ettore, Marjorana Sch. Sci. Culture, Sicily ed. Cummins, Plenum.

Dalhamn, T., 1966, Inhibition of ciliostatic effect of cigarette smoke by oxolamine citrate. (3-phenyl 5 -diethylaminoethyl-1,2, 4-oxadiazole),
     Am. Rev. resp. Dis., 94, 799.

Dalhamn, T., 1969, The anticiliostatic effect of cigarettes treated with oxolamine citrate,
     Am. Rev. resp. Dis., 99, 447.

Dalhamn, T., and Rylander, R., 1971, Reduction of cigarette smoke ciliotoxicity by certain tobacco additives,
     Am. Rev. resp. Dis., 103, 855.

Greig, N., Ayers, M., and Jeffery. P.K., 1980, The effect of Indomethacin on the response of bronchial epithelium to tobacco smoke,
     J. Path., 132, 1.

Jeffery, P.K., 1978, Structure and function of mucus-secreting cells
    of cat and goose airway epithelium,
    In Respiratory Tract Mucus (ed R.Porter, J.Rivers & M.O'Connor)
    Elsevier.
Jeffery, P.K., and Reid, L., 1975,New observations of rat airway
    epithelium: a quantitative and electron microscopic study,
    J.Anat., 120, 295.
Jeffery, P.K., and Reid, L.M., 1981, The effect of tobacco smoke,
    with or without phenylmethyloxadiazole (PMO) on rat bronchial
    epithelium: a light and electron microscopic study,
    J.Path., 133, 341.
Jones, R., 1978, The glycoproteins of secretory cells in airway
    epithelium,
    In Respiratory Tract Mucus, Ciba Foundation Symposium 54
    (new series) ed R.Porter et al.
Jones, R., Bolduc, P., and Reid, L., 1972, Protection of rat
    bronchial epithelium against tobacco smoke,
    Brit.Med.J., 2, 142.
Jones, R., Bolduc, P., and Reid, L., 1973, Goblet cell glycoprotein
    and tracheal gland hypertrophy in rat airways: the effect of
    tobacco smoke with or without the anti-inflammatory agent
    phenylmethyloxodiazole,
    Brit.J.exp.Path, 54, 229.
Jones, R., and Reid, L., 1978, Secretory cell hyperplasia and
    modification of intracellular glycoprotein in rat airways
    induced by short periods of exposure to tobacco smoke, and
    the effect of the anti-inflammatory agent phenylmethyloxodiazole,
    Lab.Invest, 39, 41.
Lamb, D., and Reid, L., 1968, Mitotic rates, goblet cell increase
    and histochemical changes in mucus in rat bronchial epithelium
    during exposure to SO_2,
    J.Path. Bact., 96, 97.
Lamb, D., and Reid, L., 1969, Goblet cell increase in rat bronchial
    epithelium after exposure to cigarette and cigar tobacco smoke,
    Brit.Med.J., 1, 33.
Reid, L., 1954, Pathology of chronic bronchitis,
    Lancet, 1, 275.
Reid, L., 1963, An experimental study of hypersecretion of mucus
    in the broncial tree,
    Brit.J. exp.Path, 44, 437.
Reid, L., 1970, Evaluation of model systems for study of airway
    epithelium, cilia and mucus,
    Arch.intern,Med., 126, 428.
Rogers, A.W., 1979, Techniques of autoradiography, 3rd edit.,
    Elsevier/North-Holland, Amsterdam, New York.
Rogers, D., and Jeffery, P.K., (in preparation),
    The effect of anti-inflammatory agents on tobacco smoke-induced
    mucous cell hyperplasia in airway epithelium.

THE ROLE OF MUCOCILIARY TRANSPORT IN THE PATHOGENESIS

OF CHRONIC OBSTRUCTIVE PULMONARY DISEASE

Donovan B. Yeates

West Side Veterans Administration and Section
of Environmental Medicine, Department of Medicine
University of Illinois, Chicago, Illinois U.S.A.

INTRODUCTION

Mucociliary clearance is responsible for removing inhaled
deposited organisms, particles and for clearing cellular debris
and waste products from the airways of the lung. Within the mucus
coating are also phagocytic, enzimatic and immunological defenses
that are, in large part responsible for keeping the lung free
from infection. Changes in the depth of mucus lining or its
rate of transport in the various airway generations could compromise any of these defense mechanisms.

In humans, the effectiveness of mucociliary transport in the
lower respiratory tract was evaluated by measuring the transport
and retention of deposited radiotagged particles (Wales, et al.,
1980) as a function of time. Tracheal mucociliary transport
rates were measured with a 6-detector scintillation probe aligned
anterior to the trachea (Yeates, et al., 1981a). The positions
of transported local concentrations of radiotagged particles
were recorded on the chart output of 6 ratemeters. Particle
retention in the right lung was monitored for 2 to 5 hours either
with a gamma camera (Wilkey, et al., 1980) or a pair of scintillation probes aligned anterior and posterior to the right lung
(Yeates, et al., 1981a). Through the study of mucociliary transport in the healthy human lung and the investigation of the
effects of drugs, disease and inhaled pollutants on this system
an hypothesis was developed of the role of mucociliary transport in the sequence of events leading to chronic obstructive
pulmonary disease, (COPD).

RESULTS

In the healthy non challenged lung mucociliary transport has been characterized by a close correlation between the tracheal mucociliary transport rate (TMTR), and the percent of bronchial clearance within the first two hours (C_{B120}), (n=22, p < 0.001). The mean tracheal mucociliary transport rate was 5.1 $\pm$ 2.4 mm/min and the percent cleared in two hours 51 + 22% (Yeates et al., 1981a). Mathematical models of mucociliary transport predict that transport rates increase proximally from 10 μm/min in the terminal bronchiols to 5 mm/min in the trachea. (Yeates, et al., 1978, Lee, et al., 1979, Yeates, et al., 1981b). This increase proximally seems to be invariant between individuals (Yeates, et al., 1981b). However, it is notable that in the converging branching system of the lung modeling predicts mucus transport rates in each generation or order distally was always equal to or faster than that predicted by the assumption that the rates are inversely proportional to the total airway circumference. As mucus is being continually secreted into the airways this is consistent with the observation that mucus depth increases proximally. Mucus velocity in the trachea is approximately lognormally distributed with a geometric mean of 4.0 mm/min (Yeates, et al., 1981b). Since the maximum TMTR is probably near 20 mm/min, the majority of people appear to have a large reserve capacity. Each person appears to have his own characteristic whole lung clearance rate (Wilkey, et al., 1980).

Having described the characteristics of normal mucociliary clearance observations of the effects of drugs, inhaled polutants and disease on this system will now be summarized.

The correlation between mucus velocity in the trachea, TMTR and clearance from the lungs, C_{B120} (n=12, r=0.63, p < 0.001) disappeared after oral administration of 20 mg of the β2 adrenergic agent, metaproterenol. There was also an increase in TMTR from 4.4 $\pm$ 2.4 mm/min to 6.9 $\pm$ 4.5 mm/min (p < 0.01, two way analysis of variance) but there was no change in the percent cleared after two hours (Yeates et al., 1981c). This indicates that mucociliary transport in different regions of the conducting airways may be independently altered.

The relationship between mucociliary transport within the intrapulmonary airways (as represented by the times for 25% (T_{25}), and (50%) T_{50}, of the activity to clear) were changed after administration of 0.5 g/kg of ethyl alcohol to 12 healthy volunteers. In this study the overall correlation between T_{25} and T_{50} did not appear to change but rather there was an increased scatter about the regression after alcohol (p < 0.005) suggesting changes in the clearance pattern. In addition alcohol caused an

increase in the variation of the retention curves about their mean
(p < 0.001) (Venezelos, et al., 1981). This is consistent with
the observation that alcohol changed mucociliary clearance in both
magnitude and direction, with the magnitude and direction of change
depending on the individual subject.

A loss in the interdependence of mucociliary transport in the
trachea and bronchial airways was seen at all levels in a dose
response study of inhaled 0.5 μm sulfuric acid mist (Leikauf, et
al., 1981). Each of 10 subjects inhaled for one hour on separate
occassions, 100, 300 and 1000 $\mu g/m^3$ of 6.5 μm sulfuric acid-mist
at 50% relative humidity immediately after the radiotagged aerosol
inhalation. The mean pulmonary retention as indicated by T_{50}
increased at 100 $\mu g/m^3$ (p < 0.02), did not change at 300 $\mu g/m^3$,
and decreased at 1000 $\mu g/m^3$ (p < 0.03). Tracheal mucociliary
transport rates did not change at any exposure level (Leikauf,
et al., 1981). When the variation of the curves about their mean
was evaluated between ½ and 4½ hours for each exposure level, then
the pooled variance after 100 $\mu g/m^3$ was 99, similar to the variance
of the control curves (113). The variance was significantly
increased after 300 $\mu g/m^3$ (432, p < 0.001) and 1000 $\mu g/m^3$
(343, p < 0.001). This is consistent with the observation that
this pollutant may change mucociliary clearance in magnitude and
direction dependent on both the individual and the dose.

In patients with chronic obstructive pulmonary disease the
clearance of particulates from the lungs may not be markedly
reduced. In two comparable control studies both 8 healthy subjects
and 15 patients with COPD had similar percent of clearance after
two hours (33% and 35% respectively). Clearance of particles in
patients with COPD was similar to healthy persons probably due to
more proximal deposition and the effect of cough. There was however marked variability in the clearance of secretions from the
subjects with COPD. The intersubject coefficient of variation
for C_{B120} in 13 healthy persons was 45% and the intrasubject
variation was 20% (Yeates, et al., 1981b). In 19 patients with
COPD and intersubject variation of 86% and intrasubject variation
of 67% was observed. In addition, Spector, et al., (1980) have
shown a loss of interdependence of mucus transport in the trachea
and lungs in persons with asthma.

CONCLUSIONS

It is hypothesized that in the unchallenged lung the
interdependence of mucociliary transport throughout the conducting
airways is responsible for the maintenance of bronchial cleansing.
Temporary loss of this coordination and/or changes in the rates
of transport resulting from changes in ciliary function and/or
airway secretions are the normal responses to drugs and inhaled

pollutants. The resultant local inhomogeneities in mucus transport could compromise the inherent defense mechanisms in the mucus
coating. Regional changes in mucociliary transport may result from
nonuniform exposure of the airways due to deposited aerosol deposition or the kinetics of particle clearance, or to regional
variations in the pharmacological and toxicological sensitivities
of the airways. Continued insult to the lung mucosa by a pollutant
or infectious agent leads to changes in mucus secreting cells
and/or the ciliated epithelium. These changes often but not always
result in increased mucus production. There is a consequent further
regional accumulation of secretions which results in chronic cough,
erratic clearance of secretions from the lung and loss of airway
patency.

     There are several implications that result from this hypothesis.
There is a initial decrease in the effectiveness of the removal of
secretions as opposed to an initial increase in transport velocities
due to compensated bronchorhea as proposed by Albert, et al., 1973;
(b) Hypersecretion is not a prerequisite for the development of
COPD although this would likely aggrevate the condition.
(c) Persons with both slow and fast mucociliary transport would
be suseptable to conditions that could lead to COPD.

ACKNOWLEDGEMENTS

     This work was supported by the Medical Research Service of
the Veterans Administration, National Institutes of Health, Heart
Lung and Blood Institute HL 13824, HL 19431 and HL 07263 and
Oak Ridge National Laboratory, Basic Agreement 7762.

REFERENCES

Albert, R.E., Lippmann, M., Peterson, H.T., Berger, J., Sandborn,
K., and Bohning, D. (1973). Bronchial deposition and clearance
of aerosols. Arch. Int. Med. 131:115-127.

Lee, P.S., Gerrity, T.R., Hass, F.J., and Lourenco, R.V. (1979).
A Model for Tracheobronchial Clearance of Inhaled Particles in Man
and a Comparison with Data. IEEE Trans. Biomed. Eng., BME-26(11)
624-630.

Leikauf, G., Yeates, D.B., Wales, K.A., Albert, R.E., and Lippmann,
M. (1981). Effects of Sulfuric Acid Aerosol on Respiratory
Mechanics and Mucociliary Particle Clearance in Healthy Nonsmoking
Adults. Arch. Environ. Health - In Press.

Spektor, D.M., Leikauf, G.D., and Lippmann, M. (1980). Tracheobronchial Mucociliary Particle Transport in Asymptomatic Asthmatics.
Amer. Rev. Resp. Dis., 121 (4 part 2), 409.

Venizelos, P.C., Gerrity, T.R., and Yeates, D.B. (1981). Human
Response of Mucociliary Clearance to Alcohol Administration.
Arch. Environ., Health - In Press.

Wales, K.A., Petrow, H.G., and Yeates, D.B. (1980). Production
of Tc-99m Labelled Iron Oxide Aerosols for Human Lung Deposition
and Clearance Studies. International Journal of Applied Radiation
and Isotopes. 13, 689-694.

Wilkey, D.D., Lee, P.S., Hass, F.J., Gerrity, T.R., Yeates, D.B.,
and Lourenco, R.V. (1980). Mucociliary Clearance of Deposited
Particles from the Human Lung. Arch. Environ. Health. 35 (5)
294-303.

Yeates, D.B., and Aspin, N. (1978). A Mathematical Description
of the airways of the Human Lungs. Respiration Physiology:
32:91-103.

Yeates, D.B., Pitt, B.R., Spektor, D.M., Karron, G.A., and Albert,
R.E. (1981a). Coordination of Mucociliary Transport in the Human
Trachea and Intrapulmonary Airways. J. Appl. Physiol. - In Press.

Yeates, D.B., Gerrity, T.R., and Garrard, C.S. (1981b). Characteristics of Tracheobronchial Deposition and Clearance in Man.
In: Inhaled Particles V, Ed. W.H. Walton Permagon Press - In Press.

Yeates, D.B., Spektor, D.M., Leikauf, C.D., and Pitt, B.R. (1981c).
Effects of Drugs on Mucociliary Transport in the Trachea and
Bronchial Airways. Chest - In Press.

REGIONAL VARIATIONS OF MUCUS CLEARANCE IN NORMAL AND IN

BRONCHITIC MAMMALIAN AIRWAYS

A. van As

Pulmonary Division, VA Medical Center and
University of New Mexico, Albuquerque, NM

Very little information is available on the relationship of
ciliary beat frequency (CBF) to mucus transport velocities (MTV)
in airways distal to the trachea in intact lungs. Ciliary beat
frequencies and mucus transport rates in airways are not uniform
at all levels of the lung, both of these increasing from the
small airways to the trachea. The purpose of this paper is to
report the relationship of CBF to MTV in normal airways and compare these relationships to abnormalities of mucus transport in
animals with spontaneous chronic respiratory disease.

Direct measurement of integrated mucociliary function at all
levels of the bronchial tree was made by observation of an intact
whole lung preparation.[1] Detailed measurement of CBF and MTV was
possible in two specific pathogen free (SPF) animals which were
regarded to be free of bronchial disease. In these normal animals
mean MTV ranged from 0.4 mm/min. in bronchioles to 20 mm/min. in
the trachea. At the same location where mucus transport was
measured CBF were recorded. The CBF ranged from 400 beats/min.
in the bronchioles to 1500 beats/min. in the mid-trachea. There
was a close correlation between CBF and MTV ($r = 0.997$) indicating that they are closely integrated.

Mucociliary activity in an animal model of chronic respiratory disease was examined using non-SPF Wistar rats with Myco-
plasma pulmonis infection. In these airways ciliary beat was
markedly disturbed as regards rate, form, amplitude and direction

1. J. Iravani and A. van As, Mucus transport in the tracheo-
       bronchial tree of normal and bronchitic rats, J. Path.
       106:81 (1972).

417

of the effective stroke. These disturbances were diffusely distributed and focal, often with closely adjacent areas exhibiting
normal ciliary function. Mucus transport was measured in a
chosen extralobar bronchus (medial cardiac lobe) and compared
with transport velocities at the same airway level in normal lungs.
In normal animals mucus was transported at 4.9 mm/min. (range 1.2-
12.0 mm/min.). Under conditions of bronchitis the transport velocity was 9.0 mm/min. (range 0.75 - 27.2 mm/min.).

This study has shown that although the gradients of ciliary frequency (3.75 fold) and mucus transport (50 fold) which are present
from the bronchioles to the trachea in normal airways are of different magnitude they correlate well. In addition, under conditions of bronchitis widespread disruption of transport occurs
focally but is preserved in areas closely adjacent to abnormal
areas. Mucus transport is generally faster in areas of normal
ciliary function in bronchitic airways then at the same level in
normal airways. This suggests that normally functioning cilia
in bronchitic airways respond to the additional load of mucus
produced under conditions of hypersecretion by beating more rapidly when this load is presented to them.

THE DEVELOPMENT OF AN IN VIVO MODEL FOR THE EVALUATION OF DRUGS

AFFECTING TRACHEAL MUCUS

Ann S. Readman, C. Marriott and K. Barrett-Bee*

Department of Pharmacy, Brighton Polytechnic
Brighton BN2 4GJ, U.K. and *Department of
Biochemistry, I.C.I. Pharmaceuticals Ltd.
Macclesfield SK10 4TG, U.K.

Previous in vivo studies on the effect of drugs, particularly
those with mucolytic activity, on mucus secretion and properties
have involved the use of either patients with obstructive airways
disease or various anaesthetised animals although it has recently
been shown that certain anaesthetic agents can modify mucus
secretion and transport (King et al, 1979). The tracheal pouch
in the conscious dog has been used to evaluate drugs affecting the
antonomic nervous system (Shih, 1978) but no studies on the effect
of mucokinetic agents have been undertaken using such a model. It
is the purpose of this investigation to evaluate the tracheal pouch
as a model for the investigation of agents which putatively alter
mucus secretion or structure.

Tracheal pouches, permanently cannulated at either end, were
established in mini-pigs (Suis Scrofa domestica, Gottingen strain)
weighing 25 kg. Samples could then be aspirated daily from each
pig over a twelve week period to establish the patency of the pouch
and provide baseline data. On separate occasions thereafter, each
animal was administered a daily dose of either S-carboxymethylcysteine (SCMC) (30 mg kg^{-1}) or oxytetracycline hydrochloride
(18 mg kg^{-1}) orally for five days. An appropriate amount of placebo
was administered separately as a control. Individual samples were
collected daily, weighed and stored at -20°C until required. On
thawing, samples were freeze dried to determine total solids
concentration. These freeze dried samples were used in the determination of protein, hexose, sulphate and DNA concentrations (Gill
and Marriott, 1979). The concentrations of the individual monosaccharides, N-acetyl-galactosamine, N-acetyl-glucosamine, galactose,
fucose, mannose and sialic acid (as N-acetyl neuraminic acid) were

determined by the method of Clamp et al (1971), and amino acids
were estimated by the method of Mayes et al (1973) using a single
column technique.

    The results after treatment with oxytetracycline and placebos
showed no significant change (p < 0.1) in any of the parameters.
The dry weight elevation which was observed with SCMC was maintained
for 9 days and a further 9 days elapsed before baseline values were
re-established. Concomitant with this marked rise, was a definite
increase in the ratio of acidic to neutral terminal sugars (fig. 1).
This trend was reproduced in subsequent replicate experiments after
an adequate recovery period. The amino acid analyses indicated
that the distribution remained unchanged and thus it would appear
that SCMC only affects the production of, or modifies the oligosaccharide sidechains.

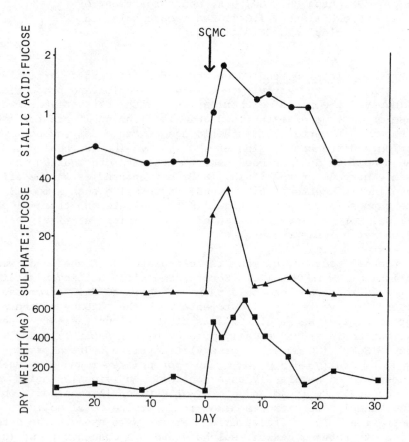

Fig. 1.   The effect of SCMC on total mucus secretion and glyco-
          protein type.

These findings support those of Havez et al (1970) who found
an increase in acidic glycoproteins from human bronchitic sputum
with the same drug. However, these workers demonstrated that the
acidic component which was seen to increase most markedly was
sialic acid whereas in this work, it was the sulphate which under-
went the most significant change. The sialic acid to fucose ratio
of the glycoprotein also increased, supporting the hypothesis that
SCMC induces an increase in acidic sugars at the expense of fucose;
the overall effect being to produce a more acidic glycoprotein.
Therefore it has been shown that, not only does the amount of
glycoprotein secreted increase under the influence of SCMC, but
also the type of molecule produced is altered.

It is concluded therefore that the porcine tracheal pouch
model is eminently suitable for the evaluation of drugs affecting
mucus synthesis and secretion.

Clamp, J.R., Bhatti, T. and Chambers, R.E., 1971, Determination
    of carbohydrate in biological materials by gas liquid
    chromatography, Methods of Biochemical Analysis, 19:229-344.
Gill, I.J. and Marriott, C., 1979, An evaluation of solubilising
    agents for sputum glycoproteins, in: "Glycoconjugates",
    542-543, ed. R. Schauer, George Thieme, Stuttgart.
Havez, R., Degand, P., Roussel, P. and Randoux, A., 1970, Mode
    d'action biochimique des derives de la cysteine sur le mucus
    bronchique, Poumon et la coeur, 26:81-90.
King, M., Engel, L.A. and Macklem, P.T., 1979, Effect of pento-
    barbital anaesthesia on rheology and transport of canine
    tracheal mucus, J.Appl.Physiol., 46:504-509.
Mayes, R.W., Mason, R.M. and Griffin, D.C., 1973, Composition of
    cartilage proteoglycans. Investigation using high and low
    ionic strength extraction procedures, Biochem.J., 131:533-541.
Shih, C.J., 1978, Ph.D. thesis, The physicochemical properties of
    mucus and their relation to secretion and mucociliary trans-
    port, University of Pennsylvania.

EFFECT OF S-CARBOXYMETHYLCYSTEINE ON THE BIOPHYSICAL AND

BIOCHEMICAL PROPERTIES OF MUCUS IN CHRONIC BRONCHITICS

A Cox, I Jabbal-Gill, C Marriott and S S Davis

Department of Pharmacy
University of Nottingham
University Park, Nottingham, NG7 2RD, UK.

INTRODUCTION

The chemical heterogeneity of sputum has resulted in confusion about the effect of mucolytic agents used in the treatment of chronic obstructive airways disease. In disease states sputum is thought to become more viscous and this is associated with the production of a less acidic mucin with less hydrophilic groups, so that the macromolecules become more entangled, with more intermolecular bonding and decreased water retention (Havez et al, 1973). After treatment with S-carboxy-methylcysteine (SCMC), some workers claim that the viscosity of sputum decreases, (Edwards et al, 1976), although increases have also been observed (Puchelle et al, 1978). It has been pointed out that elasticity and not viscosity may be the important factor controlling mucociliary transport (Richardson and Phipps, 1978; Medici and Radielovic, 1979).

The work described here is the result of an extensive study into the biophysical and biochemical properties of sputum. It was designed to identify the effects of administration of SCMC which has been claimed to be a mucolytic agent; this is a blocked thiol derivative of cysteine. Both whole sputum and purified mucin glycoprotein were investigated, with special attention paid to the ratio of acid and neutral glycoproteins, which is thought to alter after treatment with SCMC (Havez et al, 1973; Aylward et al, 1976).

METHODS

Forty patients were included in a double-blind, randomised, cross-over study lasting fifteen weeks (six patients withdrew).

Each patient provided two sputum samples every fourteen days; the
first on rising and the second later on the same day in the clinic.
Samples were stored at 4°C after collection.

To minimise disruption of the gel network, the samples were
not homogenised, but three aliquots were taken. One was used for
rheological measurements using a cone/plate creep viscometer. The
second was freeze-dried to determine the dry weight and was then
used to estimate total hexose, protein and DNA content. The third
aliquot was used to purify the mucin glycoprotein.

The mucin glycoprotein was obtained from individual sputum
samples which were washed with phosphate buffered saline to remove
the serum glycoproteins. After centrifugation the gel was
solubilised with phosphate buffer containing 0.22M potassium
thiocyanate (PB-KCNS), followed by gel filtration chromatography
on Sepharose 4B using PB-KCNS as the eluent. The material was
eluted and the excluded volume was exhaustively dialysed against
distilled water, freeze-dried and reconstituted in water. The
purified glycoprotein was analysed for protein, total hexose and
sulphate by an isotopic method (Gill and Marriott, 1979). The
individual sugars were determined by glc of trimethyl silyl
derivatives, after methanolysis with IM HCl for 16 hr at 90°C
(Clamp et al, 1971).

RESULTS

An increase in viscosity and a decrease in compliance (Fig 1)
appears to coincide with SCMC therapy. Compliance is the
reciprocal of elasticity and consequently one could conclude that
the viscosity and elasticity have increased thereby resulting in
a "thicker" gel. However, because of the wide variation between
patients, the differences were not statistically significant at
the 95% level. These results suggest that other previous studies
using smaller patient groups may be subject to criticism because
of small group size. The rheological changes for whole sputum
were not reflected in the dry weight and hexose/protein ratios,
since no significant differences were observed over the fifteen
week period. The DNA levels were low throughout (indicating no
marked infection) and therefore were not a contributing factor to
the thickening of the gel.

It was found that samples of whole sputum from the clinic
showed the greater changes in all parameters and this indicates
that sputum accumulated overnight is probably subject to
modification by enzymes.

The mucin glycoprotein is composed of the sugars: fucose,
sialic acid, galactose, N-acetyl galactosamine and N-acetyl-

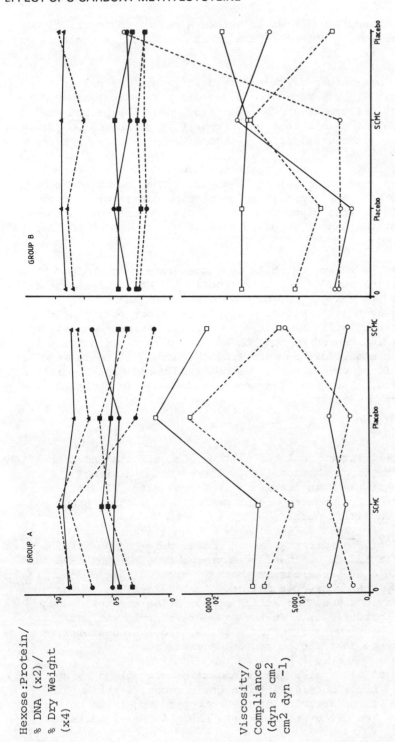

Hexose:Protein/
% DNA (x2)/
% Dry Weight
(x4)

Viscosity/
Compliance
(dyn s cm²
cm² dyn ⁻¹)

Figure 1   Comparison of Treatment Groups A and B for viscosity (O—O), compliance (□—□),
% dry weight (▲—▲), % DNA (●—●) and hexose/protein ratio (■—■). Mean values
are shown for morning samples (solid line) and clinic samples (broken line) of
whole sputum.

glucosamine. Mannose which would act as a marker for serum glycoproteins is virtually absent. Each of these sugars are calculated
as percent of dry weights.

The absolute values of the individual sugars for clinic
samples (Figure 2) both for Group A and Group B patients showed
no significant difference at the 95% level, over the period of
the trial. This suggests that the composition of the carbohydrate
side chain has not been affected by the drug.

When the individual side chain components are considered as
ratios (Figure 3 (a) and (b)) it is apparent that no significant
change in the hexose to protein ratio occurs. Therefore the overall composition of the glycoprotein molecule has not been affected.
The values for the hexose to protein ratio are typical when compared
with other studies.

The sialic acid and sulphate to fucose ratios shown in
Figure 3 (c), (d) and 3 (e), (f) respectively, do at first sight
appear to produce changes in accordance with the treatment
schedule. However, statistical analysis has shown that none of
the changes are significant at the 95% level. In most cases, the
mean value is influenced by one, or at the most, two, individual
values which are markedly removed from the mean. If these results
were to be omitted then the changes observed would be less distinct.
It is concluded, therefore, that no marked change in the acid to
neutral sugars has occurred.

CONCLUSIONS

It is still largely unknown whether SCMC affects the properties
of sputum from chronic bronchitics. There is the distinct
possibility that the response is patient dependent. If this is so,
then the presence of non-responding patients in a trial group may
well mask a beneficial effect.

The lack of significant changes observed in the purified
mucin can be explained by the fact that the drug may not have been
given for long enough periods or that any effects that did occur
were masked by the cross-over design. Seasonal variations have
been noted previously on the action of so-called mucolytic drugs
(Bryd and Sheppard, 1971). In the present trial it was necessary
to conduct studies over different seasons, and the possibility
that results were so affected cannot be excluded.

The results also raise doubts about investigations using
whole sputum. Large consistent changes in measured parameters
are necessary if the activity of a mucotropic drug is to be
demonstrated. The different responses shown for early morning

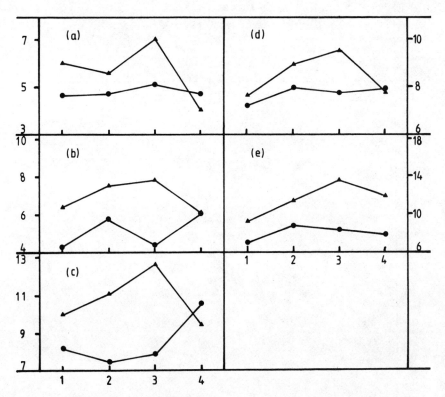

Figure 2.   Comparison of Treatment Groups A (●——●) and B (▲——▲)
for (a) fucose, (b) sialic acid, (c) galactose,
(d) N-acetylgalactosamine, (e) N-acetylglucosamine.
Therapy sequence for Group A is 1 = nil, 2 = SCMC,
3 = placebo, 4 = SCMC and Group B is: 1 = nil, 2 =
placebo, 3 = SCMC and 4 = placebo.

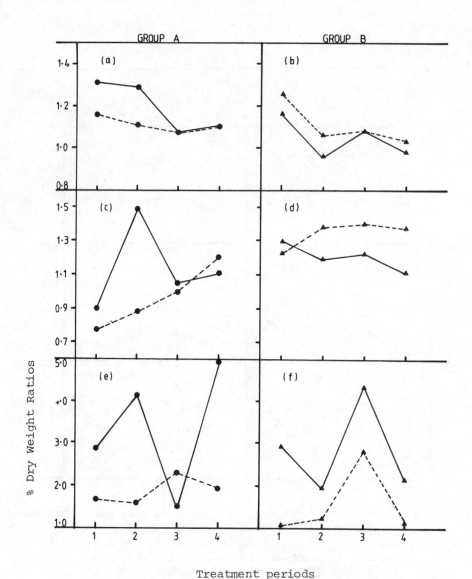

Figure 3. Comparison of Treatment Groups A and B for the ratios
Hexose/Protein (a), (b); sialic acid/fucose (c), (d);
and sulphate/fucose (e), (f), both for clinic (solid
lines) and morning (dotted lines) samples. Therapy
sequence for Groups A and B as indicated in the legend
of Figure 2.

and clinic samples are even more disturbing, since these indicate that a completely different conclusion could be drawn for the activity of a mucotropic drug, simply because of the time of day the sample was taken.

REFERENCES

Aylward, M., Edwards, G.F., Maddock, J. and Steel, A.E. (1976). Correlation between clinical status and macromolecular components of sputum in chronic bronchitis, related to the therapeutic activity of S-carboxymethylcysteine. Read at LIMA (through Berk Pharmaceuticals).

Bryd, E.M. and Sheppard, E.P. (1971). An autumn-enhanced muco-tropic action of inhaled terpenes and related volatile agents. Pharmacology 6:65.

Clamp, J.R., Bhatti, T. and Chambers, R.E. (1971). The determination of carbohydrate in biological materials by gas-liquid chromatography. Methods of Biochemical Analysism Ed. D. Glick, Vol. 19: 229.

Edwards, G.F., Steel, A.E., Scott, J.K. and Jordan, J.W. (1976). SCMC in the fluidification of sputum and treatment of chronic airway obstruction, Chest, 70:506.

Gill, I. and Marriott, C. (1979). An evaluation of solubilising agents for sputum glycoproteins, Proc. 5th Int. Symp. Glyco-conjugates, Kiel, Eds. Schauer, R., Brer, P. and Budecke, E., Stuttgart, George Thiane.

Havez, R., Degand, P., Roussel, P. and Randoux, A. (1970). Mode d'action biochemique des derives de la cysteine sur le mucus bronchique, Poumon Coeur 26, No 1:81.

Medici, T.C. and Radielovic, P. (1979). Effects of drugs on mucus glycoproteins and water in bronchial secretion, J. Int. Med. Res., 7:434.

Puchelle, E., Aug, F., and Polu, J.M. (1978). Effect of the muco-regulator SCMC in patients with chronic bronchitis, Europ. J. Clin. Pharmacol., 14:177.

Richardson, R.S. and Phipps, R.J. (1978). Tracheobronchial mucus secretion and the use of expectorant drugs in human disease, Pharmac. Ther. B., 3:441.

STIMULATION OF NASAL MUCUS SECRETION IN THE RABBIT

C. Duffett, J. Pell, R. Phipps, U. Wells, and J. Widdicombe

Department of Physiology, St. George's Hospital Medical
School, LONDON SW17 ORE, U.K.

The output of radiolabelled nasal mucus glycoprotein in
response to various stimuli has been studied in the rabbit.   The
technique is an adaptation of that used in the cat trachea
(Gallagher et al., 1975) where sodium ^{35}S-sulphate and ^{3}H-glucose
have been shown to label preferentially different glycoproteins.

Rabbits were anaesthetized with sodium pentobarbitone (36 mg/
kg).   A cannula was tied into the back of the pharynx above the
soft palate, and the nasal cavity was filled with physiological
saline solution via this cannula.   At the start of each experiment
^{3}H-glucose (0.35 mCi) was washed into the nasal cavity.   Sodium
^{35}S-sulphate (1.5-2.0 mCi) was administered either intravenously
(i.v.) or into the nasal cavity (i.n.).   After 7 hr, saline was
washed through the cannula and nasal cavity, collected at the
nostrils and discarded.   Subsequent washings were taken at intervals
of 30 min initially, and 15 min thereafter.

The nasal washings contained a mixture of unbound and glycoprotein-bound radioactivity.   Repeated dialysis in distilled water
containing 0.9% (w/v) sodium sulphate removed most unbound radioactivity.   After further dialysis against 6 M urea to prevent
aggregation of glycoproteins the radioactivity in these washings
was measured by β-scintillation counting, and expressed as becquerels
per minute of collection time for each sample.

The effect of stimuli, given at intervals during sampling, on
the output of radiolabelled mucus glycoprotein is expressed as a
percentage change over the preceding control sample.   The results
were analysed using the sign test.

In one set of experiments, the effect of sympathetic stimulation was studied. Sodium ^{35}S-sulphate was administered i.v. The peripheral cut ends of the cervical sympathetic nerves supplying the nose were stimulated with pulses of 10V, 10Hz and 1 msec. This increased the output of glycoprotein-bound ^{35}S by 106% (median value, n = 5, P< 0.05). Phenylephrine (α-adrenoceptor agonist, 0.5 mg/kg i.v.) increased the output of bound ^{35}S by 115% (n = 5, P< 0.05). Salbutanol (β_2-adrenoceptor agonist, 0.25 mg/kg i.v.) increased the output of bound ^{35}S by 41% (n = 7, P< 0.01). Dobutamine (β_1-adrenoceptor agonist, 0.25 mg/kg i.v.) gave a median increase of 32% (n = 5) in the output of glycoprotein-bound ^{35}S, but this was not significant. Preliminary experiments using thymoxamine (α-antagonist, 10μg/ml wash fluid, 2mg/kg i.v.) or L-propranolol (β-antagonist, 10μg/ml wash fluid, 2mg/kg i.v.) failed to block completely these effects. None of these drugs had a significant effect on the output of glycoprotein-bound ^{3}H.

In a second set of experiments, sodium ^{35}S-sulphate was administered either i.v. or i.n. Pilocarpine (parasympathomimetic, 1 mg/kg i.v.) and histamine (50 μg/ml i.n.) had no significant effect on the output of bound ^{35}S (i.n.), or of bound ^{3}H (i.n.). Rabbit plasma diluted 1:10 had no effect on the output of any of the bound radiolabels (^{3}H, ^{35}S i.n., ^{35}S i.v.). Ammonia vapour (3 x 100 ml i.n., 1:100), administered after draining of the nasal cavity, significantly increased the output of bound ^{35}S (^{35}S i.n., n = 5, median + 22%, P< 0.05) and bound ^{3}H (n = 5, median + 18%, P < 0.05).

Concentrated radiolabelled rabbit mucus has been separated on a molecular weight basis, using sepharose CL-2B and CL-4B columns. Two major groups of glycoproteins were found: a high molecular weight ($10^7 - 10^5$) group and one of lower molecular weight ($< 10^5$). The high MW peak was highly labelled and contained a small amount of glycoprotein (as measured by the Lowry (protein) the Winzler-Orcinol (neutral sugars), and the Alcian Blue (acid glycoprotein) assays). The lower MW peak was not highly labelled but contained a large amount of protein. The high MW peak probably represents a more rapidly synthesised group of proteins than the low MW peak.

Reference

Gallagher, J.T., Kent, P.W., Passatore, M., Phipps, R.J., & Richardson P.S. 1975, The composition of tracheal mucus, and the nervous control of its secretion in the cat, Proc. R. Soc. Lond. B, 192: 49-76.